INSTRUMENTATION AND RESEARCH PROGRAMMES FOR SMALL TELESCOPES

INTERNATIONAL ASTRONOMICAL UNION
UNION ASTRONOMIQUE INTERNATIONALE

INSTRUMENTATION AND RESEARCH PROGRAMMES FOR SMALL TELESCOPES

PROCEEDINGS OF THE 118TH SYMPOSIUM OF THE INTERNATIONAL ASTRONOMICAL UNION, HELD IN CHRISTCHURCH, NEW ZEALAND, 2–6 DECEMBER 1985

EDITED BY

J. B. HEARNSHAW

and

P. L. COTTRELL

University of Canterbury, Christchurch, New Zealand

D. REIDEL PUBLISHING COMPANY

A MEMBER OF THE KLUWER ACADEMIC PUBLISHERS GROUP

DORDRECHT / BOSTON / LANCASTER / TOKYO

Library of Congress Cataloging in Publication Data

CIP

International Astronomical Union. Symposium (118th : 1985 : Christchurch, N.Z.)
Instrumentation and research programmes for small telescopes.

At head of title: International Astronomical Union.
Includes index.
1. Telescope—Congresses. 2. Astronomical instruments—Congresses. 3. Photometry, Astronomical—Congresses. I. Hearnshaw, J. B. III. Cottrell, P. L. (Peter Ledsam), 1952– III. Title.
QB88.I583 1985 522′.2 86–17820
ISBN 90–277–2324–9
ISBN 90–277–2325–7 (pbk.)

Published on behalf of
the International Astronomical Union
by
D. Reidel Publishing Company, P. O. Box 17, 3300 AA Dordrecht, Holland

Sold and distributed in the U.S.A. and Canada
by Kluwer Academic Publishers,
101 Philip Drive, Assinippi Park, Norwell, MA 02061, U.S.A.

In all other countries, sold and distributed
by Kluwer Academic Publishers Group,
P. O. Box 322, 3300 AH Dordrecht, Holland

Printed in The Netherlands

TABLE OF CONTENTS

b) Poster Papers

III. SPECTROSCOPIC RESEARCH PROGRAMMES

a) Invited and Oral Contributed Papers

SUMMARY

PREFACE

This volume is the proceedings of IAU Symposium No. 118 on "Instrumentation and Research Programmes for Small Telescopes", where small telescopes were defined as those ground-based instruments with apertures less than 1.5m. The scientific goal of the symposium was to emphasise research programmes which were more suited to smaller telescopes, on which frequent regular observations can be made. A wide variety of topics on instrumentation, photometry, spectroscopy and polarimetry of objects in the solar system to extragalactic systems were discussed.

Each of the four scientific days of the symposium comprised a number of invited review papers, contributed oral papers and discussion sessions devoted purely to the large number (~64) of poster papers. An introductory paper on the research potential of small telescopes sets the scene for the symposium. The proceedings have then been divided into three sections.

Section I: Telescopes and instrumentation;
Section II: Photometric research programmes;
Section III: Spectroscopic research programmes.

The diversity of topics within each of these sections indicated the extent to which small telescopes have (and can) contribute greatly to astronomical research.

Dr J.A. Graham's summary of the symposium, which illustrates the opportunities available with small telescopes, concludes these proceedings.

As in all symposia, the importance of the discussion following each paper was realised. The discussion was recorded on tape (and wherever possible on questions and answer sheets), transcribed and then edited. We believe it provides a true and accurate record of the scientific comments relating to each paper. These edited comments follow each invited and oral contributed paper. However, because all of the poster papers in a given section were discussed sequentially we felt that a continuous discussion (annotated with author and paper title) following the outline of all the papers in that section, would be more useful to readers of these proceedings.

John B. Hearnshaw
Editor and Chairman,
Local Organising Committee

Peter L. Cottrell
Editor and Secretary/Treasurer,
Local Organising Committee

IAU
INTERNATIONAL ASTRONOMICAL UNION
SYMPOSIUM No. 118

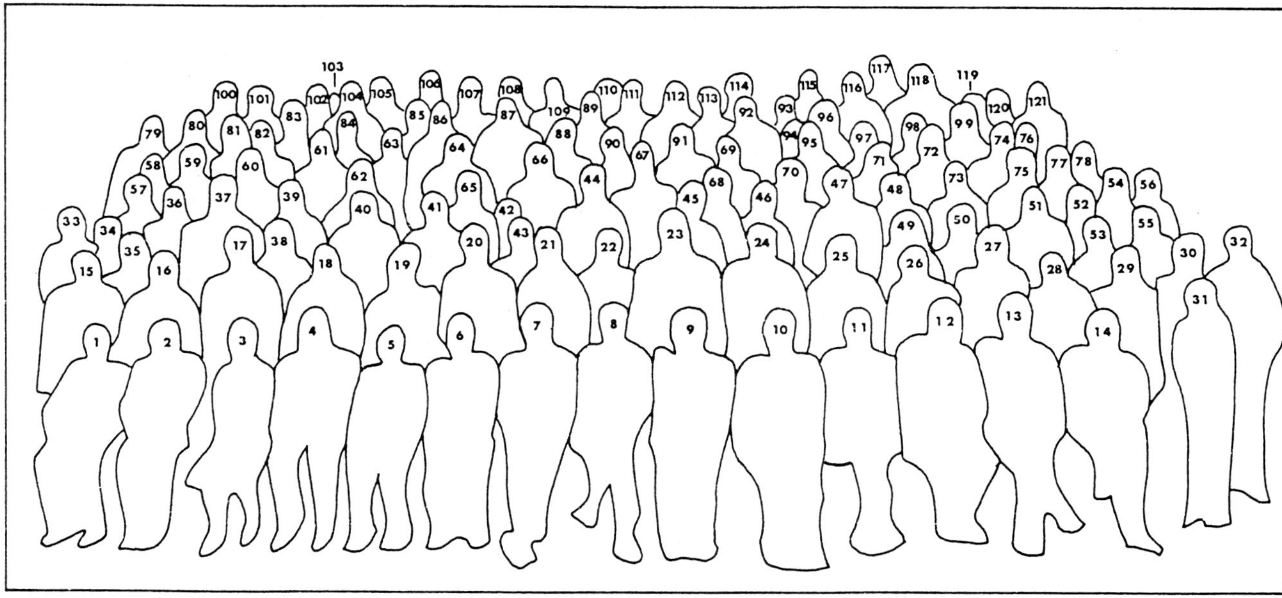

1. Yang
2. Dodd
3. Nield
4. Woodward
5. Bakos
6. Kreiner
7. Hickey
8. Genet
9. Roberts
10. Garrison
11. Hiltner
12. Warner
13. Duncan
14. Zeilik
15. Gliese
16. Storey
17. Latham
18. Barwig
19. Saxena
20. Allen
21. Millis
22. Mohin
23. Andrews
24. Crawford
25. Djorgovski
26. Hazen
27. Songsathaporn
28. Zook
29. A. Scarfe
30. Malasan
31. J. Scarfe
32. Kurtz
33. Sterken
34. Anthony-Twarog
35. Tomita
36. Jacoby
37. Ables
38. R. Loader
39. Bembrick
40. Moffett
41. Nankivell
42. Belserene
43. Angione
44. Levato
45. Ryan
46. Burki
47. A.R. Walker
48. Kitchener
49. Vanichai
50. Kilmartin
51. Osborn
52. S. Scarfe
53. Cameron
54. Page
55. C. Scarfe
56. Peery
57. Thompson
58. Innis
59. Zhu Nenghong
60. Batten
61. Gray
62. Kissling
63. Clark
64. Sandmann
65. Kulkarni
66. Warren
67. Evans
68. Weisberg
69. St. George
70. McDavid
71. Blow
72. Ianna
73. Gilmore
74. Hearnshaw
75. Beavers
76. Takagishi
77. Melsheimer
78. Jugaku
79. Costello
80. Herdman
81. Caton
82. White
83. Isobe
84. Watson
85. Wisniewski
86. MacQueen
87. Jeffers
88. Laubscher
89. Sullivan
90. Gaustad
91. B. Loader
92. Orchiston
93. Kilbey
94. Ives
95. Menzies
96. Hutcheon
97. Rumsey
98. Payne
99. Denby
100. Taylor
101. Rowe
102. Mochnacki
103. McIntyre
104. Cottrell
105. Penhallow
106. Twarog
107. Phillips
108. Redfern
109. Graham
110. Lawrence
111. Schulte in den Bäumen
112. Kviz
113. Kennedy
114. A'Hearn
115. van der Peet
116. Walkington
117. Fletcher
118. Neilson
119. Wood
120. Wynne
121. Weller

ACKNOWLEDGEMENTS

We wish to thank the financial support of the International Astronomical Union Executive Committee, the Royal Society of New Zealand, the University of Canterbury and the Kingdon-Tomlinson trust. Some of this enabled the IAU and the Scientific Organising Committee to provide travel support to this astronomical outpost in the south seas. The symposium was sponsored by International Astronomical Union Commissions 9 (astronomical instruments) and 25 (stellar photometry) and the Royal Astronomical Society of New Zealand.

The members of the Scientific and Local Orangizing Committees are listed below. The success of the symposium was due to their efforts and also to those of 133 participants from 21 countries.

Scientific Organizing Committee J.A. Graham (Chairman), D.F. Gray, J.B. Hearnshaw, M. Mayor, R.L. Millis, E.B. Newell, S. Nishimura, S.M. Runcinski, B. Warner, S.J. Yang.

Local Organizing Committee J.B. Hearnshaw (Chairman), P.L. Cottrell, A.C. Gilmore, H.M. Melville, C.H. Rowe, B.A. Sansom, D.B. Sparks, A.W.F. Thynne, S.M. Usher.

In addition to the above committee members there was a large group of helpers from the Physics Department at the University of Canterbury who helped in the final stages of the organisation of this symposium: Martin Begley, Sven Brabyn, Beverley Bristowe, Mike Clark, Geoff Costello, Gill Evans, Pam Kilmartin, Bryan Lawrence, Derek Lipyeat, Vincent McIntyre, Kirsten Nissen, Sean Ryan, Andrew Taylor and Alex Wynne. To these people we extend our thanks.

J.B.H.
P.L.C.

PARTICIPANTS

Ables, H.D.	Flagstaff, AZ, U.S.A.
A'Hearn, M.F.	College Park, MD, U.S.A.
Allen, W.H.	Blenheim, New Zealand
Andrews, F.	Wellington, New Zealand
Angione, R.J.	San Diego, CA, U.S.A.
Anthony-Twarog, B.J.	Lawrence, KS, U.S.A.
Bakos, G.A.	Waterloo, Ontario, Canada
Barwig, H.B.	Munich, F.R. Germany
Batten, A.H.	Victoria, B.C., Canada
Beavers, W.I.	Ames, IA, U.S.A.
Begley, M.J.	Christchurch, New Zealand
Belserene, E.P.	Nantucket, MA, U.S.A.
Bembrick, C.S.	Queenscliff, N.S.W., Australia
Blow, G.L.	Wellington, New Zealand
Bochonko, D.R.	Winnipeg, Manitoba, Canada
Bowers, C.L.	Bickley, W.A., Australia
Breger, M.	Vienna, Austria
Burki, G.	Sauverny, Switzerland
Cameron, A.C.	Cambridge, England
Caton, D.B.	Boone, NC, U.S.A.
Clark, M.C.	Lake Tekapo, New Zealand
Cottrell, P.L.	Christchurch, New Zealand
Crawford, D.L.	Tucson, AZ, U.S.A.
Denby, B.	Kensington, N.S.W., Australia
Djorgovski, S.G.	Berkeley, CA, U.S.A.
Dodd, R.J.	Wellington, New Zealand
Duncan, D.K.	Pasadena, CA, U.S.A.
Evans, D.S.	Austin, TX, U.S.A.
Fea, K.H.	Christchurch, New Zealand
Finkenzeller, U.	Berkeley, CA, U.S.A.
Firneis, M.G.	Vienna, Austria
Fletcher, J.M.	Victoria, B.C., Canada
Garrison, R.F.	Richmond Hill, Ontario, Canada
Gaustad, J.E.	Swarthmore, PA, U.S.A.
Genet, R.M.	Mesa, AZ, U.S.A.
Gilmore, A.C.	Lake Tekapo, New Zealand
Gliese, W.	Heidelberg, F.R. Germany
Graham, J.A.	Washington, DC, U.S.A.
Gray, D.F.	London, Ontario, Canada
Harwood, D.N.	Bickley, W.A., Australia

Hazen, M.L.	Cambridge, MA, U.S.A.
Hearnshaw, J.B.	Christchurch, New Zealand
Herdman, G.C.D.	Auckland, New Zealand
Hickey, D.	Redcliffe, Queensland, Australia
Hiltner, W.A.	Ann Arbor, MI, U.S.A.
Hutcheon, S.J.	Sheldon, Queensland, Australia
Ianna, P.A.	Charlottesville, VA, U.S.A.
Innis, J.L.	Clayton, Victoria, Australia
Isobe, S.	Mitaka, Tokyo, Japan
Ives, B.A.	Auckland, New Zealand
Jacoby, G.H.	Tucson, AZ, U.S.A.
Jeffers, S.	Downsview, Ontario, Canada
Jugaku, J.	Mitaka, Tokyo, Japan
Kennedy, H.D.	Ferny Hills, Queensland, Australia
Kershaw, G.M.	Christchurch, New Zealand
Kilbey, P.R.	Auckland, New Zealand
Kilmartin, P.M.	Lake Tekapo, New Zealand
Kissling, W.M.	Wellington, New Zealand
Kitchener, B.W.	Murarrie, Queensland, Australia
Kreiner, J.M.	Krakow, Poland
Kulkarni, P.V.	Ahmedabad, India
Kumar, C.K.	Washington, DC, U.S.A.
Kurtz, D.W.	Rondebosch, South Africa
Kviz, Z.	Kensington, N.S.W., Australia
Latham, D.W.	Cambridge, MA, U.S.A.
Laubscher, B.E.L.	Albuquerque, NM, U.S.A.
Levato, H.	San Juan, Argentina
Loader, B.R.	Blenheim, New Zealand
Mabin, D.B.	Nelson, New Zealand
McCarthy, M.F.	Vatican City State
McDavid, D.A.	San Antonio, TX, U.S.A.
MacQueen, P.J.	Christchurch, New Zealand
Malasan, H.L.	Lembang, Java, Indonesia
Melsheimer, F.M.	Longmont, CO, U.S.A.
Menzies, J.W.	Cape Town, South Africa
Millis, R.L.	Flagstaff, AZ, U.S.A.
Mochnacki, S.W.	Richmond Hill, Ontario, Canada
Moffett, T.J.	West Lafayette, IN, U.S.A.
Mohin, S.	Bangalore, India
Nankivell, G.R.	Lower Hutt, New Zealand

Neilsen, G.E.	Aspley, Queensland, Australia
Nield, K.M.	Hamilton, New Zealand
Orchiston, W.	Clayton, Victoria, Australia
Osborn, W.H.	Mt. Pleasant, MI, U.S.A.
Page, A.A.	Aspley, Queensland, Australia
Payne, P.W.	Sydney, N.S.W., Australia
Peery, B.F. Jr.	Washington, DC, U.S.A.
Penhallow, W.S.	Kingston, RI, U.S.A.
Phillips, M.M.	La Serena, Chile
Plavec, M.J.	Los Angeles, CA, U.S.A.
Redfern, R.M.	Galway, Eire
Reynolds, S.E.	Christchurch, New Zealand
Roberts, G.C.R.	Los Angeles, CA, U.S.A.
Rothermel, H.	Garching, F.R. Germany
Rowe, C.H.	Christchurch, New Zealand
Rucinski, S.M.	Richmond Hill, Ontario, Canada
Rumsey, N.J.	Lower Hutt, New Zealand
Ryan, S.G.	Christchurch, New Zealand
St. George, L.E.	Auckland, New Zealand
Sandmann, W.H.	Claremont, CA, U.S.A.
Saxena, A.K.	Bangalore, India
Scarfe, C.D.	Victoria, B.C., Canada
Schober, H.J.	Graz, Austria
Schulte in den Bäumen, J.S.	Oberkochen, F.R. Germany
Songsathaporn, R.	Chiang Mai, Thailand
Sterken, C.L.	Brussels, Belgium
Storey, J.R.	Auckland, New Zealand
Sullivan, D.J.	Wellington, New Zealand
Takagishi, K.T.	Kirishima, Japan
Thomas, M.	Dunedin, New Zealand
Thompson, K.	Clayton, Victoria, Australia
Tomita, K.	Mitaka, Tokyo, Japan
Twarog, B.A.	Lawrence, KS, U.S.A.
van der Peet, A.G.J.	Wellington, New Zealand
Vanichai, Y.	Cholburi, Thailand
Vetesnik, M.	Brno, Czechoslovakia
Walker, A.R.	Observatory, South Africa
Walker, W.S.G.	Auckland, New Zealand
Walkington, C.M.	Wellington, New Zealand
Warner, B.	Rondebosch, South Africa

Warren, W.H. Jr.	Greenbelt, MD, U.S.A.
Watson, R.D.	Hobart, Tasmania, Australia
Weisberg, J.M.	Northfield, MN, U.S.A.
Weller, W.G.	La Serena, Chile
White, N.M.	Flagstaff, AZ, U.S.A.
Wisniewski, Z.	Tuscon, AZ, U.S.A.
Wolf, G.W.	Springfield, MO, U.S.A.
Wood, F.B.	Gainesville, FL, U.S.A.
Woodward, C.E.	Rochester, NY, U.S.A.
Yang, S.J.	Nanjing, China
Zeilik, M.	Albuquerque, NM, U.S.A.
Zhu, Neng-hong	Shanghai, China
Zook, A.C.	Claremont, CA, U.S.A.

INTRODUCTION

THE RESEARCH POTENTIAL OF SMALL TELESCOPES

Brian Warner
Department of Astronomy,
University of Cape Town,
7700 Rondebosch. South Africa

ABSTRACT. The use of small telescopes for the production of significant research is examined from the points of view of cost-effectiveness, notable successes already achieved, the advantages over large telescopes in the areas of efficiency, availability, flexibility and serendipitous and speculative observation. It is concluded that small telescopes have an increasingly important role to play in modern astronomy.

1. INTRODUCTION

Interest in the design and construction of very large telescopes has never been greater than at the present time and there is no shortage of research projects for such instruments. The glamour associated with the very large can easily obscure the fact that small telescopes have a place, and an increasingly important place, in modern astrophysical research. It is vital that this assertion be understood and widely accepted: there are many Institutions, even entire countries, who cannot afford to enter the race for giant telescopes and who consequently consider themselves unable to contribute anything of significance. A study of past and present usage of small telescopes leads to a different conclusion: there is a need for more small telescopes, for better use of existing ones and for continual upgrading of the instrumentation used on them.

The purpose of this introductory article is to illustrate the truth of these statements; the remainder of the Symposium proceedings will provide a comprehensive survey by experts in their respective fields which will demonstrate better than I can the present and potential contributions made by users of small telescopes - and I choose my words with care here, for it is the users not the telescopes who do research, and the demands on comprehensive knowledge and ingenuity are if anything greater for the possessor of limited aperture than for the better endowed.

J. B. Hearnshaw and P. L. Cottrell (eds.), Instrumentation and Research Programmes for Small Telescopes, 3–16.

2. COST EFFECTIVENESS

The cost effectiveness of a telescope or collection of telescopes can be considered from several points of view. A penetrating analysis has been given by Disney (1972) in comparing a large telescope versus an array of smaller ones working on the same kind of observation. This latter condition is not appropriate here - we will show that the most effective use of small telescopes is to carry out projects that large telescopes cannot do. Nor is Disney's definition of effectiveness - as the limiting faintness to which a telescope can carry a given observation *in a given time* - appropriate for our present purpose. Instead I will present a naive domestic argument that will amuse you by its presumption.

Let us define cost effectiveness E in the way that a commercial enterprise might: as the annual return on capital investment. Then

$$E = \frac{\text{Number of Publications per annum}}{\text{Cost of Telescopes}}$$

where we count only publications that are significant!

The cost of construction of a telescope is usually expressed in the form

$$C = AD^{\alpha}$$

where D is the diameter of the mirror and the exponent α is ~ 3 for traditional designs but may be ~ 2 for designs using the recent advances in technology. We further assume that the cost of instrumentation and of maintaining a telescope scales as the cost of construction and can therefore be absorbed into the constant A.

If we have constructed n telescopes and each one produces τ^{-1} publications per year (τ is the mean time taken to complete an observational project) then

$$E = \frac{n\tau^{-1}}{A \sum_{i=1}^{n} D_i^{\alpha}}$$

For simplicity we compare the cost effectiveness of using n small telescopes of diameter D_s versus that of using one large telescope of diameter D_ℓ where the total capital cost of the alternative arrangements is the same

$$\text{i.e.} \quad nAD_s^{\alpha} = AD_\ell^{\alpha}. \tag{1}$$

Then

$$\frac{E_s}{E_\ell} = n \frac{\tau_\ell}{\tau_s} \tag{2}$$

where we recognise that it may take more observing time τ_s to achieve significant results on a small telescope than the time taken τ_ℓ on a large instrument.

Equation (2) shows that provided $n > \tau_s/\tau_\ell$ it makes sense to invest one's money in several small telescopes rather than one large one. From equation (1) we see that $n = (D_\ell/D_s)^{\alpha}$, so for the choice between one 4m telescope and $n = 4^{\alpha}$ 1m telescopes we require $\tau_s < 4^{\alpha}\tau_\ell$ which must surely always be satisfied: my own estimate is that $\tau_s \sim 4\tau_\ell$, in which case it is at least as profitable (in terms of research output) to build four half-size telescopes as one full-size telescope, and even better to build more even smaller ones!

The above analysis breaks down when we reach apertures so small that no worthwhile research can be achieved with them. In fact, the whole analysis would be worthless if nothing significant at all could be achieved with telescopes of modest size - so we interrupt further development of the case for "small is beautiful" to illustrate what has already been accomplished.

3. SOME SMALL TELESCOPE TRIUMPHS

In giving examples of what has been achieved with small telescopes I can be neither comprehensive nor unbiased. I will, however, choose principally from modern successes: if you go far enough back in time there are divers instances of important results obtained with small telescopes, but those resulted merely from the combination of the rarity of large instruments and the richness of opportunity for discovery. As an extreme case, the satellites of Jupiter and the phases of Venus were found using a refractor of only 3 or 4 cm aperture, but that was probably the world's largest telescope at the time! My examples are taken from the past thirty years or so and are intended to demonstrate that really significant achievements (and others that are important if less significant) have been made with the use of modest equipment.

3.1 Interstellar Polarization

In seeking the intrinsic linear polarization in stars predicted by Chandrasekhar (1946), Hiltner (1949) and Hall (1949) discovered that the interstellar medium impresses polarization on the light of distant galactic stars. Hall used the 40 inch reflector at the US Naval Observatory; Hiltner used the 82 inch reflector at McDonald Observatory but could have discovered the effect with a smaller telescope if one had been readily available! Much of the very extensive material now available on interstellar polarization has been gathered with modest telescopic power. The importance of these observations to studies of interstellar grains, the galactic magnetic field and to galactic structure in general needs no elaboration.

Curiously, the search for the Chandrasekhar effect has only recently met with success: in Algol, by Kemp *et al.* (1983), using a 61 cm telescope.

3.2 Circular Polarization

The first detection of circular polarization in optical astronomy was made with a 24 inch telescope (Kemp *et al.* 1971). This was the result of a deliberate search for circular polarization in white dwarf stars, stimulated by the theoretical prediction that such an effect should be present in stars with very large magnetic fields. Studies of polarization in white dwarfs have resulted in determination of magnetic field structures in isolated stars and in close binary stars (the Polars, or AM Her stars, e.g. Cropper 1985). Again much of the observational material has been obtained with modest telescopes - particularly for the Polars where circular polarizations of up to 50 per cent make worthwhile observations feasible on stars even as faint as 16 mag with telescopes of 1m or less.

3.3 The Crab Pulsar

Probably the most spectacular discovery made in recent years with the aid of a small telescope was that of the optical pulsar in the Crab Nebula (Cocke, Disney & Taylor 1969).

Soon after the discovery of radio pulsars was announced in February 1968 several groups (e.g. at Kitt Peak and at McDonald Observatory) had used large optical telescopes in attempts to produce optical identifications. After nearly a year with no success it was becoming difficult to obtain large telescope time to continue the search. Drs. Cocke and Disney "whose observing experience was virtually nil" (Cocke, Disney & Taylor 1972) found it relatively easy to obtain five nights on the Steward Observatory's 36 inch reflector. After equipment difficulties and poor weather had prevented a satisfactory search for the pulsar they managed to obtain a further two nights from the next observer (something less readily accomplished when using a large telescope) and thereby discovered the Crab Pulsar in extra time.

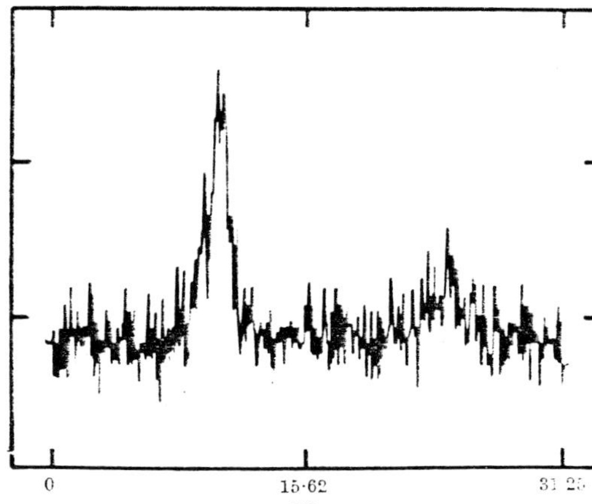

Fig. 1 *Optical pulses of the Crab Pulsar on 20 January 1969. From Cocke, Disney & Taylor (1969).*

Subsequent studies of the pulse arrival times of the Crab pulsar have been made almost exclusively with small telescopes (92 cm: Groth 1975; 60 cm: Lohsen 1981) and produce information on period changes fundamental to the understanding of the structure of neutron stars.

3.4 Double Degenerate Close Binary

The faint blue star HZ 29, recognised spectroscopically as a white dwarf, was found by Smak (1967) using a 36 inch telescope to be a variable with a period of 17½ mins. This provided the impetus that stimulated later work (Warner & Robinson 1972a; Faulkner, Flannery & Warner 1972) which showed that HZ 29, renamed AM CVn, is a short period mass transferring close binary with both components degenerate whose evolution may be controlled by the emission of gravitational waves.

3.5 DB White Dwarf Variables

The theoretical prediction that a new kind of variable star - oscillating DB white dwarfs - should exist was verified with the use of a 36 inch reflector in the discovery of pulsations in GD 358 (Winget *et al.* 1982).

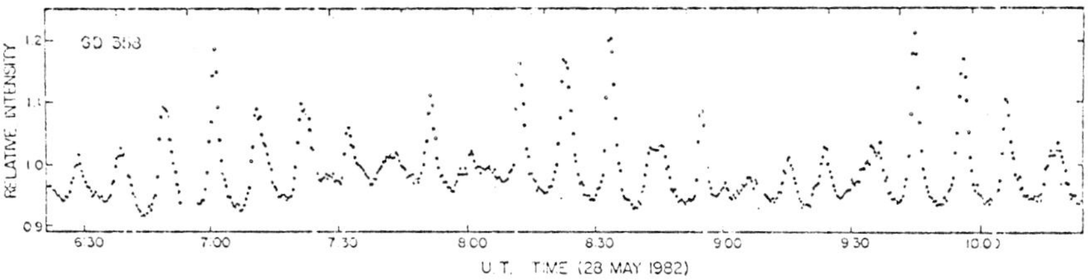

Fig. 2 *Light curve of the DB White Dwarf variable GD 358 observed with a 36 inch reflector. From Winget et al. (1982)*

3.6 Cataclysmic Variables

Because of the faintness of white dwarfs the study of their physical properties was once the almost exclusive privilege of 200 inch telescope users. However, for once nature conspires to aid the small telescope user and many unique optical phenomena associated with the white dwarfs or their accretion discs in cataclysmic variable stars occur conveniently during the eruptions of dwarf novae when they reach brightnesses (8-12 mag) within easy reach of small telescopes. The demonstration that it is the accretion disc that brightens (and not the secondary star as had been previously thought and enthusiastically modelled by theoreticians) during a dwarf nova outburst was made with a 20 inch telescope (Warner 1974). The enigmatic superhump structures present during supermaxima of SU UMa stars were discovered with 61 cm and 1 m telescopes (Vogt 1974, Warner 1975). The rapid (10-50 sec) periodic oscillations often present during eruptions of dwarf novae (Warner and Robinson 1972b) are well within reach of small telescopes.

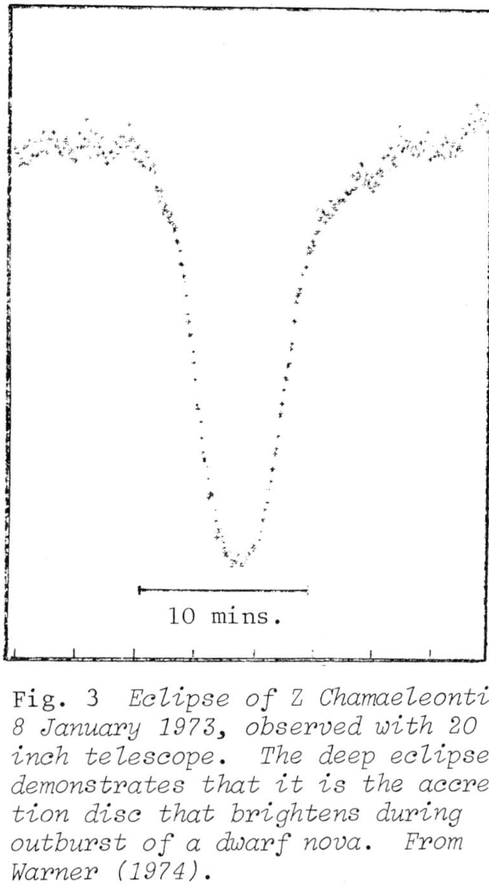

Fig. 3 *Eclipse of Z Chamaeleontis 8 January 1973, observed with 20 inch telescope. The deep eclipse demonstrates that it is the accretion disc that brightens during outburst of a dwarf nova. From Warner (1974).*

3.7 Optical Flashes from Gamma-Ray Bursters

Although not yet satisfactorily verified, the possible detection of three optical flashes from the Gamma-Ray burst source GBS0526-66 in ∿910 hours of monitoring with a high speed photometer on a 50 cm telescope (Pedersen *et al.* 1984) is an example of how a small telescope can contribute crucial information at the forefront of high energy astrophysics - and as this source is probably located in the supernova N49 in the Large Magellanic Cloud, the contribution is also one to extragalactic research.

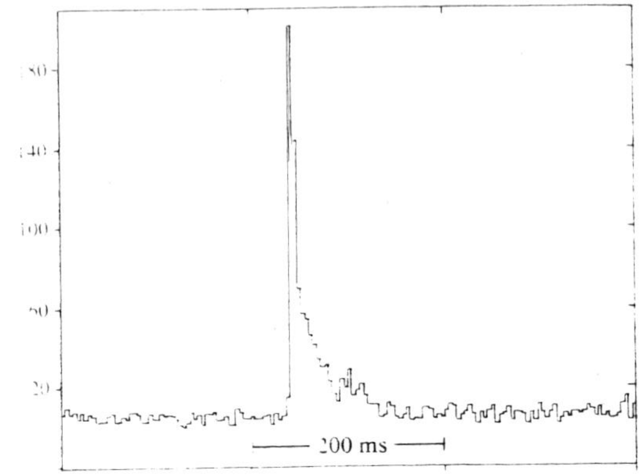

Fig. 4 *Possible optical flash from Gamma-Ray Burster GBS 0526-66 on 27 October 1983. From Pedersen et al. (1984).*

3.8 Rapidly Oscillating Ap Stars

The announcement in 1978 (Kurtz 1978) of a hitherto unsuspected class of rapid variable star, which includes among its members a star as bright as second magnitude, came as a surprise. The initial and most subsequent observations were made with a 20 inch reflector. Analysis of these low amplitude pulsators (Kurtz 1982) shows promise of providing details of the interaction between pulsations and magnetic fields and provides one of the first examples of the study of the seismology of stars other than the sun (Christensen-Dalsgaard, Gough & Toomre 1985).

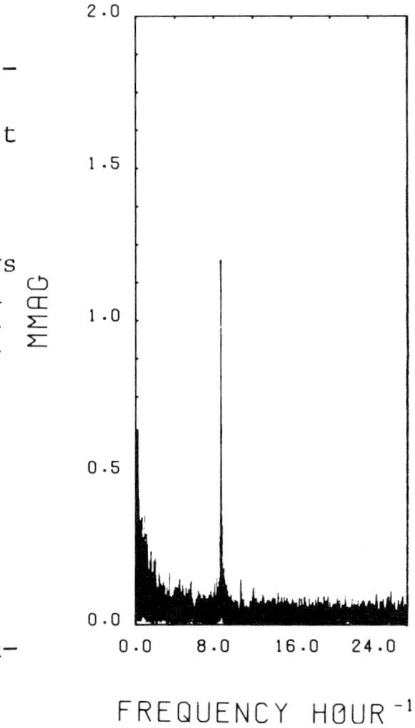

Fig. 5 *Amplitude spectrum of 70 hrs of photometry of the Oscillating Ap Star α Cir showing the presence of coherent 6.8 min pulsations. Note the very low noise (∿0.1 millimags) obtainable from extended photometry. From Kurtz & Balona (1984).*

3.9 The Rings of Uranus

The rings around the planet Uranus were discovered (Elliot *et al.* 1978) by various observers of an occultation of a star by the planet, including ones using 46, 61, 102 and 104 cm telescopes. These were the first planetary rings to be discovered since Huygens correctly interpreted the appendages of Saturn in 1655.

3.10 Charon: the Satellite of Pluto

In 1978 astronomers at the U.S. Naval Observatory claimed from photographic evidence that Pluto has a satellite, which they named Charon. This discovery was received with some scepticism until proven correct by Walker (1980) who observed an occultation of a star by Charon on 7 April 1980. The observation was made with a 1.0m telescope.

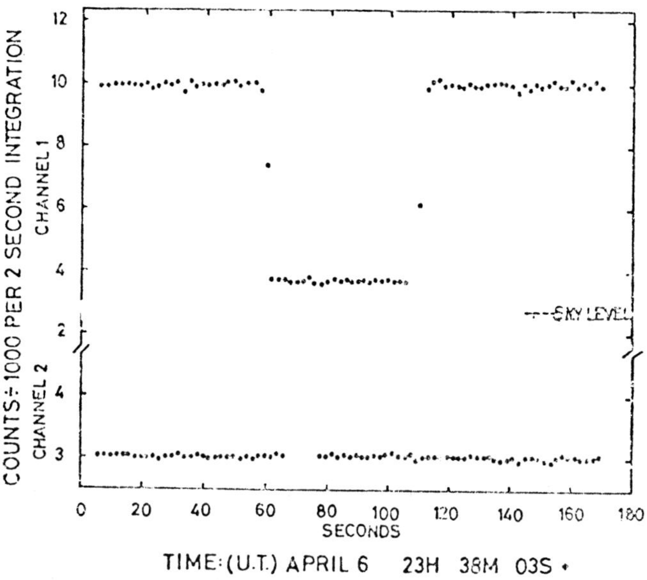

Fig. 6 *Occultation of a star by Charon (upper curve). From Report of the South African Astronomical Observatory 1980.*

3.11 Radial Velocity Measurements

Until 15 years ago almost the only source of high accuracy radial velocities was from massively designed high stability spectrographs at the Cassegrain and coudé foci of large telescopes. The quiet revolution started by Griffin's (1969) application of multiplex techniques to radial velocity measurements, using the 36 inch reflector at Cambridge, has taken much of the production away from large instruments. Apart from Griffin's own contributions, important results are being obtained with a variety of telescopes, one example of which is the radial velocity curve of the 104 day binary Capella, measured with the radial velocity meter at the coudé focus of the 61 cm reflector at the Fick Observatory (Beavers & Eitter 1977, Shen *et al.* 1985).

4. MISSED OPPORTUNITIES

In addition to the achievements already listed there are numerous phenomena well within the reach of small telescopes that by historical accident were found first with large telescopes: often much of the later work was in fact carried out with the more readily available smaller instruments. Choosing just two examples from my own field of interest, Merle Walker's discoveries (Walker 1954, 1956) that the nova remnant DQ Her is a short period (4^h39^m) eclipsing binary and a 71 sec oscillator were made with the Mount Wilson 100 inch telescope, and the first of the pulsating white dwarf stars, HL Tau 76, was discovered by Landolt (1968) with an 84 inch telescope, whereas ZZ Cet (Lasker & Hesser 1971) and several others were subsequently found with telescopes in the range 30-40 inches.

Fig. 7 *Amplitude spectrum of 38 hrs of photometry of the pulsating white dwarf (ZZ Cet type) star L19-2 showing presence of 5 coherent oscillations. From O'Donoghue & Warner (1982)*

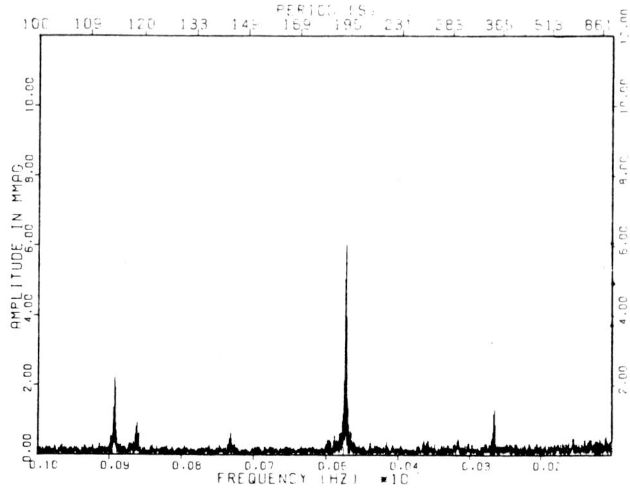

5. ADVANTAGES OF SMALL TELESCOPES

My arguments about the greater output of astronomy for a given input of funds require of course a number of caveats and qualifications. The most obvious is that for many kinds of research (photometry and spectroscopy of faint objects, high time resolution) there is simply no substitute for photons. However, claims that a given observation can always be made more efficiently or more quickly on a larger telescope need to be considered in the light of practicability.

In the remainder of this review I will look at large versus small from various perspectives, concentrating on direct research applications and omitting indirect usages such as text bedding of equipment destined for large telescopes, or aid in education and training.

5.1 Efficiency

Large telescopes are often designed to be as versatile as possible - multipurpose instruments not always efficient at their tasks. In contrast, a small dedicated telescope can be optimised for a single purpose using, for example, special optics, superreflective coatings or distinctive mounting. Examples are the DAO 48 inch reflector whose coudé spectrograph is faster than that of the 200 inch (Richardson 1971) and the 61 cm rotatable tube polarization telescopes at Yerkes and Siding Spring Observatories (Hiltner & Schild 1965).

Furthermore, the permanent installation of instrumentation on a dedicated telescope ensures stability of performance, improved monitoring of calibrations and reduced loss of time from instrument malfunctions associated with equipment changes. A site with several small telescopes, each dedicated and optimised, is an ideal arrangement; there can even be a large telescope if desired.

For some purposes, particularly in the study of rapid photometric variability in stars where only the frequency domain is of interest, it is an advantage to abandon the filters of standard photometric systems and use "white light". For very blue stars such as ZZ Cetis or cataclysmic variables, a gain in signal of 4 or 5 times that measured through a B filter is obtained, thus effectively doubling the aperture of the telescope. The use of large telescopes for such unstandardised use as "white light" photometry might not be countenanced, but there should be no such neurosis about small telescopes, which are thereby surreptitiously converted into large ones.

It is sometimes possible also to exploit the usually better seeing in small telescopes, for example with a smaller entrance aperture or slit and faster optics. This can improve the signal-to-noise for small telescopes and make larger ones less advantageous.

5.2 Adequacy

There should be a general rule not to use a large telescope where a small one will suffice. This is not to imply that all telescopes should be used near their limiting magnitudes, which would make every observation a time consuming one. Rather I have in mind that, especially for optimally designed specialised small telescopes, some extensive observing programs can be taken away from large telescopes. For example, with an efficiently designed spectrograph it would be feasible to carry out with a 40 inch telescope the surveys of radial velocities of bright galaxies currently in progress on larger telescopes (Da Costa et al. 1984). Radial velocities of stars with $m_V \lesssim 12$ are already mostly measured by the Griffin (1969) technique on small instruments. An example of the use of a 20 inch telescope in stellar population studies of elliptical galaxies is the aperture photometry performed by Wegner (1979). The adequacy of modern small telescopes is illustrated by the fact that by far the largest and most homogeneous set of integrated spectra of globular clusters (once the prerogative of large telescopes) is that obtained at 120A/mm with the 1.0m CTIO-Yale telescope by Hesser & Shawl (1985).

A further example of important extragalactic research achievable on small telescopes is the CCD photometry of clusters in the Magellanic Clouds described in more detail later in this Symposium by Alistair Walker. In the southern hemisphere a 20 inch telescope has approximately the same absolute magnitude limit in the Clouds as the 200 inch has in M31.

For rapid variables of low amplitude, non-coherent oscillations may only be detectable with the low photon and scintillation noise obtainable with large apertures, but for variables with periods in the range 5-15 mins sky transparency variations dominate the noise and large aperture has no advantage (Kurtz 1984).

5.3 Availability

Probably the dominant advantage of small telescopes is their availability - not only are there many more small than large telescopes, they are less in demand by astronomers seduced by the alure of mere size. It is common at major observatories to find that the largest telescopes are oversubscribed by factors of 5 while the smallest telescopes are underutilized. No well-equipped telescope with an aperture of 16 inches or more should find itself idle - particularly in the southern hemisphere.

The observing time available on existing small telescopes and to become available on those yet unborn constitutes a major research resource for astronomy. It is here that the small outclasses the large: projects requiring intensive use of a telescope for many weeks, or guaranteed access for a few nights a month for several years, or simply large amounts of time for the forseeable future, cannot be serviced with a large telescope. In fact, for some observers restricted to the use of large apertures, such small amounts of observing time are available that the pressure to make conspicuous use of the time generates the "temptation to do first class second-rate astronomy" (Disney 1972).

The need for generous allotments of time is obvious for photometric programs on variable stars with long periods, for photometry of standard stars or for general photometric or spectroscopic surveys. Galactic stellar population studies such as the Strömgren-Olsen project to obtain Strömgren photometry of the 15,014 HD stars brighter than V=8.3 with spectral types A5-GO (Stromgren 1976, Olsen 1979, 1980) can only be done on small telescopes.

Large amounts of observing time are also required for studies of short period variables with complicated frequency spectra. Examples are β Cep, δ Sct, PG1159, Rapidly Oscillating Ap and ZZ Cet stars, any member of which may require hundreds of hours of photometry spread over months or years in order to decipher the oscillation modes. The research rewards here are particularly exciting:- the oscillations reveal aspects of the interior structures of stars that cannot be obtained any other way. For example, using small telescopes, a measurement of dP/dt has been obtained for one of the oscillation frequencies in PG1159-035 (Winget et al 1985). Interpretation of this is currently ambiguous but should eventually test models of neutrino cooling in the cores of hot subdwarfs (Kawaler, Hansen & Winget 1985). Similar work in progress on the ZZ Cet stars will in a few years test the theory of cooling of white dwarfs (Robinson & Kepler 1980, O'Donoghue & Warner 1982).

In recent years, a new demand on small telescopes has arisen - that of monitoring (usually photometrically) objects observed simultaneously by IUE or EXOSAT. This important service will increase with the advent of the HST, AXAF, RHOSAT and other planned astronomical satellites. In order to obtain adequate longitude coverage from the ground good international cooperation and coordination is required. It may even become necessary to build new small and moderate size telescopes dedicated to such uses.

Optical monitoring of Gamma Ray Bursters, described above, is a good example of a project feasible only on small telescopes where large amounts of time are available.

5.4 Flexibility

For coordination with space observations where Target of Opportunity time has been awarded in order to catch an unpredictable event (e.g. a nova eruption) great flexibility of scheduling is required for the terrestrial telescopes. This is most easily satisfied on small telescopes, where trading of time or interruption of programs is more readily accommodated than on large telescopes where time is jealously held. Even without short lead-times, there are observations which require flexibility and exchanges of observing time: examples are planetary and lunar occultations, mutual phenomena of Jupiter's satellites etc.

As one illustration of successful coordination of observations of an unpredictable event I can cite the November 1984 supermaximum of the dwarf nova VW Hyi, which was observed (by prearrangement and only after two false alarms in the form of ordinary maxima) by Voyager 1, IUE, EXOSAT and at SAAO and ESO (Pringle *et al.* 1985).

5.5 Serendipity and Speculation

The large systematic programs, particularly of the survey type frequently undertaken with small telescopes, often disclose unusual behaviour of familiar objects or rare events or unusual objects. Many low amplitude variable stars, eclipsing binaries or spectroscopically peculiar objects have been found this way. A recent discovery, found during routine UBV photometry of hot stars, is that BD $-7^{o}3477$ is a hot subdwarf binary with a period of $2^{h}48^{m}$ and a strong reflection effect resembling the nuclei of some planetary nebulae (Menzies & Marang 1985).

Again, the low cost and relatively low demand for time on small telescopes enables speculative observations to be carried out that may be excluded from large telescopes, either because of pressures to carry out more certainly rewarding observations or because of reluctance to chance wasting time. The discovery of the Crab pulsar is a fine example of the reward of speculative observation.

6. RECOMMENDATIONS

There is no shortage of important research programs for small telescopes; more such telescopes are required, some of them can be had at no cost - they already exist but are not being used fully, sensibly, or at all. Many small telescopes at major observatories are starved of first rate instrumentation; this is criminal negligence - provision of off-set guiding facilities, two-star photometers and microcomputer data capture equipment can convert any small telescope into a major research tool - even at sites with poor weather (Warner 1971, Grauer & Bond 1981). At sites with very good seeing, on which small telescopes capitalize, limiting magnitudes are surprisingly faint - for example the 24 inch at Las Campanas produces spectra at classification dispersion (120 Å/mm) to $\sim$11 mag photographically, $\sim 13\frac{1}{2}$ mag with image tube enhancement and $\sim$16 mag for untrailed spectra. The spectral classifications produced by Corbally (1984a) with this instrument for close visual binaries - with separations 1"-5", all of which were provided with accurate UBV measurements using an area scanner on small telescopes (often operating through thin cloud - Hurly & Warner 1983), have proved useful in checking the latest isochrones derived from stellar evolution computations (Corbally 1984b).

To use a small telescope effectively requires willingness to exploit the special advantages. A basic equation in research productivity appears to be

$$\text{Current Research} = \frac{\text{Research Potential}}{\text{User Resistance}}$$

Small telescopes have the potential; it is up to us to reduce resistance to their use.

ACKNOWLEDGEMENT

This work has been supported in part by a grant from the Foundation for Research Development of the C.S.I.R.

REFERENCES

Beavers, W.I. & Eitter, J.J., 1977. *Publ. Astr. Soc. Pacific, 89,* 733.
Chandresekhar, S., 1946. *Astrophys. J., 103,* 365.
Christensen-Dalsgaard, J., Gough, D. & Toomre, J., 1985. *Science, 229,* 923.
Cocke, W.J., Disney, M.J. & Taylor, D.J., 1969. *Nature, 221,* 525,
Cocke, W.J., Disney, M.J. & Taylor, D.J., 1972. *Adventures in Experimental Physics 1972,* p. 7.
Corbally, C.J., 1984a. *Astrophys. J. Suppl. 55,* 657.
Corbally, C.J., 1984b. *Astrophys. J., 285,* 195.
Cropper, M.S., 1985. *Mon. Not. R. astr. Soc., 212,* 709.
Da Costa, L.N., Pellegrini, P.S., Nunes, M.A., Willmer, C. & Latham, D.W., 1984. *Astron. J., 89,* 1310.
Disney, M.J., 1972. *Mon. Not. R. astr. Soc., 160,* 213.
Elliot, J.L., Dunham, E., Wasserman, L.H., Millis, R.L. & Churms, J., 1978. *Astron. J., 83,* 980.
Faulkner, J., Flannery, B.P. & Warner, B., 1972. *Astrophys. J. Letts., 175,* L79.
Grauer, A.D. & Bond, H.E., 1981. *Publ. Astr. Soc. Pacific, 93,* 388.
Griffin, R.F., 1969. *Mon. Not. R. astr. Soc., 145,* 163.
Groth, E.J., 1975. *Astrophys. J. Suppl., 29,* 431.
Hall, J.S., 1949. *Science, 109,* 166.
Hesser, J.E. & Shawl, S.J., 1985. *Publ. Astr. Soc. Pacific, 97,* 465.
Hiltner, W.A., 1949. *Science, 109,* 165.
Hiltner, W.A. & Schild, R., 1965. *Sky and Telescope, 30,* 144.
Hurly, P.R. & Warner, B., 1983. *Mon. Not. R. astr. Soc., 202,* 761.
Kawaler, S.D., Hansen, C.J. & Winget, D.E., 1985. *Astrophys. J.* In press.
Kemp, J.C., Swedlund, J.B., Landstreet, J.D. & Angel, J.R.P., 1971. *Astrophys. J. Letts., 161,* L77.
Kemp, J.C., Henson, G.D., Barbour, M.S., Kraus, D.J. & Collins, G.W., 1983. *Astrophys. J. Letts., 273,* L85.
Kurtz, D.W., 1978. *Inf. Bull. Var. Stars,* No. 1436.
Kurtz, D.W., 1982. *Mon. Not. R. astr. Soc., 200,* 807.
Kurtz, D.W., 1984. Proc. *Workshop on Improvements in Photometry,* NASA Conf. Publ. 2350.
Kurtz, D.W. & Balona, C.A., 1984. *Mon. Not. R. astr. Soc., 210,* 779.
Landolt, A.U., 1968. *Astrophys. J., 153,* 151.
Lasker, B. & Hesser, J., 1971. *Astrophys. J. Letts. 163,* L89.

Lohsen, E.H.G., 1981. *Astron. Astrophys. Suppl.*, *44*, 1.
Menzies, J. & Marang, F., 1985. In press.
O'Donoghue, D.E. & Warner, B., 1982. *Mon. Not. R. astr. Soc.*, *200*, 563.
Olsen, E.H., 1979. *Astron. Astrophys. Suppl.*, *37*, 367.
Olsen, E.H., 1980. *Astron. Astrophys. Suppl.*, *39*, 205.
Pedersen, H., Danziger, J., Hurley, K., Pizzichini, G., Motch, C., Ilovaisky, S., Gradmann, N., Brinkmann, W., Kambach, G., Rieger, E., Reppin, C., Trumper, W. & Lund, N., 1984. *Nature*, *312*, 46.
Pringle, J.E., Verbunt, F., Hassall, B.J.M., Warner, B., Marang, F., Heise, J., van der Woerd, H., van Amerongen, S., van Paradijs, J., Polidan, R. and others, 1985. In preparation.
Richardson, E.H., 1971. Proc. ESO/CERN Conf. on *Large Telescope Design*, ed. R. M. West, p. 179.
Robinson, E.L. & Kepler, S.O., 1980. In *Stellar Hydrodynamics*, Colloq. No. 53, p. 343.
Shen, L.Z., Beavers, W.I., Eitter, J.J. & Salzer, J.J., 1985. *Astron. J.*, *90*, 1503.
Smak, J., 1967. *Acta Astr. 17*, 255.
Strömgren, B., 1976. *ESO Messenger 7*, 12.
Vogt, N., 1974. *Astron. Astrophys. 36*, 369.
Walker, A.R., 1980. *Mon. Not. R. astr. Soc.*, *192*, 47P.
Walker, M.F., 1954. *Publ. Astr. Soc. Pacific*, *66*, 230.
Walker, M.F., 1956. *Astrophys. J.*, *123*, 68.
Warner, B., 1971. *I.A.U. Colloq. No. 15*, p. 144.
Warner, B., 1974. *Mon. Not. R. astr. Soc.*, *168*, 235.
Warner, B., 1975. *Mon. Not. R. astr. Soc.*, *170*, 219.
Warner, B. & Robinson, E.L., 1972a. *Mon. Not. R. astr. Soc.*, *159*, 101.
Warner, B. & Robinson, E.L., 1972b. *Nature*, *239*, 2.
Wegner, G., 1979. *Astrophys. Sp. Sci.*, *60*, 15.
Winget, D.E., Robinson, E.L., Nather, R.E. & Fontaine, G., 1982. *Astrophys. J. Letts.*, *262*, L11.
Winget, D.E., Kepler, S.O., Robinson, E.L., Nather, R.E. & O'Donoghue, D., 1985. *Astrophys. J.*, *292*, 606.

DISCUSSION

Graham: I want to emphasise the importance of long term standard star programs. They can only be done with small telescopes and yet much of the work done with large telescopes depends fundamentally upon them. Work of this sort, if done well is of tremendous value. The best example is Dr Cousins in South Africa, who with care and by paying careful attention to errors has provided a set of fundamental observations. No similar set of data exists for the northern hemisphere and it is an area in which people with small telescopes could become involved.

Gaustad: Although I agree with your emphasis on the contributions of small telescopes to high energy astrophysics, which we might not otherwise think of, I should like to point out the many contributions of small telescopes in the field of astronomy. They have given us the masses of stars and distances of many stars and the basis of such fundamental knowledge as the mass-luminosity relation.

Djorgovski: Surface photometry of almost anything - galaxies, globular clusters, or comets - is now an almost exclusive province of small telescopes. It may be a misuse of large telescopes to do such things with them.

Schober: Through recent observing experience I have found that small telescopes are not well maintained and that for this reason several nights of a two week observing run may be lost. Additional improvement in observing efficiency (maybe by a factor of 3 to 5) could be attained through automation. This improvement factor relates to pointing accuracy, automatic storage of coordinates and other areas.

Warner: Dedicated, optimised and well-maintained telescopes and instrumentation, which are always in operation, have stable properties and produce an efficient system.

I. *TELESCOPES AND INSTRUMENTATION*

a) *Invited and oral contributed papers*

DETECTOR DEVELOPMENTS FOR SMALL TELESCOPES

David W. Latham and John C. Geary
Harvard-Smithsonian Center for Astrophysics
60 Garden Street
Cambridge, Massachusetts 02138

ABSTRACT. We describe the performance expected from a new generation of CCD imagers being developed by Tektronix, and we indicate the types of applications, instruments, and computer systems that can best take advantage of these devices on telescopes both large and small. Since these new detectors will be able to reach very faint limits even on small telescopes, we also describe cost-effective television viewing systems suitable for object acquisition and guiding.

1. RESEARCH PHILOSOPHY

For several types of research, small telescopes are limited only by the imagination and energy of the users, and by the detectors and instruments that are available. Since the budgets for most small telescope installations are rather limited, the instrumentation and therefore the types of scientific applications must necessarily be limited as well. In this case the following strategy makes sense: choose a research area that is suitable for the interests and abilities of the scientific users, and for the characteristics of the site. At any time, concentrate on just one or two research projects. Build an instrument system that is optimized for these projects. Support a continuing program of improvement to the instrument, the detector system, the computers, and the software. Over the long term, let the instrument system evolve to match the evolution of the particular research projects being carried out.

It is our opinion that often insufficient funds and resources are budgeted for the instrumentation compared to the telescope facility itself. Plan on spending as much (or more!) on the initial investment in the instrument, detector, and computer systems, and budget as much for the continuing development of these systems as for the operation of the facility.

At the Harvard-Smithsonian Center for Astrophysics (CfA) we have tried to practice what we are preaching here. Over the past several years the research areas that we have emphasized at our 1.5-m telescopes have involved redshifts of galaxies and precise radial velocities of stars. These areas complement each other well, because the galaxies are observed in the dark of moon, while stars are observed during bright

J. B. Hearnshaw and P. L. Cottrell (eds.), Instrumentation and Research Programmes for Small Telescopes, 19–31.

time. The two capabilities use the same detectors and computer systems, and very similar reduction software. The main difference is that an echelle spectrograph is used for the stellar work instead of the galaxy spectrograph.

A large project that has utilized this capability is the CfA Redshift Survey and extensions (*e.g.* see de Lapparent, Geller, and Huchra 1986) which is having a major impact on observational cosmology and our view of the large-scale structure of the universe. Another is a survey of halo stars, which promises to make important contributions to our understanding of the structure of our galaxy (Carney and Latham 1985).

For some observatories the above research philosophy may also make sense for large telescopes. Many of the important advances in astronomy have come from major, long-term survey projects, where the work goes much more efficiently if a large fraction of the telescope is dedicated to the program.

2. CCD DETECTORS

Optical astronomy is on the threshold of a major observational revolution. A new generation of charge coupled devices (CCDs) is being developed by Tektronix, and the first devices will see operational use at telescopes starting in 1986. To set the perspective of why we think these new detectors will prove to be revolutionary, we first review the characteristics of the CCDs now in common use in astronomy.

2.1. CCDs Now in Use

CCDs manufactured by General Electric Corporation Ltd (GEC), Radio Corporation of America (RCA), and Texas Instruments (TI) are now in fairly common use at telescopes. The characteristics of these devices are summarized in Table I.

All three are buried channel devices, an architecture which allows good charge-transfer efficiency (CTE) and low dark leakage. In practice the charge transfer on some of the GEC chips is quite clean and well behaved, but the RCA has poor transfer at low light levels, when the charge in a pixel is less than about 1000 electrons. The TI has some idiosyncracies in its design that degrades the charge transfer at low light levels unless special wave forms are used for the clocking signals.

These chips use three-phase, three-level polysilicon gate structures. If the device is unthinned, as in the case of GEC, then it must be illuminated through the gate structure, with large absorption losses as a result. This is why the GEC chip has no blue sensitivity, and rather mediocre response in the red. To get blue and near-ultraviolet response, the chip must be thinned and illuminated from the back. The amount of thinning is rather critical, because there is a strong wavelength dependence on the depth that photons penetrate into silicon before being converted to charge. Both the RCA and TI chips are thinned, but the mounting scheme is quite different. The RCAs are mounted on a glass substrate, while the TIs are stretched as a thin membrane across a frame. As a result the RCAs are quite flat, while the TIs are usually bowed and wrinkled in a way that can change as they are cooled to operating temperature. This effect is large enough to degrade the resolution in very fast cameras, simply because the depth of focus is less than the variation in depth on the chip.

Table I. Characteristics of CCDs Now in use.

Manufacturer	GEC	RCA	TI
Columns x Rows	385 x 576	320 x 512	800 x 800
Pixels, mm	0.022 x 0.022	0.030 x 0.030	0.015 x 0.015
Size, mm	8.4 x 12.7	9.6 x 15.4	12.0 x 12.0
Area, cm^2	1.1	1.5	1.4
Charge transfer	Good	1000 e^- thresh	Good with tricks
Illumination	Front, thick	Back, thin	Back, thin
QE 400 nm	0%	40%	40% (UV or NO)
QE 700 nm	40%	80%	65%
Flatness	Good	Good	Marginal
Availability	Yes, but ...	No !	No, but ...

The RCA chips win very high marks for their outstanding quantum efficiency. Two factors contribute to the extra-high sensitivity in the red. First, the chip is illuminated through the glass substrate, which gives a graded change in index of refraction from vacuum to silicon, which has an index of around 4. Second, apparently RCA has some kind of surface treatment that gives significant gains not realized by TI.

Unfortunately the RCA chips are plagued by high readout noise. This, together with the low-level charge transfer problems, makes the RCA chip a relatively poor performer for spectroscopy. However, the excellent quantum efficiency makes it a winner for deep wide-band imaging, where the sky exposure brings the charge in every pixel well above the threshold needed for good charge transfer.

On the TI chips the blue and ultraviolet sensitivity is quite low, usually well below 10%, unless the accumulation layer is charged up with a technique such as flooding with hard ultraviolet light under modest vacuum during cool down, or chemical activation with nitric oxide gas at low partial pressures. With such treatments the blue response is quite similar to that of the RCAs, and much better in the ultraviolet below about 3600 A, where the glass window on the RCA begins to attenuate seriously.

2.2. Availability

Unfortunately none of the above devices are readily available. Chips can be ordered from GEC, but very few new devices suitable for astronomy have been produced or delivered for the past two years or so. RCA recently announced that their product is discontinued.

The situation with the TI 800 x 800 chips is rather more complicated. TI developed these devices for Space Telescope (ST) but chose not to manufacture any more after the ST contract ended a few years ago. In 1982 we learned that there were several dozen packaged chips that had been rejected for the Space Telescope program. A consortium led by Don York arranged through NSF for a dozen of these reject chips to be made available to a few major observatories. About ten of these chips were finally released in 1984, and most of them are now in use. Attempts to obtain release of additional chips has been thwarted by NASA's decision to build a second Wide Field/Planetary Camera, and essentially all the finished chips have been reserved for this project. However, there exist literally thousands of chips in various stages of completion, mostly unthinned and still on their wafers. In principle it should be possible to carry these chips to completion, perhaps even with a higher yield than the very low success rate experienced in the ST effort. NSF is considering a proposal for a program to complete additional TI chips.

2.3 Tektronix CCDs

More than three years ago Tektronix (Tek), Inc., Integrated Circuits Operation, P.O. Box 500, MS 59-420, Beaverton, Oregon 97077, telephone (503) 627-6101 embarked on an ambitious program to develop a new generation of large scientific imagers. Tek had been working on integrated circuits for more than ten years, and had already demonstrated impressive capabilities for in-house applications. The decision to develop a new line of high-end CCDs for external scientific markets was one of the first examples of a fundamental shift in Tek corporate policy to support venture developments for markets outside the company.

In a key move, Tek was able to hire Morley Blouke, the engineer who had led the technical development of the 800 x 800 CCDs at TI. At Tek, Morley has been able to pick up where he had left off with the ST effort, correcting errors known to exist in the TI design and process. Morley and the team at Tek are dedicated to making imagers that will satisfy the astronomers, partly because they are interested in astronomy, and partly because they know that if they can satisfy the astronomers, they can satisfy anyone. This situation is reminiscent of the symbiosis between the astronomers and E. C. Kenneth Mees at the Kodak Research Laboratories and the development of the special photographic plates used by astronomers.

The general characteristics of the Tek CCDs are summarized in Table II, with additional comments about specific characteristics in the following sections.

2.3.1. *Pixel size.* A pixel size of 0.027 x 0.027 mm was chosen for the Tek CCDs, partly to obtain a large full-well charge, 700,000 e^-, and thus a large dynamic range, and partly with an eye towards the giant telescopes of the future, where the focal lengths will be very long even with extremely fast optics.

2.3.2. *Array size and format.* The most important thrust of the Tek project is to build CCDs far larger than any other integrated circuit yet attempted. The CCDs now in use have areas of about 1.5 cm^2, while Tek is building their standard devices in two sizes, one with an area of

1.9 cm^2, and the other with the astounding area of 30.6 cm^2. Previously the largest CCD attempt that we know about was an effort at GEC supported by the Anglo-Australian Telescope, with an area of about 9 cm^2, and although this effort has produced one or two devices that make images, they are not suitable for astronomical use. The geometries of the two standard Tek configurations are summarized below:

Pixel format	512 x 512	2048 x 2048
Chip size, mm	13.8 x 13.8	55.3 x 55.3
Chip area, cm^2	1.9	30.6

2.3.3. *Charge Transfer.* The specification on the efficiency of charge transfer is 0.99999 or better. This is barely good enough for the large chip, where the charge from a pixel in the far corner must be shifted up the parallel register 2048 rows and then across the serial register 2048 columns before it reaches the on-chip output amplifier. An efficiency of 0.99999 raised to the power 4096 is 0.96, which means there would be a multiplicative shading of 4% from the corner opposite the output amplifier.

2.3.4. *Quantum efficiency.* The Tek CCDs will be available both thick for front illumination, and thinned for back illumination. Eventually the blue-ultraviolet quantum efficiency of the thinned chips should be about 40% with no special treatment required. However, for the chips produced initially it will probably be necessary to invoke the ultraviolet flooding or NO gas treatments that have proven so effective with the TI 800 x 800 chips. In the red the quantum efficiency of early production chips may not exceed 50% by very much. Within a year or two Tek hopes to offer a coating service that should increase both red and blue sensitivity.

Table II. Summary of the general characteristics of the Tek CCDs.

Pixel size	0.027 mm
Full well charge	700,000 e^-
Charge transfer	0.99999
Illumination	Front thick, or back thin
QE at 400 nm (thin)	40% (may require UV flood or NO)
QE at 700 nm (both)	50% (higher with coatings)
Flatness, mm	0.001 local, 0.05 global

2.3.5. *Mounting.* All the Tek chips will be mounted on a kovar substrate. The original plan was to use a ceramic substrate, but tests showed that this material resulted in spurious dark events due to radioactive decays.

The mounting scheme makes it possible for the Tek chips to be quite flat, an important characteristic when the chip is illuminated by very fast optics with a shallow depth of focus. The substrate also provides an important heat conduction path during cooling. Indeed, the package is robust enough to survive extreme temperature cycling, such as immersion in liquid nitrogen.

2.3.6. *Amplifiers*. The Tek chips have a serial-parallel-serial architecture, so that they can be clocked to either of two different amplifiers located at opposite corners of the array. One of the amplifiers is meant for high-speed readout at video rates, so that an image can be read in a second or less. The other amplifier is optimized for slow-scan low-noise operation, and would normally be used to read a science image. The slow amplifier has already been demonstrated to achieve a readout noise of better than 10 e^-, and the goal is to achieve 3 e^- or less with further development in the future. The fast amplifier has a readout noise of about 20 e^-, which is quite good considering the high speed of operation.

Amplifier	Slow	Fast
Read rate	50 kHz	10 MHz
Read time 512/2048	5 s / 10 s	0.03 s / 0.4 s
Read noise	10 e^-	20 e^-

3. APPLICATIONS AND INSTRUMENTATION

The development of the large Tek CCD is such an important advance in optical detectors that we expect a whole new generation of instruments will be designed around it, both for direct imaging and for high- and low-resolution spectroscopy.

3.1. Direct Imaging

In this section we make some general remarks about the types of imaging applications that we think the large Tek CCD will be likely to see on various sizes of telescopes. As a guide to these remarks, we outline in Table III the relationship between the angular size of a pixel, the field of view covered, and the focal ratio required for the imaging optics as a function of telescope diameter.

The first, rather obvious conclusion that can be drawn from Table III is that giant telescopes will never be used to survey large areas of the sky at low resolution, since this job can be done in the same exposure time using a small telescope with the same focal ratio but wider field and lower resolution. Put differently, there are no focal ratios entered in the lower right-hand corner of the table, because such fast optics are not practical.

More generally, Table III can be used as a guide for deciding what diameter telescope is most competitive by choosing the angular resolution that is needed and then selecting the telescope diameter that has the fastest focal ratio that can be achieved. Of course, the actual resolution achieved can be no better than 2 or 3 times the pixel size. For projects at a given resolution, the telescope with the fastest possible optics, *i.e.* with the largest possible diameter, will require the shortest exposure times.

Table III. Focal ratio as a function of pixel and field size.

Pixel arc-sec	Field arc-min Chip 512	2048	F/ratio Telescope Diameter, m 1/4	1/2	1	2	4	8
1/4	2.1	8.5	100	50	25	12	6	3
1/2	4.3	17	50	25	12	6	3	1.5
1	8.5	34	25	12	6	3	1.5	
2	17	68	12	6	3	1.5		
4	34	136	6	3	1.5			
8	68	273	3	1.5				

However, smaller telescopes can be competitive whenever the exposures are short, when the elapsed time for an observation is dominated by overhead. This will normally be the case for brighter objects, especially when the objects are scattered all over the sky.

Monitoring projects also seem suitable for small telescopes, although here the reason is not so apparent from Table III but seems to depend more on the tastes of time allocation committees and the flexibility of scheduling the actual observing time.

The small Tek CCD can be competitive for isolated objects or spectroscopic studies involving isolated spectral features. The combination of a small chip on a small telescope thus seems best suited for monitoring of isolated bright objects, especially spectroscopic monitoring of isolated spectral features.

The niches that we see for CCD imaging with different diameter telescopes are summarized in Table IV.

Table IV. CCD Imaging Niches for Different Telescope Classes.

Giant	4-8 m	Very faint QSOs, stars, and other point sources. Excellent seeing is critical.
Large	2-4 m	Faint galaxies; surveys of faint QSOs and stars.
Medium	1-2 m	Surveys of galaxies; surface photometry of extended objects; monitoring.
Small	<1 m	Wide angle surveys at low resolution; surface photometry of very extended objects at low resolution; monitoring bright objects.

3.2. Spectroscopy

The potential of the large Tek CCD for spectroscopy is enormous and varied. Undoubtedly many spectrographs will be designed specifically around this detector. We mention here just two examples; a new high-resolution big-beam echelle spectrograph, and a lower-dispersion spectrograph for measuring the redshifts of galaxies. To provide a perspective on the improvement that these instruments should provide, we compare the expected performance to that of existing workhorse spectrographs now in use at Mt. Hopkins.

3.2.1. *High-resolution big-beam echelle spectrograph*. Because the large Tek CCD has so many pixels, it is feasible to build a grating cross-dispersed echelle spectrograph that gives a factor of two in wavelength coverage with a pixel size of about 2 km s^{-1} and a resolution of roughly 5 km s^{-1} more or less independent of wavelength in the range 3,100 to 10,000 A. There are even enough pixels to give some sky coverage between orders for point sources. For the largest echelle gratings now available the entrance slit can be as wide as about 1 arc sec even on a giant telescope.

For projects where wide wavelength coverage is needed, this instrument would provide a spectacular improvement over the echelle spectrographs and photon-counting Reticons now in use at CfA (Latham 1982). Instead of covering just 3000 km s^{-1} of spectrum in a single order, the Tek CCD would improve the spectral coverage by a factor of 50! In addition there is an important gain expected in the detective quantum efficiency (DQE). The measured peak detective quantum efficiency (DQE) of the photon-counting Reticons is only about 0.11 (Latham 1982), so just the higher quantum efficiency of the Tek CCDs should give at least a factor of five at all levels of signal to noise except the very lowest. Just as important, the Tek CCD would extend the wavelength coverage into the ultraviolet, an extremely important region that is not covered by our photon-counting Reticons because of the low ultraviolet transmission of the input fiber-optic face plates.

On a large telescope such as the MMT, this spectrograph could be used for detailed abundance studies of faint stars such as dwarfs in globular clusters and giants in other galaxies. An especially exciting application would be for high signal-to-noise studies of weak absorption lines in high redshift QSO's. Such studies could use much weaker lines to study the abundances of a wide variety of elements in different stages of ionization, thus giving a much clearer picture of the chemical composition in primordial intergalactic gas clouds and the halos of galaxies at much earlier times in the history of the universe.

On smaller telescopes this spectrograph would be a powerful tool for studying the chemical abundances in brighter stars and in the interstellar medium of our galaxy.

3.2.2. *Galaxy spectrograph*. We would also like to build a CCD galaxy spectrograph to replace the Z-Machine that has been used for the CfA Redshift Survey. Once again we would get at least a factor of 5 improvement in peak quantum efficiency, plus extended wavelength coverage into the ultraviolet so that the H and K lines of ionized calcium could be included. In addition, the two dimensional coverage would allow imaging along a tall slit, thus giving optimum coverage of extended objects and the nearby sky. This is worth at least a factor of root two, just by the better sky subtraction that it allows. In crowded fields of galaxies, a multiple slit could be used to observe several objects at once.

This spectrograph would allow a medium telescope such as the 1.5-m Tillinghast Reflector at the Smithsonian's Whipple Observatory to extend the CfA Redshift surveys to the limit of the Zwicky Catalog and well beyond. The tall-slit capability would also be useful for mapping velocity fields in extended galaxies and for spectrophotometry of supernova remnants.

4. COMPUTER SYSTEMS

The large Tek CCD will present a major challenge to modern computer hardware and software systems. At 16 bits per pixel, one frame will comprise 8 MBytes, and with improved readout noise the full dynamic range can potentially go as high as 17 or 18 bits. Fortunately, fast 32-bit workstations are now available, and these seem well suited for the care and feeding of the large Tek CCD at the telescope and for quick-look analysis and archiving of the data. They are also suitable for the reduction of the data later on. The investment in hardware can be similar to the cost of a large chip, when all the peripherals necessary for archiving and image display are included. The cost of software development is potentially much higher, although these costs are often hidden in the salaries of astronomers and computer programmers.

4.1. Hardware

There are many different strategies that can be adopted for the computer hardware. At CfA we have chosen a VME bus computer to provide the large amount of random-access memory needed to hold an entire frame during readout. The VME computer is in turn linked to a fully equipped VAX-Station II. In effect the VME machine serves as a memory buffer between the CCD and the VAX-Station II. The real time operating system for the control of the detector and instrument resides in the VME computer, while the astronomer works at the VAX-Station II, controlling the operation, checking the data as they come in, archiving the data, and doing some quick-look reductions.

The interface between the CCD camera and the VME computer is customized, while the interface between the VME machine and the VAX-Station II uses DRV11W boards for fast data transfer and RS232 serial lines for control.

4.2. Software

At CfA we have chosen to run the VAX-Station II under a version of UNIX, rather than under VMS. The reason is that VMS is a proprietary operating system that can not normally be transported to other computers. In contrast, UNIX is an operating system that is being implemented and supported on all the 32-bit workstations. By adopting UNIX we leave open the possibility of simple conversion to a future workstation that is not manufactured by Digital Equipment Corporation.

For image processing and data reduction we hope to run under IRAF, since this promises to be the environment of the future for most astronomical computing centers, including NOAO and STSI. Early pre-release versions of IRAF have been somewhat disappointing, but we assume that this can only get better. The goal is to be able to take advantage of applications programs developed elsewhere, and to provide the same computer environment used at the major centers so that users will be familiar with the software. All applications developed at CfA will be coded in a high level language, usually C or FORTRAN.

4.3. Low-budget approach

For many scientific applications the 512 x 512 Tek CCD should be quite adequate. In this case a low-budget approach could be adopted that would reduce the total costs substantially. The key would be to use a computer system based on a PC, both for instrument control and for image processing. The disadvantage of this approach is that IRAF and standard applications software could not be used, but the advantage would be a much lower hardware cost, perhaps 10 to 20K US dollars instead of 30 to 75K. For example, image display boards with 512 x 512 pixels and 8 grey levels are now available for PCs at a cost of about 2K, and quite large hard disks are available for similar amounts.

The real cost of this approach would be the special effort required to develop the interfaces and software systems, but this could be handled by staff members or perhaps by students.

4.4. System costs

Complete Tek CCD detector systems will not be cheap, because there are many items beyond just the chip itself. Some rough estimates of the direct costs of building a system in house are listed in Table IV, but the indirect costs for personnel such as astronomers, engineers, machinists, technicians, programmers, and students are not included.

Table IV. Rough cost estimates for Tek CCD instrument systems. In thousands of US dollars, direct costs only.

Item	512 x 512	2048 x 2048
Chip; front vs. back; quality	3 - 7	25 - 80
Dewar for CCD	4 - 8	6 - 12
Vacuum pumping station	5 - 10	10 - 15
Computer hardware	10 - 15	50 - 75
Camera electronics	2 - 4	5 - 20
TV viewing system and guider	6 - 8	10 - 15
Spectrograph	10 - 30	50 - 200
Total	40 - 82	156 - 417

If an observatory does not have the technical expertise to build a CCD camera, then it will probably be possible to purchase a system commercially. However, the direct cost of the initial purchase will be higher.

5. INTENSIFIED TELEVISION VIEWERS

With modern detectors small- and medium-sized telescopes can reach so faint that object acquisition and identification can be a major problem. An intensified television viewing system can become an essential part of the total instrument system, not only because it allows the observer to see fainter objects, but also because it brings one from the cold and dark of the dome into the warmth and light of the computer room where various aids such as finding charts and coordinates of offset stars are conveniently spread out. Just as important, a television system makes it possible to implement an automatic guider.

At CfA our intensified television viewers have evolved considerably over the years. Originally we followed the design described by researchers at the Steward Observatory (Cromwell and Angel 1979; Angel, Cromwell, and Manger 1979) and used three stages of selected Varo Gen I 25 mm tubes coupled with fiber optics to a Vikon made by Teltron, essentially a vidicon with a target whose resistivity depends on charge and therefore on light level. With this readout there is an analog integration on the target that can be as long as 10 or 20 s in areas of very low illumination. For the vidicon electronics we used standard cameras made by Cohu, being careful to provide a good ground shield between the intensifiers and the front end of the camera.

More recently we have changed both the intensifiers and the television readout. We use a single stage of Varo Gen I 25 mm intensifier in front of a Varo Gen II 25 mm MCP intensifier. We have chosen a Varo Gen I tube for the first stage in order to get the highest possible DQE. According to Cromwell *et al.* (1985) MCP tubes give away nearly a factor of three in DQE because two thirds of the photoelectrons released by the cathode never make an output flash at the phosphor. The tubes are encapsulated in GE RTV 511, and the tubes are coupled with fiber optics which are glued at each joint using Lens Bond. An important aspect of the fiber-optic boules is the transparent conducting coatings, which are tied electrically to the appropriate voltages. These coatings bleed off any high-voltage leakage in a harmless way. The coating on the boule in front of the readout also protects against noise pickup.

For the power supplies we use the standard oscillators and multipliers available on special order from Varo. A 15,000 MOhm bleed resistor is used across the high voltage for the Gen I tube, and this provides a time constant of 1 or 2 s for the front tube when the power is shut off. To hold various components in place during the potting operation, we use dabs of Dow Corning 732, since it is compatible with GE RTV 511.

For the television readout we are now using Fairchild 3000F CCD cameras. These cameras come with a fiber-optic input and a compact remote head. As a result it is possible to make a very compact viewer that we have mounted in place of the standard eyepiece on our spectrographs and offset guider stages. This system can easily see single electron events from the first cathode. The output from the Fairchild CCD is very clean, and can be fed into television monitors or directly into digital electronics for integration and/or automatic guiding.

6. INFRARED ARRAYS

The technology of infrared array detectors is experiencing very rapid evolution. One of the most significant recent developments has been the commercial availability of a 58 x 62 indium-antinomide array from Santa Barbara Research Center (SBRC). The impact of this detector will be even more dramatic for infrared imaging and spectroscopy than the Tek CCDs in the optical, because these arrays will be replacing single-element detectors, thus giving an improvement of something like 3000 for many applications. In addition, it should be much easier to get good geometrical registration and photometric accuracy because the elements of the array are all exposed simultaneously. For many applications it may not be necessary to chop.

The SBRC arrays have quantum efficiencies of about 0.5 over the spectral region 1 to 5 microns. Interestingly, the noise equivalent power of individual elements on these arrays is better than for single element detectors from the same company. The first deliveries of production devices are scheduled for late 1985, at a cost of 100K each.

ACKNOWLEDGMENT

We wish to pay tribute to Peter Crawford and Charles Hughes, the two loyal members of our instrument development team who have made possible many of the systems described in this paper. We would also like to acknowledge the fundamental contributions made by Bill Wyatt to the development of our computer systems.

REFERENCES

Angel, J. R. P., Cromwell, R. H., and Manger, J. 1979 *Adv. Electronics Electron Phys.* **52**, 347.

Carney, B. W. and Latham, D. W. 1985 in *IAU Coll. No. 88, Stellar Radial Velocities*, A. G. D. Philip and D. W. Latham, eds., L. Davis Press, Schenectady, p. 139.

Cromwell, R. H. and Angel, J. R. P. 1979 *Adv. Electronics Electron Phys.* **52**, 183.

Cromwell, R. H., Strittmatter, P. A., Allen, R. G., Hege, E. K., Kuhr, H., Marien, K.-H., Funk, H. W., and Frank, K. 1985 *Adv. Electronics and Electron Phys.* **64A**, 77.

de Lapparent, V., Geller, M. J., and Huchra, J. P. 1986 *Astrophys. J. Lett.* submitted.

Latham, D. W. 1982 in *IAU Coll. No. 67, Instrumentation for Astronomy with Large Optical Telescopes*, C. Humphries, ed., Reidel, Dordrecht p. 259.

DISCUSSION

Kulkarni: What is the availability of the new Tektronix chip?

Latham: The first 512 x 512 devices will be delivered this month (close to schedule). Larger devices will be available in March or April 1986.

Gaustad: Many of us not only have small telescopes but also come from small institutions without much engineering and/or technical support. Do you think CCD systems will ever get to the point where complete systems are available and the reliability is high enough that we can acquire and operate them without becoming engineers ourselves?

Latham: Yes, but large capital investment required (the RCA system costs about $US70,000).

Garrison: An inexpensive way to solve the acquisition problem for small telescopes is a 3-stage image tube eyepiece, which can be put together for ~$US1500. We have had such an "Ojo Magico" on our 24 inch telescope in Chile since 1973 and it allows us to reach 17-18 mag easily.

Latham: There are at least two advantages when the intensified viewer has a television readout. First, the observer does not have to go into the dome in order to guide. Instead she can sit comfortably in a warm room in front of her computer. Second, the television frame can easily be digitized and used for automatic guiding. For example, at our 0.6 m CCD imaging telescope, no-one guides by hand anymore. Indeed, if the television systems are not working, we consider the telescope to be down, and we fix the system before we proceed with observing.

CCD PHOTOMETRY WITH SMALL TELESCOPES

Alistair R. Walker
South African Astronomical Observatory
P.O. Box 9
7935 Observatory
South Africa

ABSTRACT. CCD photometry with small telescopes often complements the use of such detectors on larger telescopes and in addition it is ideally suited to projects, such as determining variable star light curves, that require considerable amounts of observing time. Following a discussion of some of the techniques found useful for producing accurate CCD photometry, the results of several projects carried out with the SAAO 1m CCD camera are presented.

1. INTRODUCTION

In the past few years CCD cameras have produced some spectacular results, both in detecting fainter objects than ever seen before and also in allowing accurate photometry, free of systematic errors, to very faint limits. Colour magnitude diagrams of galactic globular clusters, reaching far down the main sequence, is just one example of the type of observation where the use of CCDs have made a major impact by producing results accurate enough to allow detailed comparison with models (eg Harris & Hesser 1984). In general, most of this work has been done with large telescopes in the 4m class.

Concentrating on stellar photometry through broad band filters, I shall attempt to show that a CCD camera system on a 1m class telescope can do valuable astronomy and indeed complement the use of CCDs on larger instruments, much of the work on which tends to be of observations with low S/N on very faint objects. An example of this symbiosis is the colour magnitude diagram for the globular cluster NGC 6752 given by Penny (1984), where the photometric zero-points and the stars to V ≃ 19 were measured using the SAAO 1m telescope while the fainter stars to V ≃ 23 were measured on the AAT.

A 1m telescope plus CCD can do accurate broad-band (B,V,R,I) photometry (few percent or better) down to magnitude ≃ 20 given reasonable seeing with exposure times under 30 minutes. Fainter magnitudes can be reached but require much longer exposure times and/or extremely good seeing, and it is clearly a lot more efficient in these circumstances to use a larger telescope.

J. B. Hearnshaw and P. L. Cottrell (eds.), Instrumentation and Research Programmes for Small Telescopes, 33–46.

CCD photometry is free from the systematic errors that plague photographic photometry. This is partly due to the inherent linearity of the CCD itself, but is also due to standards and program stars being measured in a similar manner, without the need for sequence extending devices or measurement of standards and program stars by different techniques. The quality of photometry in the magnitude range V = 12 - 20 that can be done with a CCD camera on a small telescope is clearly superior to that achieved either by photoelectric or photographic photometry virtually independent of telescope aperture, provided sufficient attention is paid to the calibration and standardization problems. In some work that has been published the natural CCD photometric system has differed greatly from the standard system to which the photometry was reduced. This is unfortunate since the chances of (poorly determined) errors being introduced are high.

CCD systems on the (few) large telescopes tend to be used to study faint objects, while time allotments are mostly short and almost always difficult to get. For smaller telescopes time allotments are commonly scheduled in units of weeks rather than days; this is of great advantage for instance in the study of variable stars with periods in the fraction of a day - few days range and also means that colour equations can be adequately checked during the course of an observing run. Longer observing runs also mean that the astronomer can come to the telescope with several programs and select the one best suited to the combination of sky transparency, sky brightness and seeing at any given time.

The disadvantage of a CCD system on a small telescope relates mainly to its cost. Over the last few years instruments have become more complex and a larger fraction of the total cost of an installation. CCDs are no exception, particularly if the cost of an autoguider and computer system are included, while the best detectors are expensive. The computer power necessary to cope with large CCDs is considerable, particularly as some degree of on-line reduction is desirable to minimise data storage problems. A powerful computer in the VAX class is required for data reductions however software packages such as DAOPHOT and GRASP (STARLINK) have been widely distributed to astronomical institutions. Olszewski & Aaronson (1985) give an excellent description of the use of DAOPHOT.

2. SOME TECHNICAL ASPECTS

2.1. Pixel scale

CCDs in common use have pixel sizes ranging from 22μ to 30μ. The best seeing at good observing sites is generally about one arc second or a little better and therefore to cover star images with several pixels in order to fully utilize any good seeing requires a pixel size no greater than about 0.5 arc sec which implies a focal length of 10 - 15 metres. As expected, tests have shown that photometric accuracy falls off rapidly as the star images become under sampled. For work that does not require the small pixel size (eg galaxy surface photometry) shorter focal lengths can be used in order to cover a greater area of sky.

2.2. Exposure times

Exposure times are a function of brightness of the object, seeing, CCD readout noise and sky brightness. If integrations are long enough so that sky noise dominates over readout noise then multiple exposures can be made and later summed so that cosmic ray events can easily be recognized, internal photometric errors can be found and the effective dynamic range increased. For a small telescope working at f/10 to f/20 this condition can not always be achieved even with broad band filters and is usually impossible with medium and narrow band filters. The use of CCDs with as low readout noise as possible is therefore even more important for small telescopes than it is for large ones.

2.3. Cleaning CCD frames

This topic has been dealt with in great detail by Djorgovsky(1984). The SAAO practice has been to use the twilight sky for flat fields. The DC bias on each frame is estimated by recording a strip of overscan along with each frame. The preflash (if used) is calibrated by an exposure typically 100 times longer than that applied normally; the preflash is uniform to ± 10 percent across the frame. Subtraction of dark frames and defringing is carried out only if neccessary. The latter is usually only required for I band exposures of 30 minutes or more duration.

2.4. Filters

Duplication of the Johnson UBV and the Kron-Cousins RI systems is possible given that the CCD in use covers the bands (The GEC and Thomson CCDs cover VRI only and the RCA CCDs cover only part of U). Specifications for the UBVRI filters in use at SAAO with the RCA CCD and the resulting colour equations are given in the Table I. The use of methylene blue in the U filter was suggested by Dr D.D. Walker (UCL), while all the filters include WG280 glass to give a total thickness of 7mm.

TABLE I. Filter specifications and colour equations

U: 4mm 28% $CuSO_4$ soln. + 4mm 0.0025% methylene blue soln. + 2mm UG1
B: 2mm GG385 + 1mm BG18 + 1mm BG12 + 2mm KG3
V: 2mm GG495 + 2mm BG18 + 2mm KG3
R: 2mm OG570 + 2mm KG3
I: 3mm RG9

$U-B = 1.002\ (u-b)$ + small non linear term for $U-B < 0.1$
$B-V = 1.090\ (b-v)$
$V-R = 1.030\ (v-r)$
$V-I = 0.975\ (v-i)$
$V = v$

It cannot be over-emphasised enough that great care should be taken

out or the defocusing of bright stars are not in general to be recommended since one of the major advantages of the CCD is that the standard star frames and the program star frames can be measured in precisely the same way. Any departure from this must introduce a systematic error at some level.

An alternative to measuring standards singly is to have accurately measured stars in small CCD sized fields with a wide range in colour and a small range in magnitude so that zero-points and colour equations can be determined from only a small number of exposures. Such fields have been set up in the north (Christian et al. 1985, with extensions by Lindsey Davis (KPNO) in progress) but the stars are mostly faint and better suited to large telescopes and in some fields the colour range is limited. The same is generally true of the fields in the south set up by Stobie et al. (1985). A brighter set of stars with a range in B-V from -0.16 to 1.35 have been measured in M67 by Schild (1983), while for Strömgren photometry the accurate work in NGC 3293, 4755 and 6231 by Shobbrook (1983a, 1983b, 1984) is useful.

3.2. Reduction procedure

The following is a brief outline of the reduction procedure necessary in order to measure the brightness of stars on "cleaned" CCD frames (Walker 1984a): (a) Determine colour equations, zero-points and extinction from observations of standard stars, using digital aperture photometry. (b) Measure one or more bright stars on each frame, again using aperture photometry. These will be the local standards. (c) Choose a bright isolated star(s) on each frame and fit a profile to it (Gaussian, Lorentzian, spline etc.). (d) Fit this profile to all the program stars plus the local standards. (e) Find the offset between the aperture photometry and the profile fitting photometry for the local standards. (f) Apply this offset to the program star photometry.

Only steps (c,d,e,f) are necessary once local standards have been set up or if they are already available, in which case observations can be carried out in non-photometric conditions. Given sufficient attention to step (a) above, step (b) will usually limit the accuracy of the procedure at about the 0.01 - 0.02 mag level.

3.3. Faint Star Photometry

For faint star photometry the (internal) errors are not dominated by photon statistics but by the accuracy of the flat fielding, the ability to take account of CCD defects and cosmic ray events, defringing for some spectral bands and the capability of the algorithm used to fit the star profile and the background, especially in the multi-star case.

A field 8 arc min from the centre of the globular cluster ω Centauri has been observed, with BVRI magnitudes being found for 200 stars within a 68 arc sec square area down to V = 23 in order to provide faint calibrating stars for the Space Telescope Wide Field Camera and Planetary Camera. The data can also be used to construct colour magnitude diagrams for the cluster (Figures 3).

Table II gives mean errors as a function of magnitude for the summed

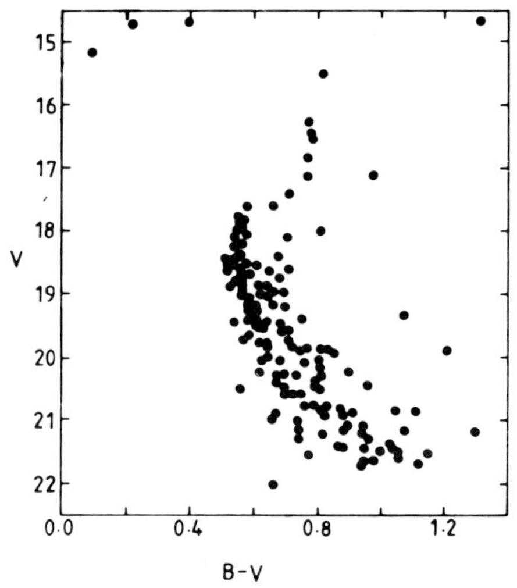

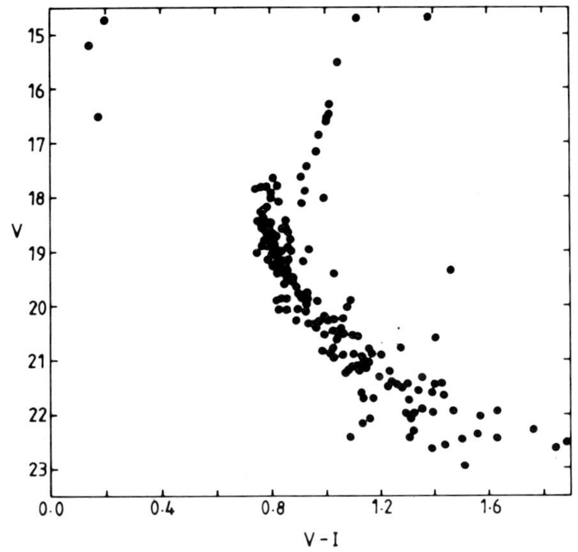

Figure 3. (a) V/(B-V) CM diagram for a field in ω Centauri. (b) V/V-I CM diagram for the same field as in (a).

frames. These were obtained for V by comparing 2500s and 3000s exposures (made on different nights) and a 3000s exposure made two months earlier while for B-V and V-I two successive 2000s exposures were compared. Comparing measurements on repeated (preferably slightly spatially displaced) frames is probably the best way of finding the error as a function of magnitude, although creating artificial stars of known brightness and measuring them along with the program stars has the advantage of also allowing estimates of completion if an automatic star finding program has been used.

The colour spread of the main sequence stars in Figure 3a must be intrinsic, considering the errors in Table II, and indeed is scarcely

Table II. V, B-V and V-I errors as a function of magnitude.

V	σ(V)	σ(B-V)	σ(V-I)
15-18	0.01	0.01	0.01
18-19	0.02	0.01	0.01
19-20	0.03	0.02	0.015
20-21	0.04	0.04	0.02
21-22	0.06	0.06	0.04
22-23	0.08		

less than that found in the photographic study by Cannon & Stewart (1981) even though the photometric errors here are at least a factor of two smaller. This result confirms that of Rodgers & Harding (1983) and is also found in a more extensive study by Dr R. D. Griffiths (Leeds) and co-workers, and shows conclusively that the ω Centauri metallicity variations are primordial. The lack of blanketing in the I band compared to the B band is responsible for the reduced colour spread in Figure 3b.

3.4. Variable Star Light Curves

3.4.1. Cepheids in NGC 1866. NGC 1866, a rich young cluster in the LMC, contains several Cepheid variables all with periods near 3 days. Field Cepheids in the LMC have been used to define the slope of the PL and PLC relation (eg Martin et al. 1978, Caldwell & Coulson 1985). These latter stars almost all have periods longer than the galactic Cepheids in the calibrating clusters and it is important to check whether or not the slope of the PL and PLC relations is the same for both the shorter and longer period stars.

Typical light curves for the NGC 1866 stars are given in Figure 4. The quality of the present results is better than the photographic light curves by Arp & Thackeray (1967) and serve to show that there are significant systematic errors in the photographic photometry. Although analysis of the light curves is not yet complete, it appears that there may be a slight change in slope of the PL and PLC relation for short period Cepheids. This is in the sense of making the LMC closer by ≃ 0.1 mag than if a linear relation is used. There is only a small spread in <B>-<V> colour, in contrast to the spread of 0.2 mag at any given period found by Martin et al. (1979) for stars with periods over 10 days.

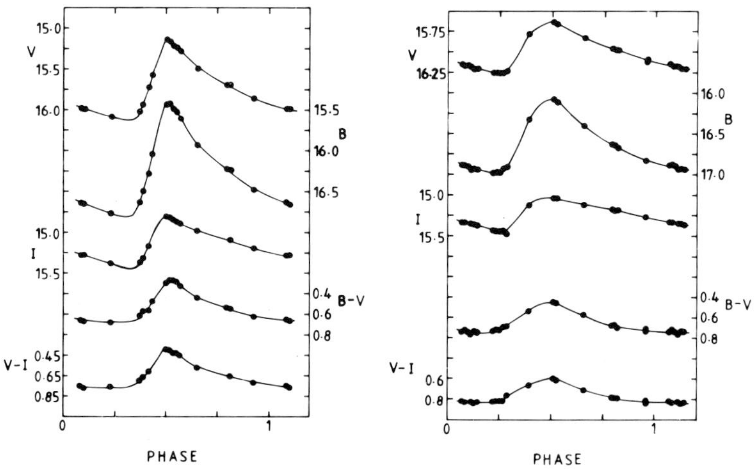

Figure 4. (a) Light curve for the Cepheid NGC 1866f (period 3.439 d), (b) light curve for NGC 1866g (period 3.523 d)

3.4.2. Light curves of RR Lyrae stars in the NGC 6522 Baade window. The star density of even relatively bright stars in this region is extremely high making any sort of photometry extraordinarily difficult. For the CCD work to be a significant advance in accuracy over other methods requires good seeing when observing and the use of a multi star fitting algorithm when reducing the data. Even so, the light curve quality is poorer than that achieved for stars of similar brightness in less crowded fields. Previous photometry in Baade's window (Gaposchkin 1956, Arp 1965, van den Berg 1971, Hartwick et al. 1972, Blanco 1984) suffers from systematic errors of varying size brought about either by the use of less than ideal techniques when setting up the photometric sequence or by difficulties calibrating the photographic plates used to measure the RR Lyraes.

CCD B and V measurements were made of a standard sequence (Blanco & Blanco 1984) and 13 variable stars in BW; light curves for Baade 118 are shown in Figure 5. The RR Lyrae colours imply that $A_V = 1.78 \pm 0.10$ near NGC 6522, in good agreement with the Mira result (M.W. Feast, private communication) and somewhat larger than usually assumed. A small correction was applied to the Blanco (1984) RR Lyrae magnitudes and hence a value of $R_0 = 8.1 \pm 0.4$ pc was derived, assuming $\langle M_V(RR)\rangle = 0.6$. The results tend to support M_V = constant rather than M_V being a function of [Fe/H] (Walker & Mack 1985).

3.4.3. V84 and V85 ω Centauri. Feast (1985) has plotted ⟨V⟩ as a function of [Fe/H] for the RR Lyrae stars in ω Centauri using the metallicity measurements of Butler et al. (1978) and Freeman & Rodgers (1975). He suggests that the data may indicate that M_V = constant for the metal poor stars but a magnitude-metallicity relation may exist for the more metal rich stars. However the most metal rich star, V84, with [Fe/H] = -0.51, is almost 0.5 mag brighter than the other metal rich stars. CCD photometry of V84 and V85, the latter having [Fe/H] = -1.16,

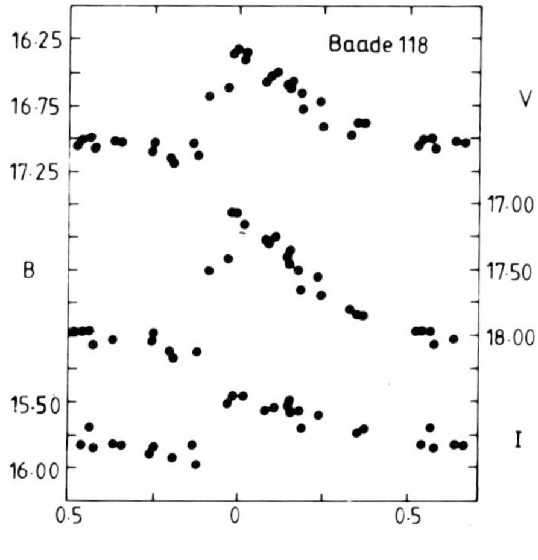

Figure 5. Light curve for the galactic bulge RR Lyrae star Baade 118.

was made in order to check on the photometry.

The light curves are given in Figure 6; the mean magnitudes are <V> = 14.25 for V84 and <V> = 14.44 for V85. These results confirm the earlier measurements, however V84 has an unusually low amplitude and a possible interpretation of the results is that it is a double star; if the variable component actually has <V> = 14.75 then a light curve is produced with normal amplitude. An alternative explanation, due to Sandage (1981), is that V84 is an overtone pulsator, but it would then be expected to have a sinusoidal light curve with very low amplitude. This is not observed. Spectroscopy of V84 should allow a definite decision to be made as to which postulate is correct since the companion star is expected not to be much fainter than the RR Lyrae and should have constant radial velocity.

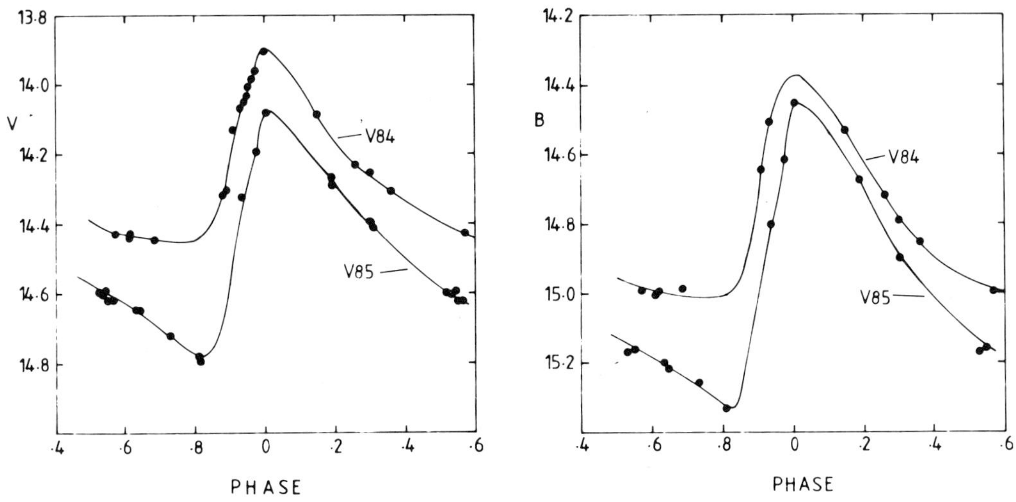

Figure 6. (a) V light curves for V84 and V85 ω Centauri, (b) B light curves for V84 and V85 ω Centauri.

3.5 High Speed Photometry

In normal operation the time resolution of a CCD is limited, by the readout time, to several seconds even if the data is prebinned into "super-pixels". Higher time resolution is possible by slowly reading the CCD out as the integration proceeds (Jacoby et al. 1985). In both cases the advantage of the CCD is that comparison stars are measured simultaneously with the program star and good data can be obtained in non-photometric conditions. In addition, stars with close companions for which normal single channel photometry is difficult can be measured. Data are usually readout noise limited because of the short exposures.

Examples of results achieved by both these methods are given by Walker (1984b) and Jacoby et al. (1985).

3.6 High Resolution Photometry

Studies of the light profiles of globular clusters are being made by P.M. Lugger (Indiana) and collaborators with the KPNO 0.9m telescope and CCD camera in order to search for light spikes at the cluster centres while D.D. Walker (UCL) and collaborators are examining the light profiles of quasars; these are but two examples of the high resolution work that is ideally suited to CCDs. The use of resolution improving techniques such as the maximum entropy method (MEM) are possible with CCD data since all stars on a given frame have the same profile. Therefore any suitably bright star can be used as a profile star, and in addition the errors are quantifiable. Walker & O'Donoghue (1984) used MEM on summed V and I frames of R136 taken in 1 arc sec seeing to achieve a final resolution of 0.4 arc sec (the pixel size) and found 22 stars with V magnitudes between 11.4 and 15.6 within 4.6 arc sec of R136a. The 0.5 arc sec central double R136a$_1$ - a$_2$ (Innes 1927) was also confirmed. The improvement achieved is shown in Figure 7. These results complement the holographic speckle interferometry of the stars within 0.7 arc sec of R136 by Weigelt & Baier (1985) who achieved 0.09 arc sec resolution. Together with the spectroscopy of the surrounding stars by Melnick (1985) and Walborn (1985), the high resolution measurements prove that R136 is merely the highly condensed centre of the very young cluster NGC 2070 and there is no need to postulate the existence of a super massive star at the cluster core.

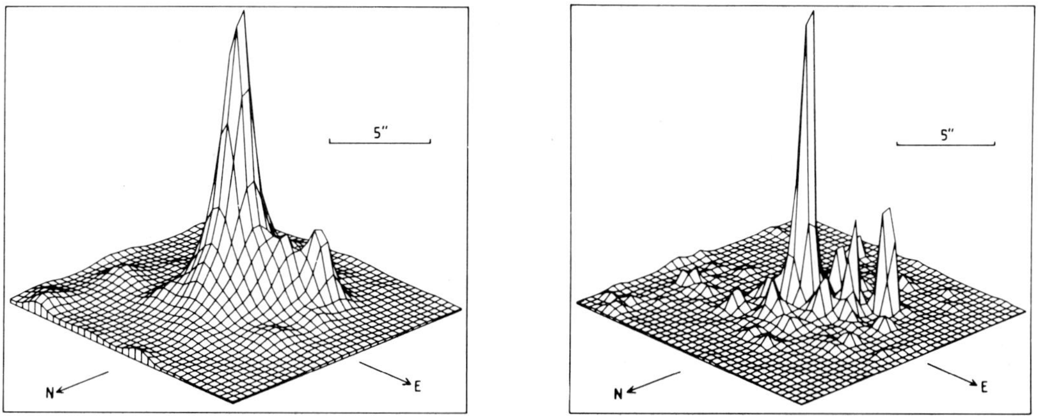

Figure 7. (a) Solid body plot of R136 prior to MEM processing, (b) Solid body plot of R136 after MEM processing, the vertical scale is half that of (a). Total exposure time was 300 seconds through an I filter.

REFERENCES

Arp, H., 1965. Ap.J., 141, 43.
Arp, H & Thackeray, A.D., 1967. Astrophys. J., 149, 79.
Blanco, B.M., 1984., A.J., 89, 1836.
Blanco, V.M. & Blanco, B.M., 1984. PASP, 96, 603.
Butler, D., Dickens, R.J. & Epps, E., 1978. Ap.J., 225, 148.
Caldwell, J.A.R. & Coulson, I.M., 1985, MNRAS, in press.
Christian, C.A., Adams, M., Barnes, J.V., Butcher, H., Hayes, D.S., Mould, J.R. & Siegel, M., 1985. PASP, 97, 363.
Cousins, A.J., 1984. SAAO Circ., No. 8, p 69.
Cousins, A.J., 1985. MNASSA, 44, 54.
Djorgovski, S., 1984. Proc. NASA/SDSU Workshop on Improvements to Photometry, ed W. Borucki & A. Young, publ. NASA, p152.
Feast, M.W., 1985. Mem. Soc. ast. Ital., 56, 213.
Freeman, K.C. & Rodgers A.W., 1975. Ap.J., 201, L71.
Gaposchkin,S.I., 1956, Variable stars, 11, 268.
Graham, J.A., 1982. PASP, 94, 244.
Harris, W.E. & Hesser, J.E., 1984. IAU Symposium No. 113, ed. J. Goodman & P. Hut, publ. Reidel:Dordrecht, p 81.
Hartwick, F.D.A., Hesser J.E. & Hill, G., 1972. Ap.J., 174, 573.
Innes, R.K.A., 1927. Southern Double Star Catalogue, publ. Union Observatory:Johannesburg.
Jacoby, G., Howell, S. & Junkkarinen, V., 1985. NOAO Newsletter, 2, 16.
Landolt, A.U., 1973. A.J., 78, 959.
Landolt, A.U., 1983. A.J., 88, 439.
Lugger, P.M., Cohn, H. & Grindlay, J.E., 1985. BAAS., 17, 602.
Lynds, B.T. & O'Neil, E.J., 1985. BAAS, 17, 596.
Martin, W., Warren, P.R & Feast, M.W., 1979. MNRAS, 188, 139.
Melnick, J., 1985. A.Ap., in press.
Menzies J.W., Banfield, R.M. & Laing, J.D., 1980. SAAO Circ. No 5, p 149.
Menzies, J.W. & Laing, J.D., 1985. MNRAS, submitted.
Olszewski, E.W. & Aaronson, M., 1985. Steward Obs. preprint no. 589.
Penny, A.J., 1984. Gemini, 12, 1.
Rodgers, A.W. & Harding, P., 1983. PASP., 95, 979.
Sandage, A., 1981. Ap.J., 248, 161.
Schild, R.E., 1983. PASP, 95, 1021.
Shobbrook, R.R., 1983a. MNRAS., 205, 1215.
Shobbrook, R.R., 1983b. MNRAS., 205, 1229.
Shobbrook, R.R., 1984. MNRAS., 206, 273.
Stobie, R.S., Sagar, R., Gilmore, G., 1985. A.Ap.Supp., 60, 503.
van den Bergh, S, 1971. A.J., 76, 1082.
Walborn, N.R., 1986. IAU Symposium No. 116, publ Reidel:Dordrecht,
Walker, A.R., 1984a. MNRAS, 209,83.
Walker, A.R., 1984b. Proc. NASA/SDSU Workshop on Improvements in Photometry, ed W. Borucki & A. Young, publ. NASA, p. 177.
Walker, A.R. & O'Donoghue, D.E., 1984. Astron. Express, 1, 45.
Walker, A.R. & Mack, P., 1985. MNRAS, submitted.
Weigelt, G. & Baier, G., 1985. A. Ap., in press.

DISCUSSION

White: A CCD could be used for lunar occultation observations of bright stars. The light is dispersed along one row. The rest of the rows serve as a buffer. The total observation consists of filling up the CCD buffer with spectra detected on the 'top' row. Clocking down a row at about 100 to 200 Hz would give sufficient time resolution.

Has a Fabry lens been applied to CCD photometry? A stationary, standard, image on the CCD could reduce reduction time by simplification of the algorithm and reduce some sources of systematic errors. As CCD dimensions increase, use of a Fabry lens may be practical for even relatively densely populated star fields.

A.R. Walker: Not that I know of.

Mochnacki: The trailed images you showed on the overhead need not in fact be elongated: Boroson and Shectman, for example, drift scan images over the CCD, reading it out at the same rate as the images drift over the CCD. This takes out a lot of the flat fielding errors.

Jacoby: With regard to scanning a CCD to improve flat fielding, two problems limit the use for photometry. 1) Clouds will affect the sky brightness, so that each row read from the chip will have different sky levels. 2) Any seeing variations affect the profile of each star differently. Both effects complicate photometric reductions.

Schober: A useful technique is to integrate at the telescope on only those stars (5-10) for which you want magnitudes. (You don't read out the whole chip). One can get differential magnitudes with respect to some reference stars. Not so much data has to be stored but good results can be obtained, with an uncertainty of a few thousandths of a magnitude.

A.R. Walker: You need good computing power at the telescope.

Ables: You showed errors in V and B-V photometry. The errors in B-V were equal to and often less than those in V. How did you accomplish this?

A.R. Walker: The contribution of photon statistics to the errors in Table II are only a minor component. It is likely that other sources of error (perhaps flat fielding) systematically affect the magnitudes as a function of position on the frame but have less effect on the colours.

Evans: What is your experience with the seeing at Sutherland?

Walker: Not often better than 1 arcsec, but usually between $1\frac{1}{2}$ - 2 arcsecs.

Phillips: Have you written your own software or do you use that of others?

A.R. Walker: For the image processing part we are using STARLINK software. Then I use my own photometry program.

Millis: How much more costly in astronomer and computer time is the reduction of CCD photometry in comparison with the reduction of conventional photoelectric photometry?

Walker: For a one week observing run one can expect to take at least this amount of time reducing the data. This lower limit applies to extracting magnitudes of only a few (~5) stars per frame in uncrowded fields. Globular cluster magnitudes would take much longer.

VARIABLE STAR OBSERVATIONS WITH AUTOMATIC TELESCOPES

Russell M. Genet, Louis J. Boyd and Douglas S. Hall
Automatic Photoelectric Telescope Service
Fairborn Observatory
1357 North 91st Place
Mesa, Arizona 85207, U.S.A.

ABSTRACT. Small automatic telescopes have been used for several years to make wide band, differential photometric observations of brighter variable stars. For example, a single automatic telescope located in Arizona has been used to study essentially the entire class of RS Canum Venaticorum binaries. These stars have changing spot structures that require once-a-night photometric observations - an ideal job for an automatic telescope located in the clear skies of Arizona. The operation of such automatic systems has become so routine that an "Automatic Photoelectric Telescope Service" now makes photometric observations for any institution so requesting. A list of stars is sent to the Arizona site of this service, and every three months the observational results are returned, with the cost typically being similar to the publication page charges. Currently the use of automatic telescopes is being extended to fainter stars and narrower bandwidths. For instance, a highly specialized one-meter telescope has been designed for automatic observations of Ca II K-line emissions from brighter active chromosphere stars. Further extensions of such automation seem likely.

1. INTRODUCTION

Every clear night for the past two years, an automatic photoelectric telescope (APT), located in Phoenix, Arizona, has been making broadband photometric (UBV) observations of variable stars. It decides itself when observations should begin, and then selects, from a list of 80 variable stars, which star should first be observed. The APT then automatically finds, centers and measures the variable star, along with appropriate comparison and check stars. This process continues all night long until the APT decides it is time to stop observations and shut down. Four times a year the data is automatically reduced, and a computer in Phoenix transmits the reduced observations to a computer at Vanderbilt University in Nashville, Tennessee. There it is added to previously transmitted observations in a computerized data base. A wide range of computerized tools, such as plotting, editing, and period

J. B. Hearnshaw and P. L. Cottrell (eds.), Instrumentation and Research Programmes for Small Telescopes, 47–55.

determination are then brought to bear on the observations by that vital non-computerized ingredient - experienced human astronomers.

Experience with the Fairborn Observatory's APT in Phoenix suggested that variable star observations with automatic telescopes could result in large amounts of high quality data being gathered with great efficiency. This seemed particularly true if: (1) routine photometric observations were required on a large number of different stars over a long period of time; and (2) the automatic telescope was located where there was a high frequency of photometric nights. As many other institutions besides Vanderbilt University would obviously benefit from automatic variable star observations, it seemed appropriate to organize such a service and offer it to astronomers world wide. To this end, the Fairborn Observatory (engineering and operations) and the Smithsonian Institution (buildings and site) have formed the Automatic Photoelectric Telescope Service.

The APT Service is located on Mt. Hopkins in southern Arizona near the well-known Multiple Mirror Telescope (MMT). The APT Service offers two basic options. With the first option, the participating institution completely owns and controls their own APT. The APT Service merely installs and maintains the APT for the participating institution. Currently 0.4-meter and 1.0-meter APT's are offered. With the second option, the participating institution places variable stars for some period of time on an existing APT owned by the APT Service. While the second option is necessarily limited to APT's and instrumentation already available (currently three 0.25-meter APT's and UBV or VRI differential photometry), it is very low in cost - typically the observations cost less than the page charges for publishing the results. Participating institutions currently include Vanderbilt University, the University of Toronto, Franklin and Marshall College, Wesleyan University, and the University of Arizona.

There is, of course, no reason why the operation of automatic telescopes should be limited to broadband photometry. For example, the Smithsonian Institution and the Fairborn Observatory are considering a light-weight and low cost 1-meter APT with a Ca II K-line photometer for dedicated observations of chromospheric activity in giant stars. The 3-Angstrom K-line filter's wavelength will be adjusted (by tilting slightly) to compensate for the earth's orbital velocity and the radial velocities of each star observed.

It will be the objective of this paper to cover in greater detail those areas summarized above. Specifically we shall discuss the operation of the Fairborn Observatory APT in Phoenix, the reasons for formulating the APT Service, the APT Service site on Mt. Hopkins, the two basic options offered by the APT Service, a proposed 1-meter APT and K-line photometer for observing chromospheric activity, and, finally, other options for full automation besides photometry.

2. VARIABLE STAR OBSERVATIONS WITH THE PHOENIX APT

The 0.25-meter Fairborn Observatory automatic photoelectric telescope (APT) located in Phoenix, Arizona, has now been in regular operation for over two years. As this system has been described in some detail elsewhere (Boyd et al 1984), we will not repeat this description here, but will instead concentrate on its operation and sample variable star observations.

The master list of stars to be observed was generated by the research astronomer (Hall), and included data on the names, positions and V-magnitudes of each of the variable, comparison, and check stars, as well as the sky measurement position. Before entry by the engineers (Boyd and Genet) onto the observing program diskette, a check was made to be certain that all requested observations were within the declination and magnitude limits of the system. If too few variable stars had been placed on the master list, then the APT would have run out of new stars to observe and would have started asking repeat observations. On the other hand, if too many variables had been placed on the list, then there would have been nights that some variables were not observed at all. We found that for UBV photometry of brighter stars that about 80 variables gave the right loading. Some thought was also given to the rough RA distribution of the stars so that they were not all bunched into one or two seasons.

Once the stars had been placed on the master list, then they were observed every clear night automatically without any human intervention or assistance. This automatic process has been described well in Boyd (1985) and Trueblood and Genet (1985), and need not be repeated here.

Every three months for the past eight quarters, the data was consolidated by star and reduced. This operation was computerized and consisted primarily of sorting the data and then reducing it using well known differential photometry procedures. The reduced data was then transmitted from Phoenix to Tennessee (Vanderbilt University) via modem at 1200 baud - a process that only takes about 15 minutes. Here the data from each star was merged with observational results from the APT for previous quarters, as well as results from nonautomatic telescopes. A very convenient set of computerized data manipulating, plotting, and analytic tools were then brought to bear on each variable star by the Vanderbilt University astronomers and graduate students.

3. FORMATION OF THE APT SERVICE

Shortly after the Phoenix APT became operational, we began examining ways of further increasing the already high efficiency of APT's. The system as originally configured and programed already spent 67% of its time making the scientific observations (the recorded integrations), and only 33% of its time moving between stars, finding and centering stars, changing filters, and deciding what to do next. Thus further efficiencies here could not be large. However, even a simple analysis showed that major increases in efficiency were possible if a number of APT's were operated together as a group.

A single engineer can readily take care of a number of APT's. The test equipment, software development equipment, and tools needed for a single APT are readily applied to a group of APT's if they are all at the same location. Much of the facilities and site support required for a single APT is applicable to a group of APT's, such as shop and office space, road construction and maintenance, utilities, etc. Also, the weather sensors can be common, and if there is a large roll-off roof, the roof and its control can be common.

With a group of APT's making observations for many different institutions in several countries, it becomes worthwhile to place the APT's at a first rate site. This not only improves the quality of the observations, but by having a high frequency of photometric nights, the continuity of the data is improved, thus reducing observational gaps.

As the Fairborn Observatory has had years of experience in the engineering development and operation of APT's, it seemed appropriate that these aspects of an APT Service should be handled by this organization. The critical questions of the site for the APT Service was studied for quite some time (Baliunas et al 1985b). The choice that clearly emerged as being best was the Mt. Hopkins site of the Smithsonian Institution's Fred L. Whipple Observatory. Not only was this site outstanding photometrically, but two buildings, including one with a very large roll-off roof, were available. The Smithsonian Institution, which has a research program of long standing on variable stars (particularly cool stars with active chromospheres), joined with the Fairborn Observatory to form the APT Service. This service is now being offered to institutions world wide that need routine photoelectric observations of variable stars.

4. APT SERVICE OPTIONS

In the first of the two basic APT Service options offered, the participating institution owns and completely controls the APT (Boyd et al 1985a). The role of the APT Service is that of installation, operation, and maintenance of the APT for the institution that owns it. While the choice of telescopes is currently limited to two (0.4-meter and 1.0-meter), there is considerable freedom of choice on filters and detectors, and limited customization of the control software to suit a unique situation is possible.

The 0.4-meter telescope is made by DFM Engineering (Melsheimer and Genet, 1984), and is a well-built Cassegrain system with a zero-backlash friction dirve and very rigid mount - characteristics very desirable for automatic telescopes. The telescope control electronics, and the computerized photometer are made by the Fairborn Observatory. The cost of an installed system is typically $35,000. Operation and maintenance of the system includes loading the list of stars sent from the institution that owns the APT, periodic maintenance, including re-aluminization of mirrors when required, repair of failed equipment (very rare), and reduction and transmittal of the data to the home institution. The annual cost of this service, which includes a share of the overall operation of the APT Service, is $15,000 per year.

The 1.0-meter telescope is made by the Fairborn Observatory, and is described later in this paper. The installed cost of this telescope including normal photometric instrumentation, is $100,000, and the annual cost for operation and maintenance is $25,000.

In the second basic APT Service option offered, the participating institution places a number of stars for some given time on one of the already existing APT's owned by the APT Service itself. Typically a number of stars might be placed on the program to supplement observations made at the home institution, observations might be requested for a limited time (a year or two) in support of a grant, or favourite stars might be followed at low cost. This second option has been described in some detail by Boyd et al (1985b). There are currently two suboptions that can be considered.

In the first suboption, stars are not observed more frequently than once per night (one differential sequence per night). It can be specified that they be observed less frequently than this, such as every other night, every fifth night, etc. The cost per observational sequence (some 33 measurements made over seven minutes) is only $2. This is so low that on a typical star if one observation is made every clear night, the cost for a year is only $400 (observations on some 200 nights). This can easily be less than the page charges to publish the results! Currently stars can be observed in UBV on a 0.25-meter APT, or in VRI on another (separate) 0.25-meter APT.

In the second suboption, a single variable star can be observed differentially for hours or even all night for several nights in a row. A single 0.25-meter APT with a VRI photometer is being set aside for these sorts of observations as they can not be readily mixed with the "once a night or less" types of observations because if mixed this would cause unfair gaps in the observations of stars supposed to be observed once a night. The cost of this service is $25 per hour.

5. ONE-METER AUTOMATIC TELESCOPE DESIGN

Having met with considerable success with the first and second generation small APT's, it was only natural that something considerably larger, say one-meter, would be considered. This is not as straight-forward as one might think because the small APT's typically make about 25,000 separate motions each night - mainly to find and center the stars to be observed. The physical smallness of these APT's is an advantage because they can be moved about quickly and settle down in a fraction of a second. If a one-meter APT had a large moment of inertia and took a long time to settle down, it would not work well with the current control system and software. Additionally, for several one-meter telescopes to fit in the roll-off-roof building on Mt. Hopkins along with the smaller APT's, they would need to be very compact in size, and should have minimal interference with each other's light paths.

Simplifying the situation somewhat is the relaxed optical requirement inherent in on-axis photometry. As the one-meter telescope is only intended for automatic photometry, no provisions need be made to accomodate a human observer in any way. For instance, the telescope can be very low to the floor, and the Dec drive can run up the middle of the back of the telescope instead of being placed on one side so as to not block the observer. Also, as only photometry will be done with a photometer permanently put in place, provisions for other instruments are not required. In short, the telescope can be a dedicated, special purpose instrument not intended for human operation.

With the above requirements and simplifications in mind, it is not surprising that the resulting design is unusually compact and light-weight yet stiff. The new mirror technology is capitalized on in terms of a lightweight, low f-ratio primary. Friction drives are utilized in both RA and Dec to achieve zero backlash without preloading. A tripod has been used to support the lightweight secondary mirror. While this increases the diffraction image somewhat above that of a spider, it has a very low moment of inertia, and it allows the telescopes to be placed close to each other without interference.

6. GIANT STAR PROJECT AND K-LINE PHOTOMETER

A good example of a specialized photometry project that is ideal for a dedicated large APT is the giant star project proposed by Sallie L. Baliunas at the Harvard Smithsonian Center for Astrophysics. In preliminary observations made at Mt. Wilson in the Ca II H and K lines, it was clear that at least some of the more rapidly rotating giants had areas of chromospheric acitivity. If a large number (several hundred) of the brightest giants visible from the northern hemisphere were monitored in the Ca II K line nearly nightly for a decade or so, not only could many rotational periods be accurately determined, but activity cycles could be determined at the lower end of the scale of chromospheric activity and perhaps lower bounds could be placed on such activity with respect to rotational speeds.

Observing the K-line in giant stars requires a large telescope to overcome the limited photons available, a very narrow filter (only 3 Angstroms), and an accurately adjustable filter center wavelength to match the rotational velocity of the earth and the radial velocities of the giant stars being observed. A photometer designed for this specific task has already been described by Baliunas et al (1985a). Briefly, the K-line narrowband filter is tuned by slight tilting from a geared stepper. To avoid the effects of changing atmospheric transparency, it is necessary to go back and forth between the K-line and continuous filters at frequent intervals. This is accomplished by two very light-weight mirrors mounted on miniature stepper motors that rotate in opposite directions (keeping the net angular momentum constant). The entire unit, including the thermoelectric cooled photomultiplier, is small enough to fit directly behind the primary mirror of the one-meter APT discussed earlier.

7. EXTENSIONS BEYOND CONVENTIONAL PHOTOMETRY

Broadband differential photometry of brighter stars provided a logical starting point for routine automated observations of variable stars. Extending this approach to very narrow band (K-line) photometry of bright giant stars requires primarily a larger automatic telescope, as discussed earlier. When some experience has been gained with one-meter-class automatic telescopes, it seems likely that besides photometric observations of fainter stars, that extensions to other forms of observation would be natural.

One of the easiest extensions would be to stellar polarimetry. One is concerned here with an on-axis image, the equipment is mainly an "add-on" to conventional photometric equipment, and the data rates are similar to photometry. One could quickly come up with many useful programs that could combine photometry and polarimetry on a dedicated automatic system that would observe almost every night for a decade.

Area photometry using CCD's is producing good results as has been suggested by others at this Symposium. There is no reason why such area photometry could not be automated, although the volume of data to be recorded and transmitted would increase by orders of magnitude over conventional photometry. A program of area photometry of the brightest 20 or so open clusters every clear night for a decade would produce a treasure trove of variable star photometry for those interested in stellar evolution.

The spectroscopists have already been asking for suggestions as to how dedicated AST's (Automatic Spectroscopic Telescopes) could take over some of their routine, long-term observations, work on catalogs, etc. Here, automatic telescopes larger than one-meter quickly enter any discussion. Such "very large" automatic telescopes and their various potential applications have been discussed recently by Boyd, et al (1986). While such speculations may be interesting, it seems likely that at least in the near future the primary contribution of automatic telescopes will continue to be in the area of photometry of variable stars.

ACKNOWLEDGEMENTS

We gratefully acknowledge the support provided by the National Science Foundation (grant AST 84-14594) for the development of automatic photoelectric telescopes and their application to RS Canum Venaticorum binaries. We also gratefully acknowledge the facilities and site support provided by the F.L. Whipple Observatory of the Smithsonian Institution. Finally, we wish to acknowledge those institutions who first used the APT Service. They were: Vanderbilt University; University of Toronto; Franklin & Marshall College, Wesleyan University; and University of Arizona.

REFERENCES

Baliunas, S.L., Boyd, L.J., Genet, R.M., Guinan, E.F. 1985a, I.A.P.P.P. Communication, 22, in press.

Baliunas, S.L., Boyd, L.J., Genet, R.M., Hall, D.S. and Criswell, S. 1985b, I.A.P.P.P. Communication, 22, in press.

Boyd, L.J. 1985, Byte, 10, 7, 227.

Boyd, L.J., Genet, R.M. and Hall, D.S. 1984, I.A.P.P.P. Communication 15, 20.
- 1985a, I.A.P.P.P. Communication, 19, 41.
- 1985b, I.A.P.P.P. Communication, 21, 59.
- 1986, Pub. A.S.P., Submitted.
Melsheimer, F. and Genet, R.M. 1984, I.A.P.P.P. Communication, 15, 33.
Trueblood, M. and Genet, R.M. 1985, Microcomputer Control of Telescopes (Richmond; Wilmann-Bell).

DISCUSSION

Osborn: What precision do you obtain from a single measure?

Genet: The system in Phoenix (10 s integration on a 10" telescope) is typically a hundredth of a magnitude. That is, no different from a non-automatic telescope.

Moffett: How do you detect cloud or rain?

Genet: This is the crucial point of the technique. There is no sense in having an automatic telescope if you have a person there to check for clouds and/or rain. We solve this in two ways. After the fact, we collect the data and look for internal consistency in three colours. If they are not consistent the data is rejected. At the telescope, if the clouds are too thick, the procedure is to make 4 tries at centering on the star. If this fails it closes itself down.

Cameron: Have you given any consideration to using red-sensitive tubes suitable for VRI photometry in these systems? If synoptic modelling of the evolution of starspot groups in RSCVn and related objects is one of your major programmes, UBV photometry has the disadvantage that problems with line-blanketing make it very difficult to resolve the ambiguity between the temperatures and the areas of the spots. Steve Vogt has shown that using the relationships between VRI colours and the visual surface brightness found by Barnes, Evans and Moffett.

Genet: The Phoenix system is just UBV. The Mt. Hopkins system uses a VRI photometer and we have ordered a Hamamatsu GaAs tube. UBVRI observations should start in January 1986.

Andrews: What is the effect of a Schmidt correcting plate of a Celestron 14 on the ultraviolet band for photoelectric observing? Have you made comparisons with conventional cassegrain scopes?

Genet: I think you just lose light (especially in U) with the corrector plate but one can transform onto the standard system using standard stars.

White: The problem of observing lunar occultations of stars showing measurable angular diameters with small telescopes is scintilation. If there exists an array of small telescopes their spatial integration will reduce the scintillation effects, as long as photons are not a limiting factor.

Schülte in den Baümen: What is the positioning accuracy of your telescope?

Genet: Long slew: within 2 arcmin of object correcting only for precession
Short slew: within 10-15 arcsec.

Page: What aperture diameters are used for the APT?

Genet: In Phoenix: 90, 60, 45 arcsec aperture, but we usually use the 60 arcsec one (10" telescope on stars brighter than 8th mag). VRI photometer: fixed aperture of ~60 arcsec.

Hickey: What is cost per observation?

Genet: $US2 per set of observations.

REPRODUCING THE UBV AND COUSINS (RI) PHOTOMETRIC SYSTEMS WITH A HAMAMATSU GA-AS PHOTOMULTIPLIER.

J. W. Menzies
South African Astronomical Observatory
P.O. Box 9
Observatory 7935
South Africa

ABSTRACT. The Hamamatsu R943-02 photomultiplier appears to be suitable for $UBV(RI)_C$ photometry. This paper discusses the tube's characteristics on the basis of one year's experience in the use of such a device on the 0.5m telescope at Sutherland.

1. INTRODUCTION

The 0.5m telescope at the Sutherland observing site of the SAAO is used extensively for photoelectric photometry. Since 1984 November a single-channel photometer equipped with a Hamamatsu R943-02 photomultiplier has been used for $UBV(RI)_C$ photometry. This tube has a 10 x 10mm² GaAs photocathode inclined at about 10° to the incoming beam.

2. PHOTOMETRIC SETUP

Because of the extended red sensitivity of the GaAs tube, it is necessary to use blocking filters with U, B and V. The filters used, with thicknesses in mm given in brackets are as follows:

U :UGI(1) + Solid $CuSO_4$(5)
B :BG12(1) + GG385(2) + BG18(1)
V :GG495(2) + BG18(1)
Rc :OG570(2) + KG3(2)
Ic : RG9(3)

All except the solid $CuSO_4$ are glass filters made by Schott. The tube is operated at 1700V and is maintained at -15C in a Products for Research thermoelectrically cooled cold box. The dark current is typically about 7 count sec^{-1} which is remarkably low for a red-sensitive tube. The overall gain of the tube appears to be sensitive to temperature and there is sometimes a progressive increase in the V zero point if the ambient temperature falls during the night.

J. B. Hearnshaw and P. L. Cottrell (eds.), Instrumentation and Research Programmes for Small Telescopes, 57–59.

3. PHOTOMETRIC PERFORMANCE

The tube is kept cooled with the operating voltage applied day and night. This leads to high photometric stability. The transformation equations are determined every three to six months by means of observations on one or two good photometric nights of numerous E-region standards (60 or more) covering a wide range of colours. The transformations are well represented by the following linear relations:

$$V = v_0 + 0.007(v - i)_0 + Zv$$

$$(B-V) = 1.008(b-v)_0 + Z_{bv} \qquad (V-R)_c = 1.043(v - r)_0 + Zvr$$

$$(U-B) = 0.987(u-b)_0 + Z_{ub} \qquad (V-I)_c = 1.006(v - i)_0 + Zvi$$

Here, the subscript zero refers to natural colours corrected for extinction and the Z's are the zero points. The coefficients are all close to 1.0 so the match to the standard systems is quite satisfactory. For (B-V) and (U-B), non-linear corrections must also be applied. The correction to (B-V) applies only redder than 1.2 and can be represented as a function of (B-V) by a straight line of slope -0.08 mag/mag. The (U-B) correction as a function of (B-V) is an S-shape curve with a maximum value of +0.02 mag at (B-V) = 0.1 and a minimum of -0.02 at (B-V) = 0.5.

The magnitudes which give 1 count sec^{-1} above sky are given in the following list, where a comparison is also made with the tubes in use immediately before the change to the Hamamatsu tube was made.

	Hamamatsu	EMI6256		Hamamatsu	EMI9658
V	20.1	19.7	V	20.1	19.7
B	20.1	20.55	R_c	20.0	19.2
U	19.6	19.7	I_c	19.3	19.4

After all linear transformations, non-linear corrections and zero points have been applied, the standard deviation for a single observation during a typical colour equation run when 85 E-region standards were observed was:

	V	(B-V)	(U-B)	$(v-R)_c$	$(v-I)_c$
S.D.(mag)	0.007	0.008	0.010	0.006	0.007

A disadvantage of the Hamamatsu tube is that there is an upper limit to the count rate measurable with it beyond which there is a signal-induced increase in dark count rate. This occurs at about 3×10^5 count sec^{-1} which corresponds to V about 6.4mag. To measure bright stars it is necessary to introduce neutral density filters.

However, the tube appears to be robust. For example, after an accidental exposure to Jupiter, the dark count rose to more than 10^4 count sec^{-1} but had returned to normal 14 hours later. There has not been any subsequent change in the colour equations or zero points.

4. ACKNOWLEDGEMENTS

I wish to thank Dr. D. Kilkenny for his assistance in the preparation of this paper and Mrs. Anne Jones for typing it.

DISCUSSION

Graham: Have you had any experience of the non-linear corrections to your U-B and B-V transformations depending on the metallicity of the star? I had a lot of trouble with this when I did my $UBV(RI)_{KC}$ photometry with an RCA 31034 tube.

Menzies: I haven't investigated this. We use Cousins standards, most of which are nearby, solar metallicity stars. Some of the (U-B) problem may be due to a metallicity effect.

Rucinski: The RCA tubes were known for frequent failures, e.g. they just stopped working. Are these tubes any better?

Menzies: I don't have any more information apart from the limiting count rate and their subsequent increased noise. The count rate effect produces a bright limiting magnitude of $\sim 6^{m}.5$ on a 0.5 m telescope. As far as durability is concerned, our current Hamamatsu tube has been in use for about one year without any deterioration in performance.

Warner: RCA tubes are much better now as they no longer leak through the pins.

Breger: Nonlinear corrections required for (B-V) transformations may be quite common. I find a similar curvature when transforming to Landolt's (GaAs) standards. Since the nonlinearity only becomes evident when stars redder than (B-V)=1.2 are measured, it can be missed when very red standard stars are not measured.

Menzies: I agree, people often don't see the effect because they don't observe the full colour range or enough standards. We have compared E region and Landolt standards and still found non-linearities. This probably means that there is no "standard" UBV system and that everybody has their own system.

Gaustad: With all the talk (see Latham) about the wonders of CCDs for photometry, and the report (see Jacoby) that KPNO is phasing out photomultipliers, I want to ask a provocative (and somewhat tongue-in-cheek) question : Why bother to develop photomultiplier systems, which seem to be a relic of the past?

Menzies: Not everybody can afford CCDs, and then there are the electronics, and maintenance requirements.

A FIBER-OPTIC THREE-CHANNEL PHOTOMETER FOR SIMULTANEOUS UBVRI MEASUREMENTS - IMPROVED DESIGN AND NEW RESULTS

H. Barwig and R. Schoembs
Universitäts-Sternwarte München
Munich
W.-Germany

A modern photometer designed for variable star observation should meet the following specifications: 1. Simultaneous measurements of object, comparison star and sky in order to allow compensation of changing atmospheric transparency and background radiation. 2. Simultaneous multicolor measurements of each channel yielding high efficiency and accurate colors. 3. Coverage of a large wavelength region (U-I) with high time resolution. A photometer, fitting these requirements has been designed at the Universitäts-Sternwarte München. In principle light from three sources in the focal plane of a telescope is guided through optical fibers to three spectrographs, each splitting the incoming signal into five wavelength regions. The resulting 15 output channels are linked to photon counting detectors. Their signals finally are processed in a microcomputer. The basic components are shown in Fig. 1. Details are given in the following:

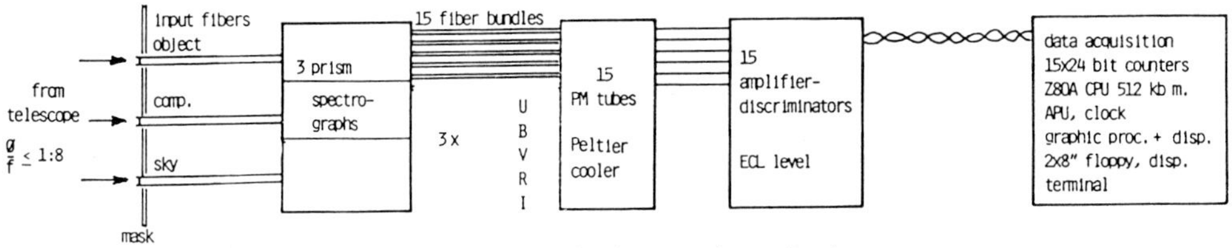

Fig. 1 Blockdiagram of the 15-channel photometer

1. FIBER INPUT

Each of the three input channels consists of a small diaphragm ($\phi \leq 1$ mm) which is projected by a lens system onto a quartz monofiber with $\overline{4}00\mu$ core diameter. The positions of the individual fibers are maintained by holes in an acrylic plate. This mask is prepared on the telescope itself just prior to observation. Star centering within the diaphragms is controlled by autoguiding a nearby offset star ($m \leq 16$ at 1 m Tel.) by means of a three-stage image tube and a Newicon TV camera.

J. B. Hearnshaw and P. L. Cottrell (eds.), Instrumentation and Research Programmes for Small Telescopes, 61–64.

2. COLOR SEPARATION

The end faces of the input fibers form the entrance slit of three identical prism spectrographs each projecting a small spectrum onto a fiber array. Proper bundling and/or shaping of its individual elements yields the separation of several slightly overlapping wavelength regions. Thus different color systems can be realized. In the present photometer version the standard UBVRI filter characteristics were approximated.

3. DETECTOR UNIT

The 15 spectrograph output fibers are connected to a very close arrangement of 15 miniature photomultipliers (PM) cooled by Peltier elements. Integrated amplifier/discriminators convert the PM pulses to normalized ECL signals.

4. DATA ACQUISITION

Photo counting is performed by 15 24-bit counters controlled from a Z80A microcomputer system. Storage of data and/or programs is possible on a 512 kB RAM disk, two 8" floppy disks (1 MB each) or a 20 MB Winchester disk. On line light curves can be displayed on a graphic screen while communications are possible from a separate terminal with display and printer. The acquisition software allows to define all observing parameters via a menu and to perform statistical analysis of all channels. The normal range of integration times covers 1s to 25.4s. However special software with less sophisticated on-line facilities allows continuous time resolution down to 5 ms. A reduction program calculates the object intensities relative to a comparison star. For this purpose the observations of a program star have to be enclosed between two calibration measurements each taking about 10 min.

5. PERFORMANCE

From various observing runs at the ESO 1 m telescope and the Calar Alto 1.2 m telescope a lot of experience was obtained with the new photometer. Compared to a conventional photometer the new instrument has gained a considerable amount of efficiency. This is mainly due to the fact, that the available observing time can completely be used for each individual channel. Furthermore the measuring principle yields an enormous gain of efficiency since observations can be performed even through cloudy skies with variations of transparency up to 80%. Of further interest in this respect is the quantum efficiency of the individual channels. A comparison of count rates obtained with the ESO 1 m single channel photometer using a Quantacon detector and with

	single-ch. photometer	15-ch. photometer
U	340 cts/2s	220 cts/2s
B	1790	2830
V	5070	5560
R	6260	2750
I	7070	820

Table I

the new instrument is given in Table I for a star with V=12^{m}.7 (U-B=.7, B-V=.9). Higher count rates are measured in B and V while the lower signal in R and I is exclusively due to the lack of red-sensitive miniature photomultipliers. The accuracy of brightness determination amounts to 1-2% if errors due to photon statistics can be neglected.

6. FUTURE DEVELOPMENTS

Improvements of the present photometer device concern the following developments:
1) computer controlled positioning of the input fibers,
2) increased spectral resolution yielding up to 30 color channels,
3) improved fiber input optics with automatic flat field control,
4) automatic calibration and continuous test of PM stability during day time,
5) improved U and R, I response,
6) increased on-line facilities (EFT, on-line reduction, etc.).

DISCUSSION

Rucinski: How long does it take to generate the aperture plate?

Barwig: Mask preparation is quite simple since we only have to center the individual stars in an eyepiece while a drilling machine, attached to it, transfers the positions to an acrylic plate.

Garrison: That is a nice piece for engineering technology, but as a user I would worry about the flexibility of the system. For example, if I wanted to work to a fixed number of counts in U for red and blue stars, I would have trouble varying the relative integration times.

Barwig: For each channel the integration time can individually be set (or counts can be summed up afterwards).

Mochnacki: Why do you use three separate spectrographs?

Barwig: In order to avoid cross-talking due to stray light.

Ryan: How well do the transmission functions of the five passbands approximate standard UBVRI passbands?

Barwig: A special array of monofibers, which have been bundled or shaped properly, can match the standard system very closely.

Caton: Do you use 3 separate Fabry lenses or do you depend on the fibre optics to mix the star and sky signals?

Barwig: We depend on the fiber optics to mix the signals. The lenses are used to project the star image in the diaphragm onto the fibre entrance.

THE USE OF WIDE ANGLE CAMERAS AND HIGH SPEED MEASURING MACHINES FOR GENERATING PROGRAMMES FOR SMALL TELESCOPES

R J Dodd
Carter Observatory
Wellington
NEW ZEALAND

H T MacGillivray
Royal Observatory
Edinburgh
SCOTLAND

ABSTRACT. The classical method of discovering objects that vary in position and/or time is to examine well-matched pairs of plates using a blink comparator. This method, being manual, is slow, subjective and, unless several matched plate pairs are used, suffers from incompleteness. The combination of wide angle Schmidt cameras, high speed measuring machines and sophisticated computer software has enabled objective methods for searching for variable objects to be devised. Using such methods it is possible to intercompare stacks of plates digitally. The deepest plate may be used as a master in position and magnitude, eliminating the need for the rest to be well matched. Following measurement and analysis it is a relatively simple matter to prepare lists of variable objects such as RR Lyrae stars, Mira long period variables, cataclysmic variables, active galaxies, comets and asteroids which may then be used as a basis for more intensive study by smaller dedicated instruments.

This, however, is only one aspect of the cooperation possible between Schmidt telescopes, measuring machines and small telescopes. In projects involving the study of large numbers of mainly non-variable objects (stars or galaxies) the small telescope has a most important role to play in providing photoelectric magnitude calibrations to convert machine measured magnitudes to a standard system. To date this has been the principal use to which the small telescope has been put, even in the study of the distribution of faint galaxies from Schmidt plates it was a small telescope which provided the necessary measurement of the background sky brightness.

A further dimension is provided by the addition of an objective prism to the Schmidt camera. In this case objects readily detected include emission line stars and galaxies. The calibrating power of the smaller instrument is now of value in providing standard galaxy radial velocities to calibrate red shifts measured from the very low dispersion prism.

TABLES.

The following tables set out areas of research which benefit from the in-

J. B. Hearnshaw and P. L. Cottrell (eds.), Instrumentation and Research Programmes for Small Telescopes, 65–67.

vidicon TV camera with low-lag time (5% after 50ms) is optically coupled by a relay lens with two dimensional photon counting tube as shown in figure 2. When an optical image is focused on the photo-

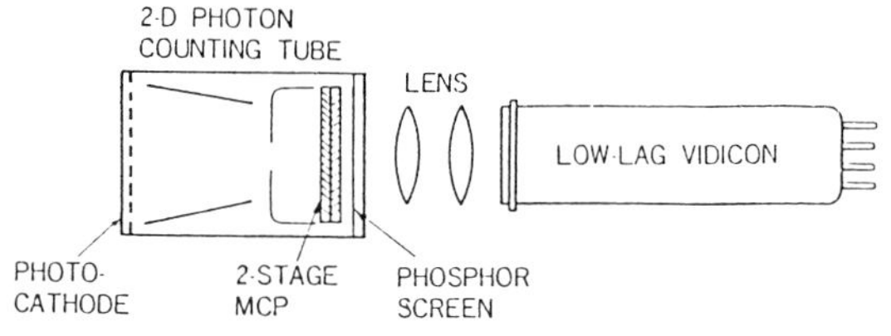

Figure 2. Schematic diagram of Hamamatsu C1960-20 VIM camera

cathode, it emits photoelectrons in accordance with the intensity of the input image. The electron image is focused on the two stage micro-channel plate (MCP) where it is amplified up to the order of 10^6. Pulse height distribution of output signal is given in figure 3. This amplified image produced on the phosphor screen is taken by the coupled video camera. When the intensity of input image becomes extremely low, one video frame of the TV scan gives only a dotted image corresponding

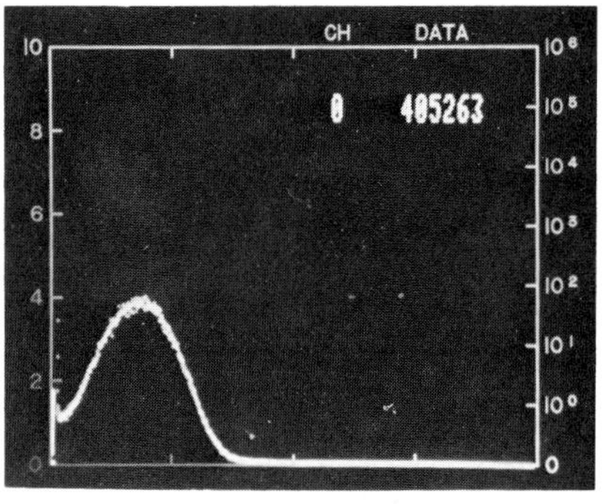

Figure 3. Pulse height distribution of camera.

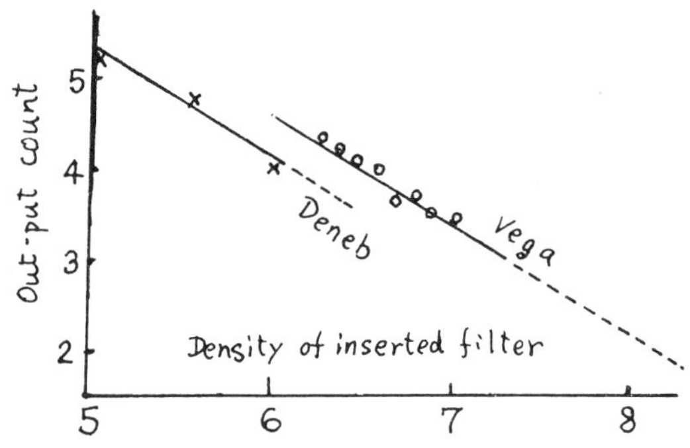

Figure 4. Relation between density and count for Vega (o) and Deneb (x).

to input photons and noise produced at the photocathode. By giving a certain discriminator level to reduce the effect of this noise, the dotted images are digitally processed and accumulated in a video frame memory over a desired number of video frames.

2. ASTRONOMICAL TEST OF THE SYSTEM

The system was fixed at the Nasmyth focus. First, the bright stars, Vega and Deneb with zero and 1.3 magnitude, were observed by adding neutral density filters in front of the detector. A good linear relation between integrated photon number in 15 seconds and total density of inserted filters is shown in figure 4 which suggests that the limiting magnitude of our system is over 21 magnitude in 15 second exposure because of low dark noise, if skybackground is dark enough. Since a star of 20 magnitude gives 10^3 photons in 15 seconds for a 75 cm telescope, quantum efficiency is about 0.1. Many objects such as the comet Halley, the Crab Nebula, M51 galaxy, and so on were observed and we obtained fine images and found this system has a wide range of applications in astronomical observations.

Table I. Tentative specification of the system

Spectral response	300 - 820 nm
Useful photocathode size	10^2 mm^2
Resolution	360 TV lines
Dark count at 25°C	2 cps
Dynamic range	10^0 - 10^9 photons/mm^2 sec
Geometric distortion	7 %

DISCUSSION

Hiltner: Has Mr. Miyaki measured the photoelectron collection efficiency of the first multiplier stage - the micro-channel plate?

Isobe: Although Mr. Miyaki may have some idea of the photo-electron collection efficiency, I have never measured it. However, quantum efficiency of the whole system was measured as 0.1 from stellar observations, which is a combined efficiency of phosphor, micro-channel plate and photocathode. Therefore, the efficiency asked by Dr. Hiltner may not be worse.

Latham: Rich Cromwell has demonstrated that for most image intensifiers, roughly half of the photoelectrons released from the cathode never make flashes of light at the output phosphor, and that therefore the true DQE is degraded seriously from the d.c. sensitivity measured for the cathode. For MCP intensifiers the counting efficiency was worse than average, typically 0.3 to 0.4 for the tubes measured by Cromwell. Only for some Varo intensifiers was a high counting efficiency found, typically 0.8 to 0.9. Magnetic intensifiers made by ITT, RCA and EMI have efficiencies of 0.5 or 0.6. When comparing photon counters with other detectors, such as CCDs, one must multiply the d.c. cathode sensitivities quoted by the manufacturers by the counting efficiency, a serious degradation.

Hiltner: ITT report a collection efficiency of 60% for their proximity focussed image tube with a micro-channel multiplier.

FICK OBSERVATORY SPECTRUM SCANNERS

W. I. Beavers and J. J. Eitter
E. W. Fick Observatory
Iowa State University
Ames, Iowa 50011
U.S.A.

ABSTRACT. Three low-dispersion (15-50 A/mm)$^{-1}$ spectrum scanners have been developed at Fick Observatory as instrumentation for small telescope research and teaching activity. They have been applied in solar eclipse, zodiacal light, stellar, and comet research programs. The two plane grating spectrometers and one holographic grating instrument have been employed on telescopes ranging from 7.6 to 61 cm aperture, generally with f/5 to f/8 configurations. Recently, one of the plane grating instruments has been modified to perform as a four-channel spectrometer for quantitative spectral classification and composite spectra studies. This instrument is also being used to make spatial-spectral studies of the coma of Comet Halley, and other recent comets.

1. HOLOGRAPHIC GRATING SCANNER

A Type IV aberration corrected concave holographic grating with 1200 grooves/mm is the only optical element in one scanner. Although capable of operation up to about f/4, this instrument is generally employed at f/8 with a single photomultiplier. However, provision is provided for a simultaneous two-cell (sky and star) mode. The reciprocal dispersion is about 40 A/mm. Scanning of the spectrum across the exit slot is performed by rotating the grating. A microcomputer is employed for data recording and managing the programmed scans over the selected wavelength regions. Recently the holographic scanner has been used most frequently with the 37 cm Cassegrain telescope in comet coma and tail studies.

2. PLANE GRATING SCANNERS

Two low-dispersion (15-50 A/mm)$^{-1}$ plane grating instruments are in use at Fick Observatory. Both are converted commercial spectrometers. One is a crossed 0.28 m Czerny-Turner configuration, and the other is a 0.25 m Ebert instrument. At times both are employed as photoelectric radial velocity spectrometers using masks with selected solar lines in

J. B. Hearnshaw and P. L. Cottrell (eds.), Instrumentation and Research Programmes for Small Telescopes, 75–76.

solar eclipse and Zodiacal light experiments with f/5 7.6 cm refracting telescopes (Beavers et al., 1980).

For stellar and comet work the scanners are used on the 61 cm and 37 cm Cassegrain telescopes at Fick Observatory. Recently, the Ebert instrument has been modified to perform as a four-channel system by providing provision for four separate photomultipliers behind the output focal plane. A mask is used to define the spectral band measured in each of the output channels. The mask may be changed quickly under computer control by displacing it perpendicular to the dispersion. Thus it is possible for each channel to be used to compare signals from the chosen spectral line region and the adjacent continuum. The four-channel system is currently being employed at f/8 on the 61 cm Cassegrain telescope with line and continuum band passes of 100 to 150 A for measurements of molecular emission bands in cometary coma. The telescope is programmed to sample a grid of positions centered on the comet for the four-channel system measurements in order to produce a combined spatial-spectral study of the comet.

A portion of this work has received support through NSF Grant 82-11126, which we gratefully acknowledge.

3. REFERENCE

Beavers, W. I., Eitter, J. J., Carr, P. H., and Cook, B. C. 1980, Ap. J. **238**, 349.

DISCUSSION

Millis: 1) What spatial resolution will you be able to get on Comet Halley? 2) How far will you have to move off the comet to reach the true sky background?

Beavers: 1) On both our telescopes, 61 cm and 37 cm, we operate at f/8. These give slit scales (arc sec/mm) of 42 and 70, respectively. 2) Not really sure, except we want to reach the background in both the solar and anti-solar directions.

I. *TELESCOPES AND INSTRUMENTATION*

b) *Poster papers*

THE INSTRUMENTATION AND OBSERVING PROGRAM AT GRINNELL COLLEGE'S GRANT O. GALE OBSERVATORY

R. R. Cadmus, Jr., F. M. Melsheimer
Department of Physics
Grinnell College
Grinnell, Iowa 50112
USA

Many small observatories are associated with undergraduate colleges at which research programs must be integrated with a broad spectrum of educational programs. The instrumentation and observing programs of Grinnell College's Grant O. Gale Observatory have been designed with this constraint in mind. Although modest by large observatory standards, the instrumentation incorporates contemporary technology such as solid state detectors and microcomputer control. Because a variety of student projects as well as a multi-faceted research program must share the telescope, often during a single night, the auxiliary instrumentation has been designed to eliminate almost all hardware changes associated with changes in observing technique (spectroscopy vs. photometry, for example). The auxiliary instruments are all interfaced to the data acquisition computer, which can also communicate with the telescope control system. Research programs have been selected both on the basis of scientific merit and compatability with our capabilities and constraints. We therefore concentrate on long-term observing programs which cannot be carried out at most large observatories and attempt to take advantage of our flexible instrumentation and computer-processing capability.

The telescope is a 61cm F/13.5 Cassegrain reflector designed and built by DFM Engineering, Inc. of Longmont, Colorado (USA). Mounted to the telescope is a guide-acquire-module which provides multiple instrument capability and guiding and acquisition. A low cost but high performance television system (detecting 16.1 magnitude stars) has been built using a multichannel plate intensifier and a Sony CCD television camera.

A computer-controlled photometer and Reticon spectrograph are under construction.

An Intel 86/380 microcomputer is used for data acquisition and instrument control. This microcomputer can also command the telescope control system to perform telescope motions such as slew to coordinates and offset.

J. B. Hearnshaw and P. L. Cottrell (eds.), Instrumentation and Research Programmes for Small Telescopes, 79.

A FOUR STAR PHOTOMETER FOR USE ON SMALL TELESCOPES

D.B. Caton and J.T. Pollock
Department of Physics and Astronomy
Appalachian State University
Boone, North Carolina, U.S.A.

ABSTRACT. An automated photoelectric four-star photometer has been designed for use on the 46-cm reflector at Appalachian State University's Dark Sky Observatory. The system will be composed of integrated detector-amplifiers, stepper-motor controlled x-y stages, and automatic counters.

1. INTRODUCTION

We propose to build a multiple-star photometer for regular use on the 46-cm reflector at ASU's Dark Sky Observatory. There are several advantages to a design allowing up to four objects to be monitored simultaneously. The instrument will be fabricated in our own department shop using readily available sub-assemblies such as mechanical X-Y stages, stepper-motor controllers, integrated detector/amplifier packages, and fiber optics.

2. THE DESIGN

The instrument is coupled to the telescope through a rotational stage, to allow rotation that may be necessary for optimizing alignment with a given stellar field. It is stepper motor driven with a step size of 0.01°, corresponding to 0.017 arc-seconds of setting accuracy (at the edge of the field of view), in the sky. An 'origin-position' detection switch and incremental encoder will allow initialization and verification that the instrument indeed is set up at the correct angle for a given object's field.

Directly attached to the rotating coupling is the upper plate, below which is a full-field filter wheel. Placement of the filters here eliminates many design problems, but adds the complication of needing to map the transparency of each filter. The filters are carried on another rotating stage which was chosen over a simple stepper motor to guarantee that the filters can be repositioned in the beam to within the stage's 0.01° rotational accuracy.

J. B. Hearnshaw and P. L. Cottrell (eds.), Instrumentation and Research Programmes for Small Telescopes, 81–82.

Each of the four stepper-motor driven X-Y stages are mounted on the middle plate below the filter wheel. Each X-Y stage carries a detector that normally services one quadrant of the field of view, although any detector can reach any point in the 80-mm (30 arc-minute) focal plane circle. The 30-arc minute field will allow one or more comparison stars to be chosen for most fields. The X-Y stages have stepper motors, home position detectors, and incremental encoders built in, and will thus provide for accurate setting to a given star field.

The light from the four beams will be brought to the detectors via four fused silica optical solid fibers held rigidly in a right angle bend. The fibers will also serve as the Fabry lens by randomizing the star+sky image before it reaches the detectors. The diameter of the fiber will limit the sky light collected. The 1-mm bundle will provide a 24 arc-second aperture, large enough to avoid spill-over of the image under adverse seeing conditions.

The detectors to be used will be compact, integrated units consisting of the photomultiplier tube, high voltage supply, output amplifier, and voltage-to-frequency converter. Such units are available in small packages that may be easily mounted on the X-Y stages with only very flexible low-voltage wires needed for hook-up. The detectors will house 3/4" diameter S-20 photomultipliers, ordered specially selected for low dark current.

Viewing will be provided by a set of four coherent fiber bundles attached at a specific offset from the fibers leading to the photomultipliers. Thus a simple computer-directed offset will bring all four viewing bundles into place; their images will be viewed in the focal plane of a single eyepiece. The computer will be able to reposition the fibers to the star pattern to such a precision that centering one star/bundle will align the entire field. Additionally, a large eyepiece at the rear plate will allow coarse initial setup on the field.

The ten stepper motors will be driven by 'smart' controller systems that are interfaced to a microcomputer via the IEEE-488 bus. Our experience is that circuits of this complexity are best bought rather than prototyped, especially with the declining prices of smart controllers. Additionally, a commercial controller will probably be more reliable, as well as more easily implemented by other observatories wishing to replicate the photometer. The output pulses from the four detectors will be fed to four programmable counters which will also be read by the computer over the IEEE-488 data bus.

By using 'smart' controllers the task of writing the control software will be greatly reduced. Software will be written to transmit to the controllers the positions for each detector for a given field. A manual control mode will aid in setup. Each mechanical stage's position will be read from the encoders to check that all stages indeed reached their destinations. The filter will be set, the counters reset, and an integration can begin. The counters can all be started simultaneously via a group command over the IEEE-488 bus. It is proposed that the software be written in Forth—an ideal control language for this type of application.

AN AUTOMATED SOUTH POLE STELLAR TELESCOPE

K-Y. Chen, J. Esper, J. D. McNeill, J. P. Oliver, G. Schneider, and F. B. Wood
Department of Astronomy
University of Florida
Gainesville, Florida 32611
U. S. A.

ABSTRACT. An 8-cm optical telescope is constructed for use at the south pole. It is designed to make photoelectric observations of selected stars continuously through an austral winter. The automated operation is controlled by a computer. The aim is to study the variability of the star γ^2 Velorum as well as the condition of the polar sky and the performance of the instrument.

The south pole may be an ideal location for astronomical observations which can be benefit from continuous monitoring of variable sources. A telescope (Schneider et al., 1984) was installed at the pole in January 1984 to evaluate the site and observe the interesting Wolf-Rayet binary γ^2 Velorum. After initial testing in that winter, the instrument has been modified and rebuilt. It will be installed at the pole again in January 1986.

The telescope is a two-mirror siderostat with an 8-cm objective and with computer control of all functions. Photoelectric photometry will be carried out with six filters. The standard B and V filters and three interference filters, centered at λ4686 (He II), λ4768, and λ5696 (C III), are selected particularly for γ^2 Velorum. Three other stars, HR 3452, α and β Gruis, are also programmed to be observed. Another interference filter centered at λ5600 is used to observe sky brightness also.

The rotating optical head, which consists of the two mirrors, is driven in the azimuthal directions by a stepping motor. The front part (the entrance mirror) can be set manually for a selected altitude (declination). The reflected light is converged through one of twenty-five 5-arc minute diaphrams at the focal plane of the objective. The diaphrams are located on a wheel. The filters are also mounted on a wheel. These wheels are driven by stepping motors for selecting proper diaphram and desired filter. An EMI9781 multiplier phototube with an electrometer is used to measure the filtered lights.

Provision is made for internal calibration of the photometric system and an internal "frost sensor" will give warning of potential frost/snow build-up on the optical surfaces. A careful thermal design

J. B. Hearnshaw and P. L. Cottrell (eds.), Instrumentation and Research Programmes for Small Telescopes, 83–84.

study has been carried out in order to deal with problems involved with the harsh south pole environment.

The computer system provides for completely automatic operation of the telescope for the entire observing season. Manual computer control is available for unanticipated changes in the observing program, and for diagnostic testing and repair. The data will be stored on magnetic disk and transmitted back to Florida from Amundsen-Scott Station via satellite.

This research is supported by the National Science Foundation grants DPP 82-17830 and DPP 84-14128.

Reference

Schneider, G., Wood, F. B., Chen, K-Y., Giovane, F., Oliver, J. P. 1984. Bull. Amer. Astron. Soc. 16, 497.

THE AUTOMATED PATROL TELESCOPE

J. W. Cochrane, P. Mitchell, P. W. Payne, J. W. V. Storey
and B. L. Webster
School of Physics
University of New South Wales
P.O. Box 1
Kensington, NSW 2033
Australia

ABSTRACT. A new astronomical system is nearing completion at U.N.S.W. in which a charge coupled device (CCD) array detector is matched to a Schmidt telescope, with the observations completely analyzed on-line. We have modified a Baker-Nunn satellite tracking camera, with its excellent 0.5m, f/1 super-Schmidt optics, converting it into a computer-controlled equatorially-mounted astronomical telescope. The telescope is expected to be located at Siding Spring Observatory in north-western New South Wales.

The Baker-Nunn telescope consists of a spherical primary mirror 0.76 m in diameter and a 3-element correcting lens 0.5 m across, giving an f/1 beam with a 55 degree corrected field. This section of the original camera remains intact, but substantial modifications were required to produce an economical drive system capable of accurate pointing and tracking.

Precision spherical roller bearings are used on both axes. The declination drive consists of the original worm gear, plus a precision harmonic drive. A separate gearing system eliminates backlash from the worm gear by exerting a constant torque to oppose or augment the main drive. The Right Ascension drive consists of a 1 metre diameter friction-driven disc, plus a precision harmonic drive. The advantages of the friction-driven disc are ease of fabrication, low cost (approximately $400), elimination of backlash, and high efficiency. One disadvantage is the limited torque that can be transmitted, which requires that the telescope be well balanced. Both axes have identical printed-armature d.c. motors. Position and velocity are sensed by incremental encoders, with a resolution of 0.12 arcsec in RA and 0.04 arcsec in Dec. These encoder signals are sent to two 6502 microprocessors which control the drive motors via 10-bit A/D converters, which cover the full range from tracking to slewing. An Apple IIe microcomputer down-loads the required positions to the microprocessors at a rate of 0.2 Hz, while sidereal rate tracking pulses are inserted at ~ 120 Hz.

J. B. Hearnshaw and P. L. Cottrell (eds.), Instrumentation and Research Programmes for Small Telescopes, 85–86.

The CCD is a GEC P8603, mounted on a thermoelectric cooler within a lightweight dewar. A field of 1.4 x 0.9 degrees results, recorded on the 385 x 576 9 arcsec pixels. Data is transferred via an IEEE-488 bus to an NEC APC III computer for processing. Transfer of a full frame takes about 11 seconds. After processing, the image can be transferred to the TV memory in about 1 second. Thereafter, the TV memory will provide a continuous display of the image.

There are three proposed modes of operating:

A) T.V. frame rate; limiting magnitude $m_V = 10$ in 20 msec.

B) Autoguiding; $m_V = 15$ in 1 second.

C) Deep Sky Search; $m_V = 19$ in 300 seconds.

(signal to noise ratio of 10:1 in each case). For photometry of variable stars, a 1% change in brightness should be detected at several sigma in an object as faint as $m_V = 15$.

The final system will be completely self-operating, slewing to any position in the sky within 80 secs. It will drive from one pre-specified position to another, observing and analysing over the one square degree field with the CCD. Several analysis modes will be possible: the field can be compared with a stored field from a previous observation and any difference recorded; a specified object could be identified and its brightness measured relative to other objects on the frame; the whole frame can be stored; or one part only of the CCD can be exposed to the sky and successive images shifted to other parts of the chip to give very high time resolution. Special features to be exploited are 1) it gives immediate feedback on transient events so that other observations can be initiated, 2) within limits, it allows searches over a fairly wide field with high time resolution, 3) by processing most of the incoming data on the spot it enables programs to be carried out that would otherwise be swamped in a mountain of data.

With this versatility we envisage a wide range of possible observing programs. These include conventional surface photometry utilizing the large field, supernova searches, monitoring interesting variable stars, and more speculative searches for the optical counterparts to γ-ray bursters.

We are particularly grateful to the Smithsonian Astrophysical Observatory for their donation of the original Baker-Nunn camera.

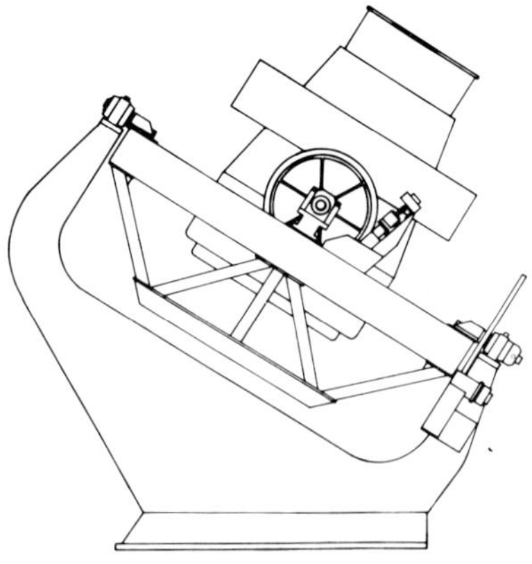

The Automated Patrol Telescope

AUTOMATIC PHOTOELECTRIC TELESCOPES: STATUS AND POTENTIALS

Dr. David L. Crawford
Kitt Peak National Observatory
National Optical Astronomy Observatories
P. O. Box 26732
Tucson, AZ 85726 USA

ABSTRACT. Automatic photoelectric telescopes now exist. They are currently doing precision photometry of variable stars, fully automatically. More such telescopes will exist in the near future, and they will be doing automatic photometry of all sorts.

New generation telescopes are not limited to the giant telescopes now being designed by major observatories. Many of the same ideas and concepts that have bred plans for these large aperture telescope programs can and are being used even for telescopes as small as 10-inch aperture. Several such small NGT facilities now exist and more are planned. There exists even a viable concept for an Automatic Photoelectric Telescope Engineering Service (APTS), to build and operate a number of such instruments, on one site, each instrument to be owned by a "customer" who funds (owns) and totally controls the research use of the telescope. The APTS acquires and puts the telescope into operation, and fully maintains and monitors the use.

Russ Genet and Lou Boyd, from the Phoenix, AZ area, are the driving forces behind this concept and supply the engineering expertise. A Scientific Advisory Panel has been appointed consisting presently of Dave Crawford (Chairman), Doug Hall, and Don Hayes. Current plans call for a battery of automatic photoelectric telescopes (APTS) to be located in an existing building on Mt. Hopkins in southern Arizona, near the MMT facility. The first telescopes should begin routine operation at this site early in 1986. They will operate routinely, being checked by a human operator less than once per week.

Programs for such APTS will consist initially of variable star light curves, all sky photometry and monitoring of bright stars and standard stars, and monitoring of special interest objects for variability. Early versions of the APTS have operated automatically for the last two years, very effectively, for variable star monitoring, in Phoenix and in Ohio. See, for example, the article by Boyd, Genet, and Hall in the July issue of Sky and Telescope for a description of the APTS and of their current use on variable stars.

J. B. Hearnshaw and P. L. Cottrell (eds.), Instrumentation and Research Programmes for Small Telescopes, 87–88.

Later extensions to somewhat larger telescopes can and will include CCD area photometry, spectrophotometry, and near infrared photometry.

Such programs are a natural growth of current new technology applied to the design and use of small telescopes. We predict a revolution in the quantity and quality of photometric data from such facilities.

UNIVERSITY OF TORONTO SOUTHERN OBSERVATORY

R. F. Garrison
David Dunlap Observatory
University of Toronto
Box 360 Richmond Hill, Ontario
Canada L4C 4Y6

Why are Canadian astronomers so keen on going to Chile to study the stars with a small telescope? First and foremost, the southern half of the celestial sphere contains scientifically exciting and unique objects which have long been neglected. Secondly the rapid growth of Canadian cities and the cloudy Canadian climate renders nearby telescopes less effective than they used to be for some (though not all) types of research.

It is on Las Campanas at the southern edge of the Atacama Desert that the U. of T. 60-cm telescope is located. Through the cooperation of the Carnegie Institution of Washington, the U. of T. was able to make arrangements to have a share in the site, which is at a latitude of 29 deg. south and a longitude of 71 deg. west. The altitude is over 2000 m.

The 60-cm telescope was purchased from Ealing in Montreal, but was built by Competition Associates in Cambridge, Mass. The f/15 Cassegrain is fitted with a camera for plates 10x13-cm in size. The scale, 22.54 arcsec/mm, is very close to that of the Newtonian focus of the 1.88-m reflector in Richmond Hill. An ITT magnetically-focussed image-tube camera, which has a useful field of about 12 arcmin, may also be used.

A grating spectrograph, designed by Garrison, may be employed with photographic plates or with a one-stage image tube at dispersions of 120 A/mm and 67 A/mm. Three photon-counting photoelectric photometers are also available. Photomultipliers on hand are of three types: S4 for UBV wavelengths, S25 extended red for RI, and an Indium-Antimonide for the infrared JHK.

There is an on-line reduction system complete with two Osborne-I microcomputers, with an HP-9815A and an HP-85 as backup. By the end of this year, an échelle spectrograph with a Reticon detector will be in use and two CCD detector systems have been applied for. The scale and availability of this small telescope make it ideal for survey programs, which are difficult, if not impossible for large telescopes.

Other auxiliary instrumentation is available to fit particular astronomical programs. A short-wave radio is available for twice-weekly (or more) communications with DDO. There is also a large tape collection to provide musical comfort for the observers.

Besides the dome, which houses the telescope, the University has built a small combination residence and operations centre, constructed of

J. B. Hearnshaw and P. L. Cottrell (eds.), Instrumentation and Research Programmes for Small Telescopes, 89–90.

local stone so as to blend into the mountain slope. The building provides comfortable accommodation for two observers, desk and work space, facilities for light meals, and a fully-equipped darkroom.

There are many special advantages to having such a telescope at a good location. One is that it allows us to carry out variable-star surveys, which are difficult with large telescopes or cloudy skies. Two large spectroscopic surveys, a large photographic survey, and several photoelectric surveys have been carried out. In addition, several new variables have been discovered serendipitously during surveys for other projects.

The most spectacular example of serendipity is that of CPD -48 1577, which is the brightest cataclysmic variable known. It was discovered during the MK spectral classification survey of southern OB stars carried out by Garrison, Hiltner and Schild (1977). Another unique discovery from that survey is that of the pure helium star CPD -31 1701; there is nothing in the spectrum other than helium. This has been confirmed with a fine analysis by the Kiel group, who found Y = 1.0.

Gauthier studied the spectra of 26 southern Cepheids around their cycles. His is the most comprehensive classification study ever carried out for these fascinating stars. His results are published in his 1983 thesis and in an article in the book "The MK Process and Stellar Classification", edited by Garrison.

Crowe extended the Mira-variable-star catalogue of Keenan, Garrison and Deutsch (1974) to include the southern Miras. Using direct and image-tube plates, he took 475 spectra of 72 variables around their cycles at a dispersion of 120 A/mm. This work represents a very considerable effort over about 6 years. Many of the stars were quite faint at minimum for such a small telescope. His results are published in his 1983 thesis and the JRASC as well as the JAAVSO and "The MK Process and Stellar Classification", edited by Garrison. The complete version will be published as an Ap. J. Supplement.

A program currently underway is being carried out by the resident observer for Bill Herbst of Wesleyan University. It consists of photoelectric observations of H-alpha in southern T Tauri stars. These fascinating stars vary irregularly and the plan is to monitor them continuously for several days of each month over a period of years.

Programs such as these can only be done with a small telescope at a great site, because larger telescopes are too much in demand. In Chile, the excellent seeing, clear skies and freedom from city lights help to make observing with a small telescope very effective in producing good and useful scientific results. It makes it worth the long trip to Chile for Canadian astronomers. The telescope time is open to all astronomers on the basis of scientific merit of the application alone.

CONSTRUCTION OF A 32CM CASSEGRAIN PHOTOELECTRIC TELESCOPE

David Hickey
27 Shields Street
Redcliffe, 4020
Queensland, Australia

ABSTRACT. This paper describes the building of a 32cm telescope which had been designed from the beginning to undertake high precision photoelectric photometry.

1. INTRODUCTION

The equipment is still under construction and has been so for the last four years. Due to the usual restrictions on time and finance the project has been slower than first anticipated. My final objective is to complete a photoelectric instrument that can be used to carry out high accuracy observations on contact and close binary systems.

2. THE TELESCOPE

The telescope is an amalgamation of two existing components; the R.A. assembly and pier and the optical tube assembly and some new parts consisting of a fork support, tube truss and a seven degree wedge. The mount is about a century old originally owned by D. Eglinton (the history of this telescope will be researched). Construction is of an old type cast iron resulting in a extremely rigid mount. The tube is an Ealing (U.K.) 32cm f16 cassegrain arrangement of aluminium construction. The tube is fitted to a cast aluminium backplate, which in turn is fitted to the truss. The photometer is also bolted to the backplate.

A wedge of 30mm steel had to be fabricated to adapt the mount to my latitude and permit finer alignment. The truss is made of 25mm hollow square section and is very sturdy. It is completely symmetrical so as not to upset the overall balance, an important factor to consider if you wish to eliminate useless counterweights. The balance of the telescope will not be perfect until the photometer is fitted, so temporary weights are mounted on the backplate.

J. B. Hearnshaw and P. L. Cottrell (eds.), Instrumentation and Research Programmes for Small Telescopes, 91–92.

2.1 The Electronics

The electronics to date consist of the following: Honeywell chart recorder, Fluke high voltage supply, HP digital printer, integrator/ amplifier (D. Thomas design). An electronic clock is the other main component yet to be constructed, this will be of the same design as the one used at Midway Observatory. All the equipment will be tested at Midway Observatory before operation begins.

2.2 The Photometer Head

The photometer head will be very similar to the ones used at Siding Spring Observatory. The cold box and as yet unconfirmed filter box are coming from there, so to make a complete unit I will have the missing parts made. The PMT will be a 1P21 or Gencom that were kindly given to me by R. Reisenweber of Rolling Ridge Observatory.

3. SUMMARY

I uphold the philosophy that, if something has been proven to work, then use it. Harold Kennedy has often pointed out to me, one does not invent the wheel twice. So this is the reason I chose to duplicate a photometric system that has been proven albeit not an easy way.

4. CONCLUSION

When the equipment is operational it will be primarily used for data acquisition on the Mass-Period relationship of eclipsing binaries (Kennedy 1986) and other related phenomena associated with contact and close binaries.

5. ACKNOWLEDGEMENTS

I would like to thank Harold Kennedy (Midway Observatory) for his endless support and advice, from Siding Spring Observatory I would like to thank Don Thomas, Bob Moone, Peter Lynstead. And to Bob Reisenweber for his generosity.

6. REFERENCE

Kennedy, H.D. 1986, IAU Symposium No. 118 (this volume)

A 75cm ALT-AZ TELESCOPE WITH SHORT TIME FOCUS EXCHANGING SYSTEM

Syuzo Isobe[1], Nobuo Shinohara[2], and Hidehiko Agata[2]
1: Tokyo Astronomical Observatory, 2-21-1, Osawa, Mitaka, Tokyo, Japan
2: Sundai Senior High School, 6-1-10, Oji, Kita-ku, Tokyo, Japan

ABSTRACT. The Sundai senior high school built a telescope with an alt-az mounting and an aperture of 75cm. A photographic camera at the prime focus, and an SIT camera, a photomultiplier, and a spectrograph at the Nasmyth focus are prepared. Time needed for the exchange between instruments is in less than a minute. Therefore, effective observations are performed by changing the instruments which are proper for the sky condition at the observed night.

The telescope of the Sundai Observatory at Kita-Karuizawa(SOK) is not only the first real alt-az telescope in Japan, but also the first working telescope with honeycomb primary mirror in the world produced by Roger Angel at the Steward Observatory. The PC-8800 micro-computer with 8 bits controls its pointing and driving in an accuracy of 0!5 and 6" in open loop, respectively. There is a flip-flop top ring between the primary and Nasmyth foci which makes a quick exchange between two foci possible. However, this system gives only tight space for an instrument of the prime focus which is a photographic camera. There are two Nasmyth foci between which one can change by tilting the tertiary mirror in 90°. Since the telescope was built for school pupils to look at planets, stars, nebulae, and clusters, one of the Nasmyth focus is used for only

J. B. Hearnshaw and P. L. Cottrell (eds.), Instrumentation and Research Programmes for Small Telescopes, 93–94.

eye observations. At the other Nasmyth focus, there is the Nasmyth platform which does not change height during the equatorial motion because of the alt-az mounting. Three instruments are permanently fixed on the platform. An SIT camera on the line of elevation axis is set on an instrumental rotator to compensate the field rotation because of alt-az mounting. We are able to obtain two dimensional images down to the 15th magnitude with 1 minute integration of a real time frame memory. In front of the Nasmyth focal plane, a dichroic mirror is inserted to give stellar light into either a photomultiplier or a spectrometer. This kind of change is done in a few seconds. The photometer contains normal U, B, V filters and the other 3 filters will be set. The spectrometer has dispersions of 20A/mm and 200A/mm. Using a detector with 3 stage microchannel plate and MOS camera, we are able to observe down to 12th magnitude in 10 minute integration. Since the observational site is not good enough, our system for exchanging observational instruments in short time makes effective observations possible.

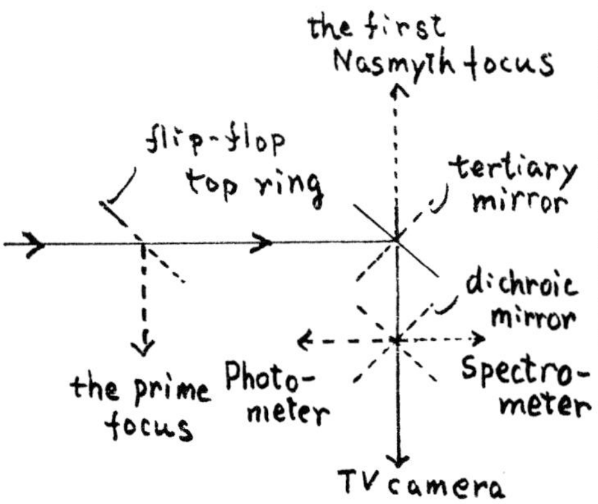

Figure 2. A flip-flop top ring between the Prime and Nasmyth foci.

Figure 3. The tertiary mirror set on the elevation axis.

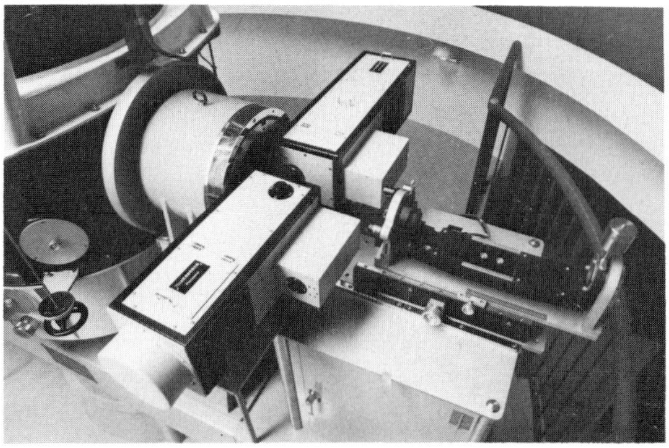

Figure 4. Arrangement of SIT, photometer, and spectrometer on the Nasmyth platform.

A SINGLE-CHANNEL POLARIMETER

S.K. Jain and J.C. Bhattacharyya
Indian Institute of Astrophysics
Bangalore 560034, India

ABSTRACT. We have developed a single-channel photopolarimeter to measure polarization ~ 1%, with an accuracy of 0.1% or so. Laboratory tests show that the performance of the instrument is as per the requirements.

1. INTRODUCTION

At Kavalur Observatory, we have started, among other programs, a program of polarimetric observations of stars with rotating extended atmospheres (Peraiah 1976), and study of members of very young ($\leqslant 10^6$ years) clusters and associations. This paper reports the design and performance of the polarimeter developed for such investigations.

2. INSTRUMENT

A schematic lay-out of the polarimeter is shown in fig. 1. The basic design is similar to that of Hall and Mikesell (1950). In this instrument the light is chopped at 125 Hz using a Pockels cell. This is, then, analyzed with an HNP'B polaroid sheet which is rotated about

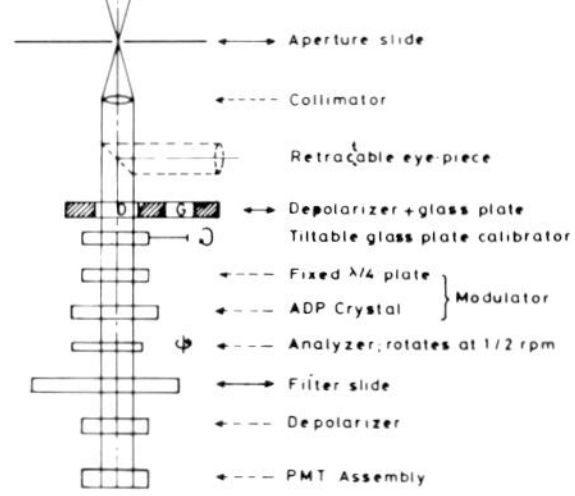

Figure 1. Optical arrangement of the polarimeter

J. B. Hearnshaw and P. L. Cottrell (eds.), Instrumentation and Research Programmes for Small Telescopes, 95–97.

CCD INSTRUMENTATION AT CAPILLA PEAK OBSERVATORY

B. E. Laubscher
Department of Physics and Astronomy
University of New Mexico
800 Yale Blvd., N. E.
Albuquerque, N. M. 87131

1. INTRODUCTION

The astronomers at the University of New Mexico are applying the finishing touches to their CCD project. We addressed many issues during the course of this project, such as whether to build or buy the CCD camera system, what upgrades our telescope and observatory were in need of, and what kind of image processing capability the instrument demanded. Our decisions and solutions are presented as a guide to other institutions which may be considering CCD instrumentation for their small telescopes, especially since most face the same constraints that we do: limited money and manpower.

2. CCD SYSTEM

Through visits to the Kitt Peak Station of NOAO, I generated rough estimates of $150,000 and two years of work to fabricate the CCD camera system at UNM. This includes no design work, but simply to "copy" the Kitt Peak System. This estimate is a lower limit and attempts to include all hardware, computers, salaries and shop time. The time estimate is based on one engineer and two technicians working full time. The other option we considered was the "off the shelf" system produced by Photometrics, Ltd. in Tucson. Their cost for the CCD system, excluding the chip, was about $70,000--very near hardware cost. We opted for the Photometrics System, in part because it comes with a complete software package to control the CCD camera and manipulate the images for preliminary processing.

3. REQUIRED UPGRADES FOR OBSERVATORY AND TELESCOPE

We determined as well that our particular observatory and telescope should be upgraded to allow more efficient use of our powerful instrumentation. Thus, the observatory "warm room" was made into a Faraday Cage; the power into the building was shielded with spark gap

J. B. Hearnshaw and P. L. Cottrell (eds.), Instrumentation and Research Programmes for Small Telescopes, 99–100.

arrestors and a capacitor bank, and the powerline to the CCD computer was put on an Uninterruptible Power Supply system. The 61 cm, Boller & Chivens, f/15.2 telescope has been outfitted with shaft encoders, filter sheel, and offset guiding system, that uses a Fairchild CCD camera and image intensifier still being developed. This ancillary equipment is all computer-controlled from the warm room.

4. IMAGE PROCESSING

The astrophysics group at UNM has purchased an I^2S Model 75 to interface with UNM's research VAX 11-785. We currently have AIPS running on the system. Both the GASP (GAlaxy Surface Photometry) software and a Dunn camera hardcopy unit are now being installed. We considered the development of a complete image processing system for data reduction and analysis to be an integral part of the CCD upgrade.

5. SCIENTIFIC GOALS

The installation of the state-of-the-art CCD instrumentation will usher in a new era for Capilla Peak Observatory. With this CCD camera system we are able to extend our observational capabilities to the realm of extragalactic imaging. Moreover, our capacity to do stellar photometry is not compromised by this new instrument. We are planning to pursue extragalactic survey work on galaxy superclustering and other large scale structure topics. Coincident with this work the stellar photometry research, primarily RS CVn stars, will be continued. Because the stellar observations can be done during bright time, we feel that the CCD allows elegant integration of two very different research programs.

6. CONCLUSION

Preliminary observations using the system have produced very encouraging results in both the CCD hardware performance and our image processing capabilities. Although some "bugs" have not yet been ironed out, we expect to produce our first research images in March of 1986. The time and effort spent on the system will result in a frontline extragalactic research tool that transforms a small telescope into one as powerful as a large telescope.

THE OPTICAL SYSTEM OF THE MT JOHN ONE METRE TELESCOPE

G.R. Nankivell and N.J. Rumsey
Physics and Engineering Laboratory, DSIR
Lower Hutt, New Zealand

The telescope will be used at the Cassegrain focal plane only. Two interchangeable systems of secondary optics are required to give relative apertures of:
(a) f/13.5 for spectroscopy and photometry over a small field.
(b) f/8 for photography over as wide a field as practicable.

Requirement (a) is met by using a Dall-Kirkham system (Kirkham, 1938), ie the primary mirror is an ellipsoid and the secondary mirror is strictly spherical. The f/8 system uses the same primary mirror, but the secondary mirror is replaced by another which, although spherical, is longer in radius of curvature and closer to the primary mirror. A pair of field correcting lens elements with spherical surfaces is placed between the secondary mirror and the focal plane. This f/8 system closely resembles one first suggested by Wynne (1949; see also Harmer & Wynne (1976) for a detailed design) for use with a paraboloid primary mirror. Schulte (1966) reported that S Rosin had worked out that if the primary mirror is an ellipsoid the correcting lens elements move down to a much more convenient location than in the original Wynne-type system. The Mount John telescope is probably the first to use this modification.

Because the secondary mirrors are spherical, they are much quicker and easier to make and align than those of classical Cassegrain or Ritchey-Chrétien telescopes. The f/13.5 system has a relatively large amount of coma in the images away from the centre of the field. This will restrict the field of apparently good definition to about 12 minutes of arc when the seeing is 4 seconds of arc (very common at Mount John), or to a smaller field when the seeing is unusually good. The photographic f/8 system has fairly uniform definition over a field of 1.6 degrees diameter. Most of the light in each star image falls within a circle of one second of arc diameter. This is not as good an image sharpness as other systems have achieved, but is entirely adequate under the conditions usually found at Mount John and permits the wide field considered of importance in this case.

The material used for the mirrors is "Zerodur" which has a coefficient of thermal expansion very close to zero. This speeds up manufacture by eliminating waiting time for thermal equilibrium before

J. B. Hearnshaw and P. L. Cottrell (eds.), Instrumentation and Research Programmes for Small Telescopes, 101–102.

testing, and prevents warping of the mirror from temperature changes during observations.

The primary mirror was worked on the machine built by Waland for working the optics of the 15 inch and 30 inch Schmidt-Cassegrain cameras of the University of St Andrews, Dundee, (King, 1955). The grinding tools were faced with square ceramic tile facets stuck down on aluminium alloy bodies with a beeswax and rosin mixture. The fillet of wax around each facet was given a rounded concave profile so that there would be no re-entrant angles where abrasive grains could lodge. The tools performed extremely well.

After coarse grinding, during which the mirror blank was brought to its final shape (apart from figuring), the hole in the primary mirror was trepanned out. The material so removed has been sliced up and test plates for the correcting lenses are being made from it.

During the early stages of figuring, the profile of the mirror was checked by the conventional zonal Foucault test. Once the profile was close to its correct shape a null-testing Offner compensator (Offner, 1978) was inserted. This permitted small irregularities to be seen easily that would have been difficult to detect with the conventional test. The compensator was made from two identical, centrally perforated, concave spherical mirrors facing each other, with their separation established by carefully machined rods.

The working of the primary mirror is described in detail by Nankivell (in press).

REFERENCES

Harmer, G.F.W. & Wynne, C.G. (1976), *Mon. Not. Roy. Astron. Soc.* **177**, 25P-30P, 'A simple wide-field Cassegrain telescope'.

King, H.C. (1955), 'The History of the Telescope', (Griffin, London). See pages 374 & 377.

Kirkham, A.R. (1938), *Scientific American* **158**, 374, 'The Cassegrain telescope with spherical secondary mirror'.

Nankivell, G.R. (in press). PEL Report: 'The primary mirror for the Mount John one metre telescope'.

Offner, A. (1978), 'Optical Shop Testing'. (Malacara, D., Ed.; Wiley New York), pp. 444-452, 'The Offner compensator'. (See especially pp. 449-452.)

Schulte, D.H., (1966), I.A.U. Symposium No. 27: 'The Construction of Large Telescopes.' (Crawford, D.L., Ed.; Academic Press, London & New York), p. 37.

Wynne, C.G. (1949), *Proc. Phys. Soc.* **B62**, 772-778, 'Field correctors for parabolic mirrors'.

BLACK BIRCH ASTROMETRIC OBSERVATORY

D. Robinson, Dr. S. Dick, Dr. R. Stone, B. Loader
United States Naval Observatory
Washington DC
U.S.A

The seven inch transit circle near Blenheim, New Zealand is intended in conjunction with the six inch circle at Washington, to produce a fundamental catalogue and to re-observe the Southern Reference Stars.

The transit circle is designed to set and automatically acquire a star on input of its reference number. After the observation the data gathered is stored on disk. Subsequently following reduction it is stored on magnetic tape.

Photographs and diagrams showed the Black Birch site, the telescope, computer and clock system, image dissector and the setting circle with scanning microscopes.

J. B. Hearnshaw and P. L. Cottrell (eds.), Instrumentation and Research Programmes for Small Telescopes, 103.

A NEW POLARIMETER FOR STELLAR POLARISATION MEASUREMENT

A.K. Saxena
Indian Institute of Astrophysics
Bangalore 560034
India

ABSTRACT. A polarimeter for the measurement of Stokes' parameters for stellar and quasi stellar objects in specific wavelength regions has been reported. The variable retardation and dispersive property of a small angle Babinet Compensator has been exploited for such measurement. The theoretical basis of the method is given in detail.

1. INTRODUCTION

Multi element detector arrays like CCD, IPCS, IDS, etc., have given a new dimension to astronomical instrumentation. CCDs with high quantum efficiency, low read out noise and small pixel size (~10-15μ) have become available. Such CCDs are finding their place at the detector end in many astronomical instruments. These have been successfully used in polarimeters enabling many spectral and spatial elements to be observed simultaneously. The Babinet Compensator has been known as a sensitive device for measurement of small phase changes in polarised light. It has been shown that with suitable precautions[1] the minimum detectable phase change[2] may be made as low as 2π milli radian. CCDs coupled to a Babinet Compensator-polariser combination in a specified manner can provide a convenient means of measuring four Stokes' parameters separately in very narrow wavelength intervals simultaneously for a considerably large wavelength range.

2. THEORY

Fig. 1 shows the optical set up for the polarimeter. When a babinet Compensator having a small wedge angle is placed in the path of a polarised collimated beam followed by an

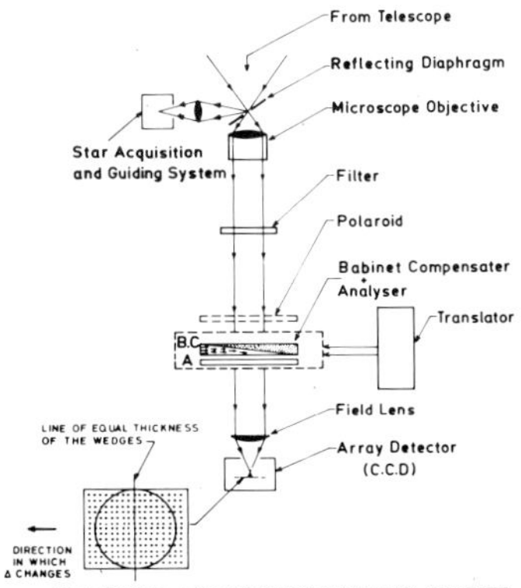

Fig.1. OPTICAL ARRANGEMENT OF THE POLARIMETER

J. B. Hearnshaw and P. L. Cottrell (eds.), Instrumentation and Research Programmes for Small Telescopes, 105–107.

analyser, a varying phase retardation is introduced between the E & O components of the wave along one diameter of the beam. But a constant phase difference remains along the perpendicular direction parallel to the line of equal thickness of the wedges. If this beam of light is imaged with a suitable field lens to a detector array in such a way that, say the columns of the detector coincide with the direction of constant phase, then the change in intensity of the beam in the perpendicular direction due to the difference in phase change introduced by the wedges can be obtained from the corresponding column average. The value of phase change introduced by the compensator along any column at a distance χ from the line of equal thickness of the wedge is given by

$$\Delta = \delta_1 - \delta_2 = \frac{2\pi}{\lambda} (n_e - n_o)\, 2\chi \tan \alpha \tag{1}$$

where n_e and n_o are the extraordinary and ordinary refractive indices of wedge material and α is the wedge angle. If further $I_{\lambda up} + I_{\lambda P}$ Q_λ, U_λ, and V_λ are the Stokes' parameters of the incident beam of wavelength λ then modulation of the intensity at any point in the perpendicular direction for this wavelength can be obtained by the Muller matrix product of the corresponding elements.

If λ_1 to λ_2 is the range of wavelength getting transmitted through the filter and this wavelength range is divided into n small intervals, then the total intensity at any point will be the summation for all wavelengths.

$$I_m = \frac{1}{2}\left[\sum_{n=1}^{n} I_{\lambda_n up} + \sum_{n=1}^{n}\left(I_{\lambda_n P} + U_{\lambda_n} \cos\Delta + V_{\lambda_n} \sin\Delta\right)\right] \tag{2}$$

I_m is the intensity which is measured from the column average. A quantity over all average intensity per pixel can be defined for the purpose of normalisation from equation (2) by taking an average over a large number of cycles of Δ i.e. integration over the values 0 to $2K\pi$.

For the spectral response of the detector and filter and other components proper weighting factors can be introduced and the equation (2) can be written in expanded form as

$$\begin{aligned}(I_{m,\Delta}) = I_{uptot} &+ a_1 I_{\lambda_1 P} + a_1 U_{\lambda_1} \cos \Delta_{\lambda_1} + a_1 V_{\lambda_1} \sin \Delta_{\lambda_1} \\ &+ a_2 I_{\lambda_2 P} + a_2 U_{\lambda_2} \cos \Delta_{\lambda_n} + a_2 V_{\lambda_2} \sin \Delta_{\lambda_2} \\ &+ - - - - - - - \\ &+ a_n I_{\lambda_n P} + a_n U_{\lambda_n} \cos \Delta_{\lambda_n} + a_n V_{\lambda_n} \sin \Delta_{\lambda_n}\end{aligned} \tag{3}$$

One needs to do at least (3n+1) measurements of $(I_{m,\Delta})$ and the corresponding (3n+1) equations so obtained can be solved to give I_λ, U_λ, V_λ and Q_λ.

3. REMARKS

1. Approach is simple with no movement required for any component during observation

2. Simultaneously it can give all four Stokes' parameters for a reasonably large wavelength range.

3. Correction for defects in the optical polarising elements can be done in the reduction process.

REFERENCES

1. Pandya, T.P. and Saxena, A.K. *KOB Ser A* 2, pp 107 (1978)
2. Jerrard, H.G. *J. Opt. Soc. Am.* 39, pp 1031, (1949)

A SIMULTANEOUS TWO CHANNEL PHOTOMETRIC SYSTEM

D. J. Sullivan and A. van der Peet
Physics Department
Victoria University of Wellington
Wellington, NEW ZEALAND

ABSTRACT. A photometric system capable of simultaneous dual channel operation has been developed for use in NZ. The photometer head unit, developed and constructed by the University of Texas at Austin, permits simultaneous observation of two stars using suitable optics and mechanically coupled photomultiplier systems. The instrumentation and control system is centred around a DEC micro/PDP-11 computer and incorporates general purpose hardware and software developed specifically for the system.

1. INTRODUCTION

There is no doubt that a research programme for a small telescope should include photometry of variable stars. Monitoring stellar variability using ground-based telescopes is, of course, limited by atmospheric disturbances and, in particular the subtle changes in atmospheric transparency that sometimes accompany an otherwise good night. Simultaneous observation of a comparison star along with the star of interest is an obvious and now well established approach to this problem. A number of years observing experience with New Zealand's island climate has convinced us that the two channel approach is a virtual necessity for accurate and reliable photometry, especially where relatively short integration times are involved. For our two channel system, we have chosen the option of using a divided light path and two photomultiplier systems. Another approach combines a single photomultiplier with a fast chopping method, enabling effective simultaneous observation of two stars for time scales long relative to the chopping period. This has the advantage of avoiding any differential drifts between two photomultiplier tubes, but photometric efficiency is halved. Although recent developments in high efficiency 2D detectors (such as the charge coupled devices) now make this a preferred option, the current high capital costs involved will not fit within our small telescope budget.

J. B. Hearnshaw and P. L. Cottrell (eds.), Instrumentation and Research Programmes for Small Telescopes, 109–110.

2. HARDWARE

The Texas photometer-telescope unit employs a standard light path for the primary channel: light reaches the main photomultiplier via an aperture, a filter wheel driven by a stepper motor, and a fabry lens for imaging the telescope mirror on the photocathode. A 45 degree mirror containing a hole for the primary channel is positioned at the entrance to the photometer, and an adjustable 2D slide arrangement permits accurate location of a second stellar image in a narrow annulus in the field surrounding the primary object. This offset guiding mechanism can be used directly, or a flip mirror allows redirection of the light to a second smaller photomultiplier, via an aperture, stepping motor-controlled filter wheel and fabry lens.

Data acquisition and control for the system is accomplished using a DEC micro/PDP-11, featuring a 10 Mbyte hard disk and 5¼ inch floppy disks. A number of special purpose boards have been constructed. Single photoelectron pulses from the photomultipliers are amplified and shaped using units similar to Taylor (Taylor, 1980), and these enter the computer environment via counter/buffer hardware on one interface board. The other interface unit features items for analogue display, stepper motor control and time synchronisation with an external crystal-controlled clock.

3. SOFTWARE

A general purpose software package for the computer system has been produced. It follows the framework of a package produced previously for another system (Sullivan, 1982), but incorporates new features provided by the micro/PDP-11. Control of, and responses to, external hardware are handled by the multiple priority interrupt system, while background routines perform a range of tasks that are not time-critical. An important component in the software package (as with any modern instrument) are routines which allow real-time display of the incomming data streams in a variety of ways. The software has been kept as general as possible in order to allow flexible development for a number of proposed programmes ranging from high-speed photometry to multiple-filter photometry of eclipsing binary stars.

4. ACKNOWLEDGEMENTS

The authors would like to acknowledge the generous financial support of the NZ Lottery Board and University Research Grant Committee.

5. REFERENCES

Sullivan, D.J. (1982), Southern Stars **30**, 19
Taylor, D.J. (1980), Pub.Astr.Soc.Pac. **92**, 108

CCD CAMERA FOR THE 60CM TELESCOPE AT KAGOSHIMA SPACE CENTER

K.Takagishi
Faculty of Engineering, Miyazaki University
1-1-1 Kirishima, Miyazaki 880, Japan

M.Matsuoka
Institute of Space and Astronautical Science
4-6-1 Komaba, Meguro-ku, Tokyo 153, Japan

T.Omodaka
College of Liberal Arts, Kagoshima University
Kohrimoto, Kagoshima 890, Japan

ABSTRACT. A CCD detector has been developed for photometry and image detection with the 60cm reflector at Kagoshima Space Center. The structure and the manipulation of the instrument were simpified by the use of Peltier devices for cooling the CCD in order to eliminate thermal noise. A micro-computer system is used to control the instrument and process the data. A test observing run has demonstrated that a photometric sensitivity of 20th magnitude in the W band(3800-7000 Å) can be achieved in a 3600sec exposure.

1. GENERAL DESCRIPTION OF THE CCD CAMERA

As an astronomical detector, Charge Coupled Devices(CCDs) have the excellent properties of high quantum efficiency and broad dynamic range with good linearity and stability. We have developed a CCD camera designed for photometry and image detection with the 60cm reflector at Kagoshima Space Center(KSC). The CCD we have used is NEC μPD3511D, a front surface incident and interline transfer type CCD with 384(H) × 490 (V) pixels whose unit size is 23(H)× 13.5(V)μm.

Our CCD camera is composed of three units; the vacuum chamber, the circuit box and the micro-computer system. The first two units are connected together and mounted at the prime focus of the 60cm reflector, at which the image size on the CCD is 11.8' × 8.8'.

The configuration of the vacuum chamber is shown in Fig.1. The chamber consists of a cylindrical vessel of stainless steel with a quartz window and a copper disk on which the CCD and 3 layers of Peltier elements are compressed by a plate spring. The heat transfered from the hot side of the Peltier element to the disk is removed by the circulation of cooled water. The chamber is pumped to 10^{-3} torr by a rotary vacuum pump.

J. B. Hearnshaw and P. L. Cottrell (eds.), Instrumentation and Research Programmes for Small Telescopes, 111–112.

By using Peltier devices to cool the CCD, the structure and the manipulation of the instrument were simplified rather than the standard liquid nitrogen cooling system.

The upstair electronics in the circuit box consists of three cards; the clock card produces the required TTL level clocking from a master clock in the downstair micro-computer; the driver card converts the TTL level clock to the proper CCD driving levels; the analog card converts the CCD output to 12bits digital data. The control signals to and image data from the upstair circuit are carried along a 15m multi-wire cable from/to the micro-computer. The image data of the CCD is read synchronously with the master clock into the micro-computer memory at a rate of 80k pixels/sec, and recorded onto the floppy disks along with the time and date, integration time and other housekeeping information.

2. THE PERFORMANCE OF THE CCD CAMERA

The operating temperature of -70°C can be maintained with sufficient stability by our cooling method. The thermal noise of the CCD we have used is ~1 electron/sec/pixel at an operating temperature of -60°C, roughly equivalent to the sky brightness of KSC, and so the cooling ability of our instrument is sufficient for practical use.

The results of laboratory measuremants showed the total readout noise to be ∿50 electrons/pixel with no incident light, and the non-uniformity in sensitivity between pixels was found to be small, not exceeding 5% over the field.

A test observing run for photometric calibration using several standard stars demonstrated that in a 1000sec exposure a photometric sensitivity of up to 17th magnitude extending over a 5 magnitude range can be achieved with better than 5% accuracy in the V band, while a sensitivity of 20th magnitude in the W band (3800-7000 Å) can be achieved within a 3600sec exposure.

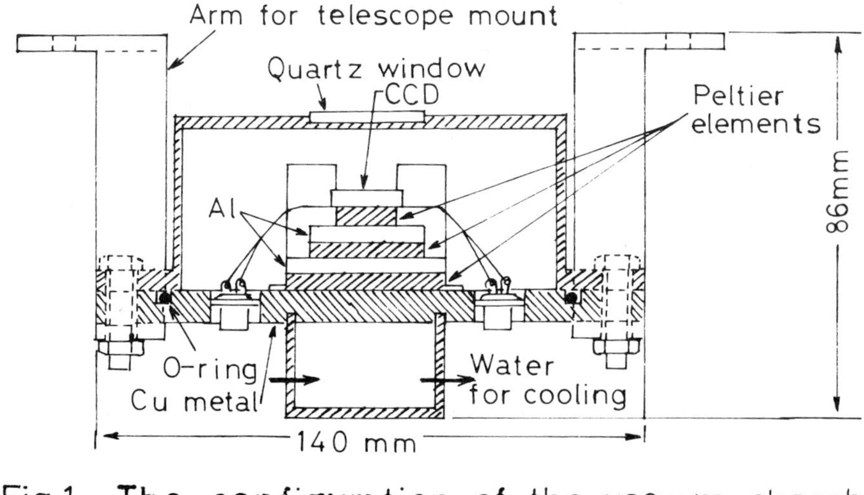

Fig.1 The configuration of the vacuum chamber

A TWIN REFLECTING ASTROGRAPH

K. Tomita
Astronomical Society of Japan
4-11-20 Yoga Setagaya, Tokyo
Japan

K. Takahashi
Takahashi Optical Co.
Ohara Nerima, Tokyo
Japan

ABSTRACT. A newly developed catadioptric optical system and a modified york-type equatorial mounting were designed and constructed. This telescope will be installed at Saipan Island for observing Comet Halley in Feb.-Apr. 1986.

1. OPTICAL SYSTEM

An optical system consisting of a hyperbolic mirror and a simple lens corrector was designed and manufactured featuring a wide field and a small focal ratio.

The reflecting system, which is 30 cm in aperture and 113 cm in focal length (F/3.76), has an unvignetted flat field of 3° in diameter, using 70 mm roll-films at the Newtonian focus.

The image quality at the field margin is 20 μm in the wavelength region of 380-680 nm. The image shape is symmetric so that it is suitable for precise position measurement.

2. MOUNTING

A modified yoke-type equatorial mounting supports two optical systems, one of which can be offset up to 5 degrees both in right ascension and declination.

The tracking around the polar axis is performed by means of two ball screws, which are rotating oppositely to each other and are employed alternately. They are driven by a pulse motor controlled by a micro computer. There is also a ball screw around the declination axis in order to track a high-velocity object.

Applications of this twin telescope include photographic observations of extended tails of comets with an offset telescope and simultaneous two color or polarization observations of the same field. A set of two photoelectric photometers is available, and so simultaneous continuous observations of an object and a comparison star are possible.

J. B. Hearnshaw and P. L. Cottrell (eds.), Instrumentation and Research Programmes for Small Telescopes, 113–114.

3. OBSERVATION OF HALLEY'S COMET

When Halley's Comet approaches the Earth in February to April, 1986, the twin equatorial telescope will be instralled at Saipan Island (latitude 15°N, longitude 146°E) to observe the comet. The weather on the island is reported to be very fine in that season.

4. REMARKS

Similar systems with smaller apertures (20, 16 and 13 cm) are commercially available as TAKAHASHI EPSILON SERIES.

The authors are greatly indebted to the support by a Grant from the MAINICHI NEWS PAPER CO. and to the Shinwa Optical Co. and Norizuki Giken Co. for the construction of the mounting.

AN OPTICAL FIBER FEED FOR SMALL TELESCOPES

W.G. Weller, T.E. Ingerson
Cerro Tololo Interamerican Observatory
Casilla 603
La Serena, Chile

ABSTRACT. We have constructed an optical fiber link between the CTIO 1.5 meter telescope and the instrument development laboratory located in the 4 meter telescope building. We have successfully observed a variety of objects using the 4 meter échelle spectrograph fed with light from the 1.5 meter telescope. These first observations indicate that the system is extremely stable, both in its spectral response and in wavelength calibration. The principal aim of the project was to provide the capability of recording high resolution spectra at the 1.5 meter, using linear detectors. Even with the loss of light associated with the use of the fiber, a conservative estimate of the speed of the system indicates a factor of 20 increase in speed over the existing coudé spectrograph.

1. INTRODUCTION

Optical fibers have been used for multiple object spectroscopy (Hill, Angel and Scott, 1983) and as a pseudo-coudé feed (Gray, 1983). In the latter use, light is collected by the telescope and conducted to a remote location for analysis and recording. There are several advantages to be gained. Solidly mounted stationary instruments are not subject to flexure effects. Since they do not move with the telescope they can be built at a considerably reduced cost. The environment of such instruments can be controlled to eliminate changes in calibration caused by thermal variations. Also, and very important for radial velocity measurements, errors caused by variations in slit illumination caused by guiding errors, and misalignments of the comparison optics can be virtually eliminated.

There are also disadvantages. The most obvious of these is the loss of light at the fiber. This occurs almost entirely at the input to the fiber, and with care in matching can be reduced to acceptable levels. One also loses the capability of observing extended objects. This can be partially overcome by sampling the long slit with several fibers. Absolute spectrophotometry cannot be done using this type of technique.

J. B. Hearnshaw and P. L. Cottrell (eds.), Instrumentation and Research Programmes for Small Telescopes, 115–116.

2. THE EXPERIMENTAL SETUP

The fiber is positioned at the telescope end of the feed in an adaptor which resembles a standard CTIO photometer. The fiber projects through a mirror in the focal plane of the telescope. This mirror reflects the field into the optics of an ISIT television camera used for acquisition and guiding of the star image onto the fiber. The adaptor also contains a quartz envelope tungsten filament lamp and a Thorium-Argon hollow cathode lamp used as calibration sources.

At the spectrograph end of the feed the fiber is simply clamped in position at the focus of the collimator. This simple system will soon be replaced by a focal ratio matching lens.

The telescope and spectrograph controls are both located in the console room of the 1.5 meter telescope and the observer is not concerned that the spectrograph is located in another building.

3. THE FIBER

There are two important parameters to consider when selecting a fiber. The first of these is the spectral transmission. All-silica fibres are available with good transmission from the ultra violet to the infra-red. We have chosen to use Spectran type SG840. This is a step index fiber with core diameter of 200 microns. At the casegrain focus this translates to 4 arc seconds on the sky. All light in the stellar image is accepted by the fiber.

The second important parameter is the focal ratio degradation. We feed the fiber with an f13.5 beam, and find that the output focal ratio is f4.5. This over fills the spectrograph collimator, and our next improvement of the system will be the inclusion of focal ratio matching optics.

4. FIRST RESULTS

By intercomparing spectra of the Thorium-Argon lamp taken on different nights we have determined that the zero point of the wavelength calibration drifts by no more than .03A (1.3 km/sec at 6500A) from night to night. This was determined from only 23 spectral lines in 3 orders. This performance is expected to improve considerably as our data reduction techniques improve.

5. CONCLUSIONS

The échelle system and fiber feed have proven to be a factor of 20 faster than the existing coudé.

References:
Gray, P.M., 1983, SPIE v.374, p.160-164.
Hill, J.M., 1983, Prepr. Steward Obs. no. 445.

TO UPGRADE A TELESCOPE?

Nathaniel M. White
Lowell Observatory
Mars Hill Road, 1400 West
Flagstaff, Arizona 86001
U.S.A.

ABSTRACT. The effectiveness of existing telescopes can often be greatly enhanced through upgrading. Two important questions which should be answered in the planning stage are discussed, based on recent experience in upgrading two Lowell Observatory telescopes.

The scientific contributions that result from the use of small telescopes are well-documented in this symposium. With high quantum efficiency, solid-state detector arrays, small telescopes have become faint object instruments, in addition to their more classical regular, time-intensive usage.

Telescopes should operate with a degree of efficiency commensurate with the new instrumentation and demands. Present computer and integrated-circuit technology has made possible their inexpensive upgrading at a level of sophistication previously available only to the larger observatories. New mirror fabrication technology may have a similar effect. But, where does one start?

The first question is to decide whether to maintain, upgrade, or replace the existing telescope. If maintenance has become too expensive, or new capabilities are required, often the only consideration is replacement. However, excluding greater aperture as the primary requirement, upgrading the old telescope can be a viable alternative at a few percent of the replacement cost.

Upgrading may be concerned with any part of the telescope, i.e, optical, mechanical, electro-mechanical, electronics, and building or site. A decision to upgrade must be based on very careful scrutiny of what needs to be improved. If it can be clearly shown that only one or two categories need attention, upgrading may be a good choice.

As an example, a careless decision to upgrade a telescope's drives and control system can be a costly mistake when it is <u>later</u> found that no improvement was gained because the support structure, bearings, or gearing were faulty. It may have been wiser to maintain this telescope at its present capability while acquiring a new one.

If the decision is to upgrade, should the work be done within the organization, by contract with a commercial firm, or through some

J. B. Hearnshaw and P. L. Cottrell (eds.), Instrumentation and Research Programmes for Small Telescopes, 117–118.

combination of these? The recent upgrading of two Lowell telescopes was done within the Observatory because of available expertise. This allowed flexibility in choosing computer and interface electronics compatible with existing equipment; and, since the upgrading was of our design, long-term maintenance and future modifications can easily be managed. Eventual duplication of the improvements on other telescopes at the Observatory also provides a potential savings. Without sufficient in-house expertise, a commercial contract would have been the most satisfactory approach.

Thus, the two questions that should be answered are: Is the area of the proposed upgrading well defined? Is the work to be done commercially? The former question is the most important and is the most difficult to answer. It is often neglected in one's eagerness to cure some obvious problem.

AN AUTOMATED PHOTOMETER FOR A ONE METER TELESCOPE

Andrew T. Young, Frank L. Beale, and Ronald J. Angione
Mount Laguna Observatory
Astronomy Department, San Diego State University
San Diego, CA 92182
U.S.A.

ABSTRACT. A new photometer is described which has some unique optical features. Emphasis is given to those features that improve the measurement precision and efficiency of use.

We have designed and are constructing an automated photometer with some unique features, primarily in the viewing and measuring optics. Upon completion this photometer will be installed on the one-meter telescope at Mount Laguna Observatory, which is operated by San Diego State University in partnership with the University of Illinois. The basic philosophy of this instrument is through the use of good design practices and high quality optical images to produce precise measurements and achieve efficiency in its operation by the astronomer.

The photometer can be operated either at the telescope using an eyepiece, or from the control room, in which case viewing is done using an intensified CID system. In both cases all component movements (e.g. filter wheel) are motor driven. The optical design allows a single eyepiece to be used for both wide field viewing and centering in the focal plane aperture. There is also an even number of reflections to produce a correct view of the sky. All lens surfaces have anti-reflection coatings to minimize reflection losses. A flip mirror allows the observer to select either a blue (EMI 6256) or red (RCA C31034) sensitive photomultiplier housed in thermoelectrically cooled boxes.

Although temperature effects on photometric measurements, particularly on filters, have been known for twenty years, few photometers control this important variable. In addition, most of the electronics of the pulse counter will be inside the photometer, and these components will produce more precise results if they are held at a constant temperature. In our photometer the temperature will be actively controlled to better than one degree Celsius using platinum RTD's as temperature sensors and resistors as the heating elements. The photometer will be held at a temperature of 26 degrees Celsius year round. Such a system is already working in our solar radiometer. The RTD's will also give us a readout of the temperature, which will be

J. B. Hearnshaw and P. L. Cottrell (eds.), Instrumentation and Research Programmes for Small Telescopes, 119–120.

recorded periodically with the photometric data. Temperature control will necessitate a quartz window at the optical entrance to the photometer, but this additional light loss will be more than compensated for by the increase in precision. It will also keep dust and moths out of the photometer. In order to minimize the heat dumped into the dome the entire photometer will have an inch of insulation around it; this is what we do with the solar radiometer.

The optical design increases precision in four ways. First, good optics produce a sharp image of the focal plane diaphram, and this, coupled with proper diaphram illumination, allowsprecise centering of the object to be measured. This is particularly important when trying to measure faint stars or faint, diffuse galaxies. Second, one needs a large diameter beam on the filters to minimize the effects of filter nonuniformity. Third, many filters in use are interference filters, and the central wavelength depends on the cosine of the angle of incidence. It is necessary therefore to minimize the angular divergence of the rays passing through the filter. Our solution is to pass parallel light through the filters. Last, the diameter and placement of the lenses is such that there is no vignetting of the measuring beam.

A standard light source will be added, which will be useful in the measuring process and as a diagnostic in case of problems. The microprocessor that controls the instrument will also use the measurement technique of F. Rufener (Geneva Observatory) of calculating various statistical quantities to evaluate each measurement.

Several features of the photometer that allow convenience are not unique, such as being automated, or in the use of a remote, intensified viewing system. Our choice of a CID with two microchannel plates of intensification is based on economy. First, it has economy of size. Second, the microchannel plates are separate and can be replaced at a cost of only $2000 if the device is "zapped" by a bright source. If our current ISIT goes, then the entire device must be replaced at a cost of $14,000. A related convenience is that the interface between the telescope and photometer contains an ISIT and optical components allowing the change in just 15 minutes from the photometer to the Observatories' fiber optics coupled spectrograph.

It is all too common to find in astronomical photometers serious problems of vignetting and/or optical aberrations, particularly in the viewing optics. Our optical system stresses proper design, such as the addition of field lenses, and the use of high quality components, such as a Clave eyepiece and commercial enlarger lenses for the relay optics, to avoid these problems. This allows the astronomer to see and center fainter objects, and also to center all objects faster. Thus more objects can be observed per night.

Most photometers have two sets of viewing optics, one before the focal plane aperture with a wide field for finding objects, and one after with high magnification for centering. This is both awkward for the observer and would require two, costly remote viewing devices. Our system allows both wide and magnified views through the same eyepiece. In addition, we have added a roof mirror to give us an even number of reflections so that the view through the eyepiece has the same orientation as most published finding charts.

THE MAIN FEATURES OF THE 1.56-M TELESCOPE AND ITS STATUS QUO

Wan Lai, Zhu Neng-hong, Wang Lan-juan,
Yang Zheng-hua and Zheng Yi-jin
Shanghai Observatory, Academia Sinica
80 Nandan Road, Shanghai
China

ABSTRACT. This telescope was designed for determining the trigonometric parallaxes of faint stars. The otpical configuration is a Ritchey-Chrétien system and has only a f/10 Cassegrain focus for photographic work.

The mounting is a combination of fork and yoke frame which can reduce the flexure due to gravity. A high speed digital servo-system was adopted for controlling and driving the telescope. At present some photographs of Comet Halley have been obtained.

1. OPTICAL SYSTEM

(1) Main optical system.

* Configuration: Ritchey-Chrétien system.
  Concave hyperbolic primary mirror: Diameter - 1560 mm. Thickness - 240 mm. Focal ratio - f/3.3.
  Convex hyperbolic secondary mirror: Diameter - 530 mm. Thickness - 80 mm.
* Focal ratio: f/10. Focal length: 15600mm. Free coma field: ± 15' or ±30'.
* Image quality: Image disc within 20' is less than 1". 85% of the light is gathered in 0."6.

(2) Both the guiding and finding telescope are refractors.
Guiding telescope: Diameter 220 mm; focal length 2860mm; field ±1.°0.
Finding telescope: Diameter 220 mm; focal length 800mm; field ±1.°5.

(3) Offset guiding system.
Offset guiding can be accomplished visually or photoelectrically. The pick-off prism (26x20 mm) can be rotated 270° around the axis of the main tube.

2. MECHANICAL ASSEMBLY

(1) Mounting
The mounting is a combination of fork and yoke frame. Its characteristics are (1) the optical axis has the minimum tilt angle resulting

J. B. Hearnshaw and P. L. Cottrell (eds.), Instrumentation and Research Programmes for Small Telescopes, 121–122.

from the flexure of the mounting, (2) the polar region can be observed with the full aperture, (3) when the declination axis is above the polar axis, the visual eyepieces are very near the polar axis and (4) there are two spherical oil bearings on the end of the mounting. Thus the alignments of the azimuth and the elevation are very easy to do.

(2) Tube

The tube consists of (1) a primary mirror cell - the primary mirror is supported both in axial and lateral direction by an 18 lever counter-weight mechanical system, (2) secondary mirror cell - the secondary mirror is supported in the axial direction by evacuating the air behind the mirror and in the lateral direction by a mercury tube and (3) multiple exposure camera - for saving observing time a multiple camera is provided, which contains four plates to be used automatically one by one.

(3) Automatic focusing device

This device contains four parts: (1) Two Cervit rods used for keeping a constant distance between the back surface of the primary mirror and the vertex of the secondary mirror, (2) a slit illuminated by a light source is fixed on the upper end of one rod, (3) a thin lens is fixed onto the centre of the secondary mirror and (4) a microscope objective and a CCD are fixed onto the upper end of another rod.

The image of the slit is formed on the CCD by the thin lens and the microscope. Thus the image will move on the CCD when the tube expands. As a result, the CCD will feed a signal to drive a stepper motor which moves the secondary mirror to the optimum position.

3. CONTROL SYSTEM

(1) Drive control

A high speed D.C. motor and several resolvers constitute a digital high speed servo-system to accomplish the slewing, setting, guiding and tracking rates.

(2) Photoelectric autoguide

By reversing a prism in the offset guide component the guiding mode can be changed from visual observation to the photoelectric autoguide mode. By means of a semicircular knife disc and a photon counting technique the guiding accuracy could reach $0\overset{''}{.}2$.

4. DOME BUILDING

Site - western suburb of Shanghai: Diameter - 17 m. Height to the centre of the dome - 21.7 m.

5. THE STATUS QUO

At present the telescope has been assembled and aligned and had its "first light" in November 1985 in the workshop.

It will be disassembled and transported to the new dome next year.

Our aim is to put the telescope into normal operation between the end of 1986 and early in 1987.

DISCUSSION OF POSTER PAPERS : SECTION I

Cadmus and Melsheimer - The instrumentation and observing program... .

Gaustad: Is it easy to add new instrumentation to the guide-acquire module?

Melsheimer: The only requirement is that they have the same focal plane position. Some flexibility is obtained by having the N, E, S and W ports of the guide acquire module with a slightly different focal plane position. A translating rotating diagonal mirror brings the light out at the required port.

Hiltner: My comment does not apply so much to small telescopes as to larger ones where the cost of light loss becomes quite significant. I'm always concerned about the loss of light at the flat mirror needed for directing light to the various instruments. This mirror may "cost" two hours of observing for each 10 hour night.

Melsheimer: In a small college where one doesn't have the manpower, or one doesn't want students changing (and possibly dropping) instruments in the middle of the night, it's a convenient and versatile package.

Evans: It's OK in a teaching institution, but it can be a curse in a working observatory to have all types of instrumentation on every telescope.

Melsheimer: The television system which Bob Cadmus has made works very well, is very inexpensive and uses a SONY CCD TV camera which costs $US750. He bought an image intensifying system and can now see $16^{m}.5$ at 30 frames/second.

Caton and Pollock - A four-star photometer for use on small telescopes

Latham: What is the status of this project?

Caton: We are awaiting NSF grant funding, as it is well beyond the normal equipment budget at our university.

Barwig: Do you use photon counting techniques?

Caton: We are not currently using pulse-counting, but rather we use a voltage-to-frequency converter and integrate this signal.

J. B. Hearnshaw and P. L. Cottrell (eds.), Instrumentation and Research Programmes for Small Telescopes, 123–131.

Genet: Norman Walker at Royal Greenwich Observatory has a 4-channel fibre optic feed photometer in operation. It chops between the four channels.

Graham: Is fibre-coupling routine and fairly cheap to do now? I presume it is also convenient.

Caton: Yes, and they cost about $US100. For our application fibres are essential. We may want to access close pairs of stars and we couldn't get two detectors into this small area whereas we can get two fibres in.

Latham: There are some subtle effects of light loss through the fibre as one changes the angles at which the fibres bend.

Caton: The angles are kept fixed as the detectors travel on the plate.

Chen et al - An automated South Pole stellar telescope

Latham: Advantages of observing at the South Pole are a long night (~18 weeks) and constant airmass for a given object. There is also a very low precipitable water vapour content, which could make it ideal for IR work. Is any planned?

Wood: Yes we do but not during the present winter.

Crawford: Can you quantize the amount of water vapour content?

Wood: The Pole is in the middle of one of the biggest deserts on earth. The Pole is equivalent to 11,000 feet in ordinary conditions.

Gaustad: I believe water vapour at the South Pole is significantly less than 1mm of precipitable H_2O, sometimes as low as 50 microns.

Scarfe: What are the observing conditions like at the South Pole, with regard to length of cloudless periods, seeing, wind (and blown snow)?

Wood: Seeing is apparently not too hard, but wind and blown snow are real problems. We have had a continuous run of ~400 hours, but high cirrus and aurora can be a problem. These haven't bothered us to date because we are working on a bright object (γ^2 Velorum).

Latham: How important are continuous (in time) data sets (20 - 50 hours) for the kind of work you have been doing with different periodicities?

Kurtz: It's essential! The sun, pulsating white dwarfs, Ap stars and cataclysmic variables all need continuous data sets. We need to use observers at different longitudes or polar observations (although I'm not convinced about this) to acquire these data sets.

Cochrane et al - The automated patrol telescope

Latham: What is the schedule for the completion of this telescope?

Payne: The telescope is now moving but the observatory (on Siding Spring mountain) is yet to be built or financed. We hope to have this completed by mid-1986.

Schulte in den Bäumen: Are there special reasons for using 2 different drive systems on this telescope?

Payne: For the declination axis we used the original worm gear on the telescope. We replaced the RA gear with a friction drive, principally because it was cheap (~$A400).

Graham: Have you given any thought about the way you search for supernovae? The Colgate experiment didn't turn up many more and most SN have been found the old way.

Payne: Not yet. My involvement has been with the servo-system and the software, and I haven't got around to the details of the astronomy yet.

Crawford - Automatic Photoelectric telescopes

No discussion of this paper.

Garrison - University of Toronto southern observatory

No discussion of this paper.

Hickey - Construction of a 32cm Cassegrain photoelectric telescope

Garrison: I am most impressed with the idea, the quality and the dedication of amateurs who are putting together photoelectric photometers for work on variable stars.

Crawford: I want to second Garrison's comments on the quality of some of the current "amateur" telescopes and observations. They are excellent, much better than most or even all of the "professional" small telescopes. They are truly new generation telescopes. Bart Bok in his talks, often called himself an amateur astronomer - a lover of astronomy. That trait is shared by amateurs and professionals.

Isobe et al - A 75cm alt-az telescope with short time focus... .

Latham: How was the mirror made? Spin-casting or caste in a mould?

Isobe: A caste, where the moulds were removed by sand- blasting.

Graham: Is there a problem with field rotation?

Isobe: We have an instrument rotator on both the prime and Nasmyth foci. For the photometer and the spectrograph we don't worry about image rotation.

Latham: What kind of bearing do you use for the main azimuth bearing?

Isobe: High angular contact ball bearings.

Jain and Battacharyya - A single-channel polarimeter

No discussion of this paper.

Laubscher - CCD instrumentation at Capilla Peak Observatory

Zeilik: There is a hidden cost in the cost estimate. This is for the image processing system. We use a VAX and various peripherals. The cost in ~$US200,000 excluding the computer and the 4-5 man-years of software.

Latham: Have you had enough operation yet to see what the reliability of the CCD system is or are you coping with problems as they arise?

Laubscher: We have had only one problem so far and that was due to connectors and cables. The system has exhibited excellent photometric properties from four short runs.

Graham: What are the weather statistics like? Is it dark? How does it compare with southern Arizona?

Laubscher: It used to be good, but the number of photometric nights has dropped off over the last three years.

Zeilik: New Mexico is, in my opinion, an underrated area for doing optical observations. The percentage of clear nights in the southern part of the state is equivalent to the Tucson area. The number of photometric nights has been ~30 - 40% over the last 5 - 7 years. The sky is very dark : V-band is 22^m/square arcsec. We have a good site at high elevation (9000 feet).

Jacoby: Now that you have worked with putting this system together and used it, I'd like to ask if you would work with developing this sort of system again?

Laubscher: Yes, I'm a graduate student! But overall, although it has been a lot of hard work, it has certainly been worth it.

<u>Nankivell and Rumsey</u> - The optical system of the Mount John 1m......

Latham: What is the status of the telescope?

Nankivell: The optics for the f/13.5 configuration (for on-axis work - photometry and spectroscopy) are complete, but the f/8 optics (for wide-field work) are still being worked upon. We hope to take the telescope down to Mount John in early February and have it operational in mid- to late-February 1986. (As at end February the telescope has been installed at Mount John - ed.).

Saxena: How, and with what methods, was the optical quality of the mirror tested? Are any quantitative values as to the quality of the surface available?

Nankivell: The mirror was ground and polished spherical to establish an axial symmetric system. It was then deepened and tested using a Foucault test. The final testing and figuring was done using an Offner compensator. This removed the overall asphericity of the mirror, and any deformations on the surface can be easily seen and refigured. When using any compensator one has to ensure that it is rigorously built and that dimensions and spacing of mirrors are accurately measured. No quantitative work on the wavefronts has been done. We will do that (e.g. a Hartmann test) once the mirror has been installed in the telescope at Mount John.

Robinson et al - Black Birch Astrometric Observatory

No discussion of this paper.

Saxena - A new polarimeter for stellar polarisation measurement

Latham: Polarisation measurements can be very precise, which means you need to have extremely high S/N, that is you need to be able to register a very large number of photons. In most CCDs, the individual pixels cannot hold that many charged (a few x 10^5). Are you going to combine pixels or are you confident that you can get the required S/N for polarisation measurements? The maximum S/N ratios that I've seen for CCDs are 300:1 or 400:1.

Saxena: Yes. We will need to add several integrations to improve our S/N ratio.

Duncan: There has been some work (Meyer at Chicago and Jura at UCLA) on weak interstellar lines, which achieved S/N ~ 1000:1 with a CCD. They have done a lot of processing (many frames with shifting and adding) to achieve this so I guess one can't directly compare this.

Graham: When will this system be working?

Saxena: In about a year or so, using the new Tektronix CCD chips.

Sullivan and van der Peet - A simultaneous two channel photometric... .

Latham: What is the schedule for completion of this instrument?

Sullivan: The photometer head has been completed by UT and is now in Wellington (NZ). The computer and electronic systems are underway and we hope it will all be completed and operational by February/March 1986. We rely on other institution's telescopes and we have been working closely with Carter Observatory. The total cost of the system is ~$NZ 75,000 for somewhat "old technology", although it is a two channel system. However the cost of a CCD is beyond the funding capabilities of NZ institutions, at least in the short term.

<u>Takagishi et al</u> - CCD camera for the 60cm telescope at Kagashima... .

Latham: The temperature that you operate at, -70°C, is quite warm compared to other CCDs. Do you have any trouble with dark leakage building up at this warm temperature? What is the longest exposure that you have made?

Takagishi: Our CCD has small pixel size compared to TI and RCA chips and we feel that -70°C is sufficient. Our longest exposure is 60 minutes.

Latham: That certainly answers my question!

Djorgovski: Can any of our Japanese colleagues tell us anything about the development of astronomical CCDs by Japanese industry.

Takagishi: The Japanese manufacturers of CCDs are very consumer orientated (e.g. the video camera). We are trying to get them to manufacture astronomical quality ones but we have had no success yet. We have to show them that there is a real need.

Melsheimer: How much did it cost?

Takagishi: $US15,000.

<u>Tomita and Takahashi</u> - A twin reflecting astrograph

Latham: One should always be looking for industrial funding rather than government sources. How did you convince a newspaper to fund your work?

Tomita: Halley's Comet fever has arrived in Japan, and the newspaper wanted a good photograph of the comet.

Andrews: You have a hyperbolic primary mirror and a Newtonian flat. Whereabouts are the correcting lenses in this system?

Tomita: A four element lens is used.

Weller and Ingerson - An optical fibre feed for small telescopes

Hearnshaw: Is the feed from the Cassegrain or prime focus of the 1.5m? What is the focal ratio? And what is the beam size of the échelle?

Weller: Cassegrain, f/13.5. This is not optimum but it was a compromise. We would like a faster input, close to the numerical aperture of the fibre, that is, using the prime focus. The other consideration however is that the faster the input beam, the faster it comes out and since we are feeding the échelle collimator focal plane a fast beam would overfill the collimator. Our échelle has a 10cm beam size. What we need is a custom built échelle spectrograph matched to the fibre.

Rowe: Intrinsic light losses in the fibre are very small, most of it occurs at the ends. One can match the fibre quite well using a microscope objective.

Weller: This is true. We use a sapphire ball on the end of the fibre. The major advantage of fibre feeding is the stability for radial velocity work. (No flexure and the temperature can be kept constant). The fibre thus provides a good mixer for the stellar spectrum, comparison spectrum and flat fielding lamp.

Mochnacki: What size fibre core are you using? DDO has 2 people working on fibres > 50μm but I am interested in thin fibres (core diameters < 50μm). Do you have any information on this?

Weller: 200μm. Jacques Beckers has been testing some thin fibres in a single mode operation.

Hearnshaw: Schiffer at Heidelberg is using a 34μm fibre to feed their spectrograph. The degradation from input to output is 10% (f/3.5 to f/3.2). Information on this system has been published in Mitt. A.G. 54, 182 (1981).

Rowe: Most of the parameters for these fibres have been published in the engineering literature.

White - To upgrade a telescope?

White: We intend to send a description of the upgrade of the 42" and 72" telescopes at Lowell to P.A.S.P., but we thought this conference was an appropriate place to raise this question of upgrading telescopes with, for example, computer controls.

Melsheimer: What kind of pointing accuracy are you getting now?

White: Two answers:
- 72", a 50 year old, off-axis English mount telescope is of the order of 10 or 20 arcsec if we make all the corrections.
- 42", a rigid fork mounted telescope sets to better than a few arcsecs.

Young et al - An automated photometer for a 1m telescope

Latham: What is the status of this project?

Angione: It is under construction at the moment and we expect to have it operational by June or July 1986. We hope to be able to do the most precise photometry possible which was one of the initial philosophies behind this project. It has more optical components than one normally sees in a single channel photometer but we wanted high quality images, maximum throughput and also to avoid vignetting.

Crawford: The kinds of precept which you have described are important for all photometric work.

Angione: Andy Young is actually writing another paper which is detailing all the considerations which went into our design.

Wan Lai et al - The main features of the 1.56m telescope and... .

Hearnshaw: What is the mirror material? What is the expansion coefficient of the material? Are mirrors blanks able to be purchased from this company?

Zhu Neng-hong: Both the primary and secondary mirrors were made by the Shanghai Yao Hua Glass Factory. The material is just like Zerodur or Cervit, and has an expansion coefficient of $1-2 \times 10^{-7} K^{-1}$. Mirror blanks can be purchased from this factory with diameters up to 2m.

Kulkarni: I met a Chinese engineer in Delhi and he told me how they measure the thermal expansion of the glass. They take two one inch, $\lambda/20$ finished pieces of this glass. They leave one at room temperature and put the other one into boiling water. Then they put the two together and count the interference fringes. In addition, I would like to say that there are some very novel and clever features in this telescope.

II. PHOTOMETRIC RESEARCH PROGRAMMES

a) Invited and oral contributed papers.

STELLAR PHOTOMETRY WITH SMALL TELESCOPES

B. A. Twarog
Department of Physics and Astronomy
University of Kansas
Lawrence, Kansas 66045-2151
U.S.A.

ABSTRACT. The dominant trend within stellar photometry in recent years has been an increase in the number of intermediate and narrow-band photometric observations, particularly with small telescopes. Recent examples illustrating the scope and emphasis of such programs are reviewed. It is concluded that with continuing improvement in detector technology and telescope automation, small telescope photometry will grow in importance.

1. INTRODUCTION

The goal of stellar photometry is simple. By observing the energy output of a star through a series of well-chosen passbands, one hopes to measure the cumulative distortion of the stellar black-body spectrum by the absorption (emission) features of the stellar atmosphere to obtain the big three of intrinsic stellar properties, temperature, luminosity/surface gravity, and metallicity. The data can provide a direct probe of stellar structure and evolution and, when combined with extrinsic parameters such as distance, position, and space velocity, allow delineation of the evolutionary history of the ensemble of gas, dust, and stars known as our Galaxy.

Since the postwar period when availability of cheap photomultipliers made photoelectric photometry feasible for the general community, the field has followed a pattern of development familiar among new techniques in astronomy. Initially, broad-band UBV photometry appeared to provide the big three with an accuracy comparable to high dispersion spectroscopy, without the magnitude restrictions and the complex uncertainties of spectroscopic analysis; it helped make the 50's and 60's the golden age of stellar photometry. With time, however, photometry began a steady decline and an increasing shift to smaller telescopes. The related reasons for this are fourfold: (1) there is a sense in the literature that the broad outline of our understanding of stars was complete, and that the remaining work would concentrate on filling in the modest details; (2) it was recognized that broad-band, two-color data were incapable of providing reliable independent information on the big

J. B. Hearnshaw and P. L. Cottrell (eds.), Instrumentation and Research Programmes for Small Telescopes, 135–147.

three; (3) the use of broad-bands oversimplified too much and hid the fine, structural differences among stars which were now important, particularly in the area of metallicity; and (4) improved observational and computational techniques made high dispersion spectroscopy a logical, feasible, and attractive alternative for the study of larger, fainter, stellar samples. It is the purpose of this review to demonstrate that (1) was exaggerated, while (2) through (4) have been overcome, in great measure, by the growing trend toward intermediate and narrow band photometric systems, almost exclusively in use on small telescopes. The latter point is illustrated in Table 1, which shows the distribution of papers published in 1974 and 1984 as a function of photometric system used on small telescopes for nonvariable stars. Between 1974 and 1984, the total number of papers increased by about 30%, the largest group for both years was the UBVRI sample, but intermediate band systems, particularly the uvby system, garnered a growing share of the total, and the total number of photometric systems went from 9 to 14.

Rather than attempt a comprehensive review of all systems and the published work of the last five years, I will concentrate in detail on a handful of programs done predominantly with small telescopes which exemplify the fine, quality research which can be done with smaller instruments, in contradiction with assumption (1) above; section 2 will deal with the broad area of stellar structure and evolution, while section 3 will cover stellar populations. Section 4 will close with a summary, and a discussion of possible future directions for small telescope photometry. It should be emphasized that the choice of which programs to discuss is only partially a reflection of my personal bias; the statistics of Table I partially dictated the emphasis on specific systems. Even so, I apologize to the dozens of photometrists whose excellent work will go unmentioned in the short time available.

Table I

Distribution of Papers by Photometric System

System	1974	1984	System		1974		1984
UVBRI	54%	42%	DDO		4%		2%
IR	11%	9%	Geneva		1%		4%
uvby Hβ	10%	23%	Others	(2)	3%	(7)	8%
Vilnius	9%	5%					
RGU	8%	7%	N TOTAL		109		138

2. STELLAR STRUCTURE AND EVOLUTION

2.1 Star Formation

One of the least understood stages of a star's life is the way it forms. While the earliest stages of the process when the protostar is shrouded in a thick cloud of gas and dust are studied optimally with IR and longer wavelength observations, some crucial insight can be gained by optical observations made as the star emerges from its opaque cocoon. Of primary interest are the related questions of the large scale relationship among individual star forming regions within an extended cloud complex, and the factors controlling the formation timescale and mass function within the clouds.

An ideal testing ground for current theories is the region of the CarOB1 complex containing the open clusters Tr14, Tr15, Tr16, Cr228, NGC3324, NGC3293, and IC2581. From UBVRI, uvby Hβ, and Walraven photometry, and classification spectroscopy of individual clusters (Turner et al. 1980 and references therein--hereafter TGHH), there is little doubt that all the clusters lie at the same distance. When uniform age estimation is applied to all the clusters, as in TGHH (see Fig. 10 of their paper), a very obvious pattern emerges. The clusters are strung along a line parallel to the galactic plane over a distance of 150 pc., with an age gradient where the older subgroups are at lower galactic longitude. While this qualitatively agrees with the Elmegreen and Lada (1977) theoretical picture of sequential star formation and the observations of nearby associations such as Orion, the quantitative analysis shows several nagging discrepancies. The age spread (5×10^6 years) appears to be far too small, and the individual subgroups are much richer in members than expected. A closer look at a member of the oldest subgroup (NGC3293) reveals even more problems.

The reddening corrected CMD for NGC3293 (Fig. 6, TGHH) shows a well-defined upper main sequence with a turnoff in the early B star range, but is discrepant in two ways. Four stars near the tip of the main sequence are in anomalous positions for the cluster age, while three faint, cool stars are located near the main sequence with a predicted age that's three times larger than the cluster turnoff age. While it is easy to dismiss the discrepant stars as statistical aberrations, the follow-up work by Herbst and Miller (1982,HM) has emphasized just how important they are. With photographic plates from a 0.6m, a CMD to V=16 was constructed of the cluster environs. It shows a main sequence populated to, at least, Mv=2.8, or M=1.4 M$_\odot$ with a potential pre-main sequence at fainter magnitudes. The pre-main sequence contraction age implies a cluster age of 20×10^6 years, three times larger than that from the main sequence turnoff. When taken in the context of the earlier work on NGC2264 (Warner et al. 1977), h and χ Persei (Vogt 1971), and the Pleiades (Landolt 1979), the implication is inescapable. Coeval star formation in clusters is a myth. Either the theoretical timescales for pre-main sequence evolution are seriously in error, or low mass stars in clusters form first. From the work of Stauffer (1982) on the Pleiades, where the age spread is about 3×10^8 years, it can be inferred that the formation of low and high mass stars occurs in two distinct phases, or the stars which form at any given time cover a specific mass range, with the mean mass increasing with time; in either case, at the low mass end, star formation is not a continuous process. At the high mass end, there is some evidence for continuous star formation. In agreement with the above results on NGC3293, Jackobsen (1984), from uvby Hβ data of hundreds of B stars covering the mass range of 2.5 to 13 M$_\odot$ in nearby open clusters, has found age spreads within individual clusters of between 10^7 and 10^8 years. Similar conclusions have been drawn from the composite CMD studies of Mermilliod and Maeder (1984).

Two questions immediately come to mind. What is the mechanism for mass-dependent, star formation delay in molecular clouds, and, if star formation is a discontinuous process, where does the break or over-lap in mass for the two phases occur?

Two relevant observational points are that many of the younger, well-studied open clusters exhibit extended halos populated predominantly by low mass stars (HM and references therein) and mass functions deficient in low mass stars compared with the field (Van den Bergh and Sher 1960; HM). While similar results for older clusters are invariably interpreted as the effect of dynamical mass segregation, in younger clusters the distribution is explained best as a fossil record of the initial star formation distribution. From the analysis by HM, one can conclude that the classical fragmenting protocloud picture of star formation, controlled by the Jeans' mass, is no longer supported by the observations, in agreement with recent theoretical approaches. The paucity of low mass stars removes the model of Norman and Silk (1980), but is consistent with the theoretical predictions of Larson (1982, 1985). After compressing into sheets, cloud fragmentation occurs because of the gravitational instability of the sheet, with the minimum protostellar cloud size controlled by the temperature of the gas. This naturally leads to sequential mass formation, with a lower mass cutoff growing with time as the temperature of the gas rises due to the appearance of higher mass stars. The significance of these models is difficult to overestimate. Larson (1985) outlines the effects of making star formation a bimodal process, analogous to a picture presented by Eggen (1976), where the peaks of the two distributions in the mass function are allowed to change with time and position in the Galaxy. As also noted by Gusten and Mezger (1982), an appropriately parameterized mass function can explain the dip in the luminosity function of nearby stars at $M=0.7\ M_{\odot}$, the unseen mass in the solar neighborhood, chemical evolution of the disk without infall, the steep Galactic abundance gradient, and the mass-to-light ratios, colors, and gas contents of spiral galaxies, along with a variety of other galaxy parameters.

If even partially correct, the above would imply that the notion of a universal mass function over the entire mass range is invalid. One should be able to find star-forming regions where only higher mass stars have formed, as in the core of Orion, while in others, such as Taurus, the mean mass of the stars in formation will be significantly lower. Additional observations of the type noted above are needed to clarify such questions as what variables control the relative strength of the two modes in associations vs. clusters, what mass range of stars determines the sequential star formation initiation process across an extended cloud complex, and over what range of scale does the pancake picture of the cloud collapse extend?

2.2 Up, Down, and In Between

Besides demonstrating the effects of time-dependent star formation on the structure of the CMD, precise photometry can be used as a probe of the effects of varying internal structure on a star's surface properties. Comparisons between theory and precision observations provide exquisite tests of just how realistic and complete our picture of stellar astrophysics and the variables which influence it are.

The first such confrontation of theory with observation has already been partially discussed in 2.1. Studies of luminous, high mass stars within galactic associations (Humphreys 1978) and within the field (Garmany, Conti, and Chiosi 1982) reval that these stars exhibit the same spread in the CMD as found by Jakobsen (1984). In contradiction with the predictions of stellar evolution theory (Maeder 1984), these stars populate the CMD redward of the termination point of the hydrogen core burning evolutionary models in large numbers. Because of the important role played by these stars in galactic nucleosynthesis, their apparent sensitivity to local initial conditions, and the ease with which they are observed in other galaxies (Conti 1984, Maeder 1981), the observational constraint imposed on current theory by the existence of a wide main sequence has both disturbing and far-reaching impact. Bertelli et al. (1984) find that only limited combinations of mass loss, convective overshoot, and opacity enhancement can reproduce the effect to date.

Moving to slightly cooler, late type B stars, we find our first example of a down point in the stellar distribution, a gap region where the CMD is underpopulated either because of some internal structural discontinuity and/or a rapid evolutionary phase. The gap is easily seen in the two-color diagram on the Geneva system in Mermilliod (1976) where it is also demonstrated that the position of the gap changes as the stars evolve. The gap was emphasized by Eggen (1976) who felt it represented the breakpoint between the two modes of star formation. He also noted, in support of his suggestion, that the brightest stars below the gap had a high rate of spectroscopic peculiarity of the type associated with circumstellar shells, presumably because these stars were still evolving toward the main sequence (see also Turner 1982). To date, no satisfactory explanation has arisen for the source of this feature.

Moving to even cooler regions of the F stars, we find further evidence for main sequence widening with a twist. uvby observations of the intermediate age cluster NGC 752 by Twarog (1983) show the existence of a bimodal main sequence separated in color by 0.03 to 0.04 mag in b-y. The availability of c1 rules out binaries, leaving rotation or two episodes of star formation as logical explanations. The age difference required is 3 x 10^8 years. Evidence that this is not an isolated case comes from a uvby study of NGC 3680, by Nissen (1985, private communication) as shown in Fig. 1. Preliminary results on IC4651 using a CCD show only weak evidence for the effect, if any, despite the fact that it should be the same age. Likewise, a uvbyHβ study of M67, twice the age of NGC752, by Nissen, Twarog, and Crawford (1985) shows the main sequence below the hydrogen exhaustion phase gap to be no wider than 0.015 mag in b-y, (Fig. 2). What is found is that the stars above the gap delineate a blueward hook which is between 0.04 and 0.07 mag wide in b-y; the shape of the turnoff does not agree well with

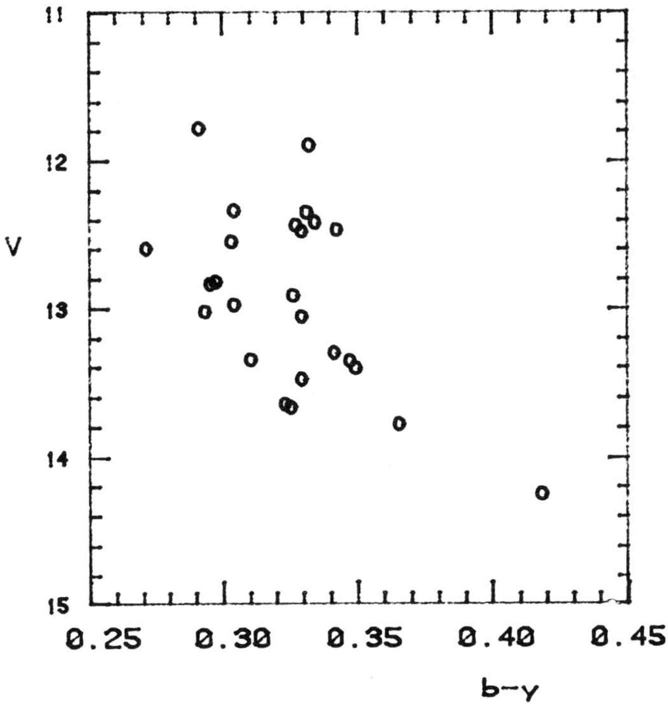

Fig. 1
uvby CMD of
NGC 3680
(Nissen 1985)

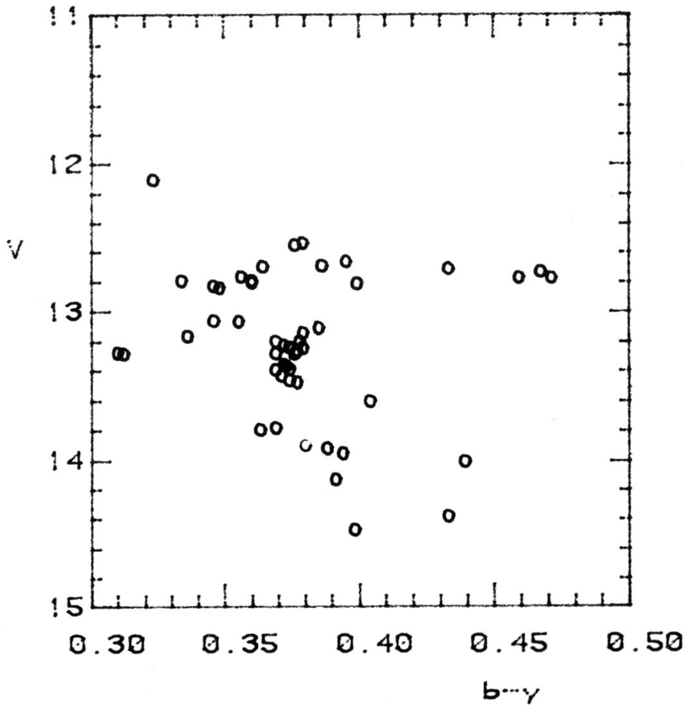

Fig. 2
uvby CMD of
M67 turnoff

the recent isochrone fits by VandenBerg (1985). The nature of the bimodal sequence and the blueward hood in M67 remain unexplained at present, though it has been noted that the Bohm-Vitense-Canterna gap does occur in the temperature range of the former. It is also worth noting that the significance of these effects would not have been recognized without a means of testing for composite systems. The work of Carney (1982, 1983) on photometric detection of unequal component binary systems deserves wider application for field star analysis.

Finally, we turn to low mass stars in the post main sequence phase. Photographic surveys of the giant branches of globular clusters show the presence of a 0.2 mag gap in the giant branch one magnitude above the turnoff in NGC 6752 (Cannon and Lee 1976), ωCen (Da Costa and Villumsen 1982), and NGC 288 (Buonanno et al. 1984). The existence of this gap is not predicted by standard theory, though an ad hoc method of inducing a gap has been proposed by Armandroff and Demarque (1984). The picture has become even more confused with the luminosity function study of the giant branch of 47 Tuc by King, Da Costa, and Demarque (1985). Their photographic work shows that the above gap does not exist in 47 Tuc, but an expected luminosity peak does. The peak reflects the passage of the H-burning shell through a composition discontinuity, and its luminosity location is dependent upon metallicity, helium content, and stellar mass. As in NGC 362, the peak in 47 Tuc is fainter than predicted, indicating that convective overshoot on the subgiant branch is not an insignificant effect as is often assumed in standard models. Further observations are needed.

3. STELLAR POPULATIONS AND GALAXY EVOLUTION

3.1 The Solar Neighborhood

Mould (1982) has defined a stellar population as "a collection of stars of similar age, composition, and kinematics." Implicit in such a definition is the hope that the three properties are uniquely correlated in a way which provides information about the evolution of the Galaxy, rather than in a random distribution, which tells little. Because of the need for reliable space velocities, distances, and stellar parameters, there is no better sample for such a program than the stars nearest the sun, and no better way to study them than with a small telescope. Over the last 40 years, no individual has devoted more of his time to just this approach than Olin Eggen. It would be an easy matter to spend not only the entire talk, but the entire conference discussing just his work; in the last five years alone he has published over 50 papers on the topic. For the sake of time, however, let me discuss only one area of recent research with which he is undoubtedly most associated, and apologize to Olin for mentioning only a minute fraction of his contributions to small telescope photometry.

A key concept in Eggen's analysis of the solar neighborhood is that of moving groups (Eggen 1970 and ref. therein). As star clusters evolve with time, they slowly dissolve into an annular distribution because of the combination of internal velocity dispersion and interaction with the galactic disk. Though difficult to recognize because of their low density, such groups should have essentially the same V velocity at a given galactocentric location and be recognizable from their common motion. From the group V, proper motion measures, radial velocity data, and position, one can derive a distance. In this way, Eggen has identified over a half dozen young stellar groups, and a smaller number of older groups extending through the old disk and into the halo. There is little doubt that the younger groups are real, and the more extreme characteristics of the others have made membership likely, e.g. the HR 1614 group should contain stars significantly above the Hyades in metallicity, a prediction confirmed with DDO photometry by Smith (1984). Others, like the Wolf 630 group, may be questionable (McDonald and Hearnshaw 1983).

Eggen (1983, 1985 and references therein) recently has expanded this concept in two ways. One of the groups (Pleiades), through the extensive use of intermediate band photometry and all available space velocity information, has grown to the scale of a local association containing α Persei, IC2602, NGC2516, Sco-Cen and Lup-Cen associations, and possibly M34. Though comparable in age, there is evidence for an age spread among the luminous stars on a scale of 5 x 10^7 years, while the fainter main sequence stars show ages closer to 2 x 10^8 years. The size of the association is about 100 pc.

A similar expansion has occurred for two of the older groups, where Eggen has found evidence for a Hyades supercluster and a Sirius supercluster. We can use the results for the Hyades supercluster to emphasize the importance of this concept, if it is correct. First, assuming group membership, one can obtain reliable absolute magnitude information for stars which are normally beyond the limits of trig parallax. Second, the expanded sample allows one to more accurately map out the pattern of post-main sequence evolution compared to poorly populated open clusters, providing luminosity and mass information on stars rarely found in nearby open clusters. For the Hyades, Eggen (1985) has extended the red giant branch to Mbol = -5, identified the AGB, and found two C star members of the supercluster. A survey of nearby white dwarfs showed 20% to be members of the Hyades, including two very hot stars, F24A and V417 Tau B, the nucleus of a planetary nebula, and a half-dozen cool white dwarfs with cooling times greater than the age of the cluster, among others. A group of stars with very active chromospheres and white dwarf companions has been shown to lie about 0.8 mag above the main sequence. Third, the paucity of old disk clusters and nearby globulars makes the study of old disk and halo groups like Arcturus and Kapteyn's star invaluable, especially on the lower main sequence where age estimation for random field stars is normally impossible. Finally, the surveys of these groups complement the more traditional approach for finding nearby members of the Hyades by Upgren, Weis, and Hanson (1985 and references therein) and emphasizes the fact that the solar neighborhood may not be a representative sample of the Galactic disk chemically or kinematically. It is estimated that one-third of the bright B stars may be members of the Local Association, while 15% of the cooler stars could be members of the Hyades supercluster. Such considerations are rarely, if ever, included in analyses of specific stellar types based upon nearby stars. Further evidence for local anomaly is provided by the reddening studies of Perry et al. (1982), who find the sun to be located in a dust free region at least 75 pc. in radius.

3.2 Beyond the Solar Neighborhood

Over the last ten years, it has been increasingly obvious that our picture of galactic chemical and dynamical evolution is grossly distorted through oversimplification and biased data samples (e.g. Norris, Bessell, and Pickles 1985). The chemical history of the Galaxy is constrained uniquely by the age-metallicity relation, the star formation history, and the observed stellar metallicity distribution. Attempts at the first and third constraints in the 70's were dominated by too small samples (e.g. Pagel and Patchett 1975), inadequate knowledge of the scale height history of the disk, and large uncertainties in individual age determination. Recent attempts using larger samples and uvbyHβ photometry (Mayor 1976, Twarog 1980 and ref. therein) have produced more reliable estimates of the AMR, but the remaining two constraints are still uncertain. Ongoing programs, hold the promise of an order of magnitude improvement in both cases. Olsen (1983) has completed his awesome survey of almost 15,000 stars on the uvby system earlier than G0 down to V=8.3, and age analysis should provide hundreds of stars per billion years within the old disk sample. Radial velocity observations are in progress for over 4,000 of these. Likewise, Ardeberg and Lindgren (1985) are completing a comparable program on cool stars involving photometric observations of 4,000 stars and radial velocity observations of over 2,000. The final constraint, the scale height history, is currently under attack by a variety of groups including Adamson et al. (1985), Twarog (1984), and the Danish group using uvby data on A and F stars, and Grenon and Mayor (1985, private communication) using Geneva photometry of late type stars. These programs should clarify the questions concerning the existence of a thick disk, the abundance gradient, and the nature of the high velocity metal-rich population found in the K-giant survey of Hartkopf, Yoss, and Neese (1985, private communication). The relationship between the high-velocity early type stars and the thick disk remains uncertain (Pier 1982, Stetson 1981). The large radial velocity and photometric surveys by a number of groups (e.g. Carney and Latham 1985) have already produced significant revisions of our knowledge of halo kinematics with more surprises expected.

Finally, we close with a discussion of a direct link beyond the Galaxy, the extragalactic distance scale as defined by the Cepheids. Two programs deserve special note because of their scope and precision. The traditional technique of finding distances to Cepheid clusters through main sequence fitting has long been plagued by field star contamination, binaries, variable reddening, inadequate numbers of stars, and evolutionary effects. Because of the almost vertical nature of the main sequence in the B star range, small errors in photometry led to large discrepancies in distance moduli. These difficulties have been greatly reduced through the use of uvbyHβ photometry by Schmidt (1984 and ref. therein) to observe the Cepheid clusters, and by Balona and Shobbrook (1984 and ref. therein), using the extensive cluster observations of Shobbrook to recalibrate the uvbyHβ system for early type stars, avoiding the use of association stars. The result is inescapable; for the most reliable cluster data, the standard Cepheid scale is shown to be off by 0.4 to 0.6 mag in the sense of providing too bright a luminosity for the Cepheids.

4. SUMMARY, CONCLUSIONS, AND SPECULATION

Having completed our selective survey of the recent photoelectric literature, one might ask what it is that makes these programs exemplary of the type of research best done with a small telescope?

First, there is an implicit assumption that our picture of stellar astronomy is woefully incomplete, and that a good deal of our currently accepted wisdom will end up in ten years as scientific mythology. Within the magnitude limits of a 36 in. telescope, every star in the HD Catalog is observable with any intermediate or broad band photometric system. Since fewer than 10% of these stars have been observed on any system but UBV, it appears difficult to accept that nothing important remains to be learned.

Second, the programs discussed were science driven. They were well-defined and had specific goals in mind which, when obtained, would have an immediate impact upon one or more areas of stellar or galactic evolution. In the area of stellar photometry, too often programs are designed around the need of filling telescope time with the only readily available, easily operated and maintained equipment, a photometer.

Third, the photometric aspect of the programs represents only one component of a more comprehensive project involving high or low dispersion spectroscopy, radial velocities, proper motions, and every other technique in observational astronomy. Photometry is a tool, one tool, which has its specific areas of application and advantage, whether it be on a small or a large telescope. Its speed, efficiency, and precision make it too valuable an approach to exclude from any observational arsenal.

Fourth, if there has been one unifying theme of this talk, it's that the type of project that small telescope photometry serves best is one where crucial trends are hidden in the noise of previously inadequate statistical samples. Often small telescopes are the only option available for such large demands on telescope time.

With these points in mind, what are the likely future directions for small telescope photoelectric research? Let me close by emphasizing only one area of fundamental research and two technical changes. Because of the growing trend toward intermediate and narrow band photometry, and the likelihood of dramatic results from many of the ongoing programs over the next few years, the strong possibility exists that stellar photometry will undergo a modest renaissance, with the concommitant increased demand for larger telescope time, especially in support of space astronomy from Hipparchos (Grenon 1985) and Space

Telescope. One of the greatest shortcomings of the newer systems has been the lack of a large, consistent set of fainter standards for use with larger apertures. Over the years the community has been fortunate to have the dedicated services of individuals such as Landolt, Graham, Cousins, and their coworkers in expanding the availability of broad band standards. It is work for which they have received far too little credit and only modest support. If the newer systems are to have a greater impact, they require the same dedication from their promoters to expand their use.

The need for improved standards is also linked to two technical advances which should change the face of photometry over the next ten years. Since they are both topics of other talks at this meeting, I will only mention them as a plug for the speakers. The first is the development of automated photometric telescope systems (APTS - Genet, this volume). The increased speed, efficiency, and reliability of this approach are too great for it to be ignored as an amateur endeavor, especially in view of the extraordinarily low cost of these systems. The second is the changeover to CCD's (Walker-this volume). They represent such a quantum improvement over photomultiplier photometry that, with a modest decrease in cost, they will make conventional photometry obsolete for most applications. As an example, if these two advances were combined, one person with a few weeks of small telescope time would have the capability of entirely redefining a standard photometric system in terms of CCD observations, single-handedly superceding years of work by a large number of observers using single channel systems. In short, there is every reason to view the future of small telescope photometry with both optimism and excitement. The best is yet to come.

It is a pleasure to acknowledge travel support to this meeting from the International Travel Grant Program of the AAS, and from the IAU. Part of the research discussed was carried out by the author under NSF grant AST-8302091.

REFERENCES

Adamson, A. J.. Hilditch, R. W., Hill, G., and Fisher, W. A. 1985, I.A.U. Coll. No. 88, "Stellar Radial Velocities," p. 355.
Ardeberg, A., and Lindgren, H. 1985, I.A.U. Coll. No. 88, p. 151.
Armandroff, T., and Demarque, P. 1984, Astr. Ap. **139**, 305.
Balona, L. A., and Shobbrook, R. R. 1984, Astr. Ap. **130**, 279.
Bertelli, G., Bressan, A., Chiosi, C. 1984, Astr. Ap. **130**, 279.
Buonanno, R., Corsi, C. E., Fusi-Pecci, F., Alcaino, G., Liller, W. 1984 Ap. J. **277**, 220.
Cannon, R., and Lee, S. 1981, I.A.U. Coll. No. 68, p. 501.
Carney, B. 1982, A.J. **87**, 1527.
Carney, B. 1983, A.J. **88**, 623.
Carney, B., and Latham, D. 1985, I.A.U. Coll. No. 88, p. 139.
Conti, P. 1984, I.A.U. Symp. No. 105, p. 233.
Da Costa, G. and Villumsen, J. 1981, I.A.U. Coll. No. 68, p. 527.

Eggen, O. J. 1970, Vistas in Astr. **12**, 367.
Eggen, O. J. 1976, Q.J.R.A.S. **17**, 472.
Eggen, O. J. 1983, M.N.R.A.S. **204**, 405.
Eggen, O. J. 1985, A.J. **90**, 74.
Elmegreen, B., and Lada, C. 1977, Ap. J. **214**, 725.
Garmany, C., Conti, P., and Chiosi, C. 1982, Ap. J. **263**, 277.
Grenon, M. 1985, E.S.A. Coll. SP-234, p. 113.
Gusten, R., and Mezger, P. 1982, Vistas in Astr. **26**, 159.
Herbst, W., and Miller, D., A. J. **87**, 1478 (HM).
Humphreys, R. M. 1978, Ap. J. Suppl. **38**, 389.
Jakobsen, A. M. 1984, I.A.U. Symp. No. 105, p. 345.
King, C. Da Costa, G., and Demarque, P. 1985, preprint.
Landolt, A. 1979, Ap. J. **231**, 468.
Larson, R. B. 1982, M.N.R.A.S. **200**, 159.
Larson, R. B. 1985, M.N.R.A.S. in press.
Maeder, A. 1981, Astr. Ap. **101**, 385.
Mayor, M. 1976, Astr. Ap. **48**, 301.
McDonald, A. and Hearnshaw, J. 1983, M.N.R.A.S. **204**, 841.
Mermilliod, J. 1976, Astr. Ap. **53**, 289.
Mermilliod, J. and Maeder, M. 1984, I.A.U. Symp. **105**, p. 349.
Mould, J. 1982, Ann. Rev. Astr. Ap. **20**, 91.
Nissen, P., Twarog, B., and Crawford, D. 1986, in preparation.
Norman, C., and Silk, J. 1980, Ap. J. **238**, 158.
Norris, J., Bessel, M., and Pickles, A. 1985, Ap. J. Suppl. **58**, 463.
Olsen, E. 1983, Astr. Ap. Suppl. **54**, 55.
Pagel, B., and Patchett, B. 1975, M.N.R.A.S. **172**, 13.
Pier, J. 1982, A.J. **87**, 1515.
Perry, C., Johnston, L., and Crawford, D. 1982, A.J. **87**, 1751.
Schmidt, E. 1984, Ap. J. **285**, 501.
Smith, G. E. 1984, A. J. **88**, 1775.
Stauffer, J. 1982, A. J. **87**, 1507.
Stetson, P. 1981, A. J. **86**, 1882.
Turner, D. 1982, P.A.S.P. **94**, 655.
Turner, D., Gueve, G., Herbst, W., and Harris, W. 1980, A. J. **85**, 1193.
Twarog, B. 1980, Ap. J. **242**, 242.
Twarog, B. 1983, Ap. J. **267**, 207.
Twarog, B. 1984, A. J. **89**, 523.
Upgren, A., Weis, E., and Hanson, R. 1985, A. J. **90**, 2039.
van den Bergh, S., and Sher, D. 1960, Publ. D.D.O. **2**, 203.
VandenBerg, D. 1985, Ap. J. Suppl. **58**, 532.
Vogt, N. 1971, Astr. Ap. **11**, 359.
Warner, J., Strom, S., and Strom, K. 1977. Ap. J. **213**, 427.

DISCUSSION

Garrison: In your diagram of halo stars, were those mostly giants?

Twarog: Yes.

Garrison: In Delhi, we managed to push through a resolution requesting time assignment committees to allocate time for the establishment of faint standards so that large telescopes can be used more effectively in cases where neutral density filters are not appropriate - e.g. space telescope.

Graham: Do you have any comment about the relative merits of the various intermediate-band photometric systems for detecting such quantities as the metallicity of a star?

Twarog: While I have a bias toward the Strömgren system, I feel that it is important for the originators of a system to promote and explain its application to potential users by publication of first-class scientific results. The best way to attract users is by showing how much high quality research one can do with a system.

McCarthy: Would you please justify your statement about the number of objects observed in the Strömgren and in the Geneva systems? Could you give some numbers please.

Twarog: I feel that the importance of a system is not measured by the number of stars observed on it, but by the application and interpretation of the data. The Strömgren system has produced far more valuable scientific interpretation than the same number of stars observed in the Geneva system.

OBSERVATIONAL POLARIMETRY PROGRAMS FOR SMALL TELESCOPES

Michel Breger
University of Vienna
Türkenschanzstraße 17
A 1180 Vienna
Austria

ABSTRACT. A number of promising observational programs involving linear polarization measurements are discussed after consideration of the observational limits based on telescope size and instrumental accuracy. These projects include studies of interstellar extinction, circumstellar shells and extended envelopes, inhomogeneities on stellar surfaces such as nonradial pulsation, magnetic stars, and interactive binary systems.

1. INSTRUMENTAL AND OBSERVATIONAL LIMITS

The high observational accuracy of modern polarimeters has opened the door to many exciting, new astronomical applications. The choice of suitable polarimetric projects is limited by telescope size (through the number of available photons) and the accuracy of the equipment. Since a surprisingly large number of polarimetric papers in the literature contain doubtful results, considerable attention should be given to the size of the expected, as well as actual uncertainties of measurement. For a bright star measured with a modern polarimeter an accuracy in the range of $\pm$ 0.01% should be possible with repeated measurements and at least 2×10^8 photons. With our single-channel polarimeter (Breger 1979), for a 200-second integration the limit is set by scintillation noise, which results in $\pm$ 0.02% on a 2-meter telescope. In this talk we would like to mention a few promising linear-polarization projects and emphasize broadband measurements.

The type of polarimeter used may well be of secondary importance in determining the observational feasibility of the projects. The use of rapid modulation or multiple channels permits high accuracy to be achieved even under nonoptimum weather conditions. Observations of standard stars (Serkowski 1974, Hsu and Breger 1982) are strongly recommended. The unpolarized standard stars allow the determination of the size of, and possible small changes in instrumental polarization. On the other hand, the observations of polarized standards (or another suitable method of calibration) fix the zero-point of the position angle and provide

J. B. Hearnshaw and P. L. Cottrell (eds.), Instrumentation and Research Programmes for Small Telescopes, 149–157.

TABLE I

LIMITING MAGNITUDES WITH A 1-METER TELESCOPE
(Single-beam, background not included)

Photon-Statistics Accuracy	Filter	10-minute integration	60-minute integration
0.01%	Open	$8^m.0$	$9^m.9$
	B	$6^m.6$	$8^m.5$
	V	$5^m.8$	$7^m.7$
0.1%	Open	$13^m.0$	$14^m.9$
	B	$11^m.6$	$13^m.5$
	V	$10^m.8$	$12^m.7$
0.5%	Open	$16^m.5$	...
	B	$15^m.1$	...
	V	$14^m.3$	$16^m.2$

a check on the size of the polarization value.

Table I lists typical magnitude limits (photon statistics) for different values of desired accuracy. In this table, the filter "Open" refers to filterless measurements with an efficient EMI 9658A photomultiplier tube. The actual observations require more time than listed because of background measurements and other observational activities such as recentering of the star in the aperture.

2. INTERSTELLAR POLARIZATION

The presence of interstellar dust leads to extinction as well as polarization of light passing through the interstellar medium. With modern polarimeters the resulting value of polarization is often more easily and accurately measured than the corresponding extinction or reddening value. An interstellar reddening of $E(B-V) = 0^m.01$ is difficult to determine accurately through conventional photometric or spectroscopic means. A reddening of $E(B-V) = 0^m.01$ is accompanied by a relatively easily measured polarization value near 0.06%. (The exact value depends on the direction of the interstellar magnetic field, among other factors). In order to use polarization as a probe for the interstellar medium, several properties of the interstellar medium need to be considered:

(i) Interstellar polarization follows a standard wavelength dependence of polarization, shown by Serkowski, Mathewson and Ford (1975,hereafter referred to as SMF) to be of the form

$$p(\lambda)/p(\max) = \exp\left[-1.15 \ln^2 (\lambda(\max)/\lambda)\right],$$

where $\lambda(\max)$ is the wavelength at which the polarization reaches its maximum value, p(max). Recent discussions (Wilking, Lebofsky and Rieke 1982) of a possible dependence of the constant, -1.15, on $\lambda(\max)$, and the presence of structure in the wavelength dependence (Wolstencroft and Smith 1984) indicate fascinating new areas of investigation. Nevertheless, applications of the general formula still appear meaningful and the wavelength dependence of polarization is a powerful tool to separate interstellar from other sources of polarization.

(ii) In clusters with constant values of foreground extinction from star to star, the amounts of polarization are usually also constant. Since cluster nonmembers are at different distances relative to cluster members, the nonmembers often show different amounts of extinction and polarization. For these clusters polarization can be used to separate members from nonmembers. This argument was successfully applied to NGC 7789 (Breger 1982), where it was shown that the blue-straggler candidates cooler than the main sequence do not belong to the cluster. In star clusters with variable reddening and polarization, the comparison of reddening and polarization values may permit a modeling of the different interstellar cloud patches. This is based on the fact that reddening is purely additive, while polarization is a vectorial quantity. The presence of a second cloud with a different magnetic-field direction, for example, increases the reddening value, causes a rotation of the position angles of polarization, and could lead to depolarization. The Pleiades cluster is a good example. Our model of a uniform foreground cloud and a patchy cloud moving through the western part of the cluster is also consistent with atomic-line data as well as radio and millimeter measurements.

(iii) The value of $\lambda(\max)$ is related to the ratio of total to selective extinction by $R = 5.5\,\lambda(\max)$, as pointed out by SMF. Polarization measurements are, therefore, important to locate unusual values of R in the interstellar medium.

For small telescopes a promising application seems to be the study of systematic regional versus local origins of abnormal values of R. The historical question whether or not values of R greater than the normal value of 3 (i.e. $\lambda(\max)$ greater than 0.55 μm) show a systematic dependence on galactic longitude and latitude has not been settled (e.g. see SMF). Our speculation is that both local and large-scale systematic regional variations are involved. The Orion cluster is a good example for local effects near the stars, with variations in the value of R from 3 to 5.

On the other hand, the stars in the Scorpius-Ophiuchus region of the sky may be an example of unusually large R values covering a large region of the sky. For this region, Meyers et al. (1985) propose a weak shock in the interstellar medium in the Scorpius-Ophiuchus direction. The rotation of the position-angle of polarization up to a few degrees with wavelength, determined so far for a few stars, may be related to these processes in the interstellar medium.

To answer these questions, accurate determinations of λ(max) and position angles are required for many more relatively bright stars.

3. CIRCUMSTELLAR SHELLS AND EXTENDED ENVELOPES

Scattering in aspherical circumstellar envelopes surrounding stars leads to polarization which often exceeds 1%. Be, T Tauri and hot pre-main-sequence stars are good examples. Time variability and a noninterstellar wavelength dependence of polarization are important clues.

The variability of the polarization of the polarization in γ Cas led to the discovery of the intrinsic polarization in Be stars (Behr 1959). The polarization does not exceed 2%. Striking features of the wavelength dependence of polarization of Be stars are the rapid decrease in polarization shortward of the Balmer limit at 0.365 μm and the decrease of polarization in the Balmer emission lines. A good example of the wavelength dependence is shown in the paper by Poeckert, Bastien and Landstreet (1979), and a very thorough discussion can be found in the monograph on B stars (Underhill and Doazan 1982).

The extreme brightness of many Be stars encourages a time variability study with small telescopes. Due to the "discontinuous" wavelength dependence of polarization, the standard UBV filters would not be an optimum choice for such a study. The standard U filter, for example, does not avoid the Balmer jump, while the u filter of the uvby system would be more suitable for a cleaner ultraviolet measurement.

The polarization of T Tauri stars can be as high as 12% (Bastien 1983). A standard wavelength dependence of polarization does not seem to exist. The question of the existence of periodicities, or better, timescales, of these variations still needs to be settled. The hot pre-main-sequence stars also deserve considerably more photometric and polarimetric study. Many of these stars are brighter than 12th magnitude and can be observed with small telescopes.

4. INHOMOGENEITIES ON THE STELLAR SURFACE

Inhomogeneities, asymmetry and distortions of the stellar disk can, at those wavelengths where scattering dominates as an opacity source, lead to a small amount of linear polarization. The size of this polarization is near the present limit of detectability of 0.01%. Some origins of inhomogeneities could be nonradial pulsation, spots, or large convection cells. The small size of the resulting polarization makes it difficult to separate these intrinsic effects from interstellar polarization. However, the different wavelength dependence of polarization and the expected time variability are important clues.

4.1 Nonradial Pulsation

Among hot stars, in which electron scattering is an important opacity source, the distortions of the visible disk could lead to polarization variability with small amplitudes. The size and phasing of the linear polarization as well as the position angles are dependent on the particular pulsation mode(s). This was suggested as a test for mode-typing by Odell (1979) as well as Stamford and Watson (1980). It was also shown in these papers that the amplitudes of polarization variability should be larger in the ultraviolet and blue regions than in the visible. The most promising objects for study appear to be hot pulsators with low orders of nonradial pulsation, ℓ and m, as well as objects in which one pulsation mode is dominant. A (possibly large) number of pulsation cycles could be averaged for better signal-to-noise ratios. Prior knowledge of the photometric frequencies and epochs would be useful, but not an absolute requirement.

The β Cephei variables appear well-suited for such polarimetric studies since many members of the class fulfil the conditions listed above and are also bright enough for a high count rate. In a pioneering paper, Odell (1981) measured the star BW Vulpeculae during four nights. He found a statistically significant (4 sigma) change of polarization during the $0^{d}.201$ cycles and a peak-to-peak amplitude of 0.008%. The phase of maximum polarization occurred near maximum light.

Such a small polarization amplitude below the accuracy of a single observation must naturally be treated with some caution. According to a private communication by Odell, Kemp has repeated the measurements of BW Vulpeculae and finds a similar amplitude, although uncertainties due to possible large phase disagreements are suspected.

A polarimetric project to remeasure BW Vulpeculae over a larger number of cycles would be important. Other bright β Cephei stars with a strong nonradial mode (such as β CMa and ν Eri) also appear promising.

4.2 Convection Cells in the Photosphere

It has been known for a long time (e.g. Dyck 1968) that cool giants and supergiants show polarization, which is intrinsic and variable with time. A plausible physical explanation is the scattering from a photosphere with brightness variations across the photospheric surface. According to Schwarzschild (1975), large convection cells appear and disappear on the surface over a few months. The resulting linear polarization would be less than 1%.

A well-studied example is the M supergiant α Ori (Hayes 1981, Schwarz and Clarke 1984). In another recent study (White, Shawl and Coyne 1984), 15 out of 39 red giants in the globular clusters M3 and M13 were found to show intrinsic polarization with sizes generally less than 2%. This shows that relatively high intrinsic polarization may well be a common property of cool giants and supergiants. More polarimetric surveys with attention given to the subtraction of the interstellar component, if any, are needed.

5. MAGNETIC STARS

The study of linear polarization variability in magnetic CP2 stars has a long and contradictory observational history. The presence of intrinsic linear polarization and its variability (caused by the rotation of the star) has been predicted many times. Demkina and Obridko (1973), for example, have pointed out that the transverse component of the magnetic field causes linear polarization over a wide spectral range. In a field strong enough for the Zeeman components to be completely resolved,

$$p \sim W_{\pi}(\eta_0) - 2\, W_{\sigma}(\eta_0/2),$$

where p is the degree of linear polarization in the line; W_{π} and W_{σ} are the equivalent widths of the π and σ components, respectively; and η_0 is the ratio of the line-center opacity relative to the continuum opacity. For weak lines the polarization cancels out. For partially saturated lines the polarization reaches a maximum value. Rough calculations show that the net broad-band linear polarization in the typical CP2 star should be less than 1%. A careful choice of wavelength regions (through examination of the lines in the regions) could maximize the measurable amounts of polarization. As is the case for other intrinsic sources of polarization, time variability and a nonstandard wavelength dependence of polarization would allow us to distinguish between interstellar and intrinsic polarization.

The early reports of polarization variability were regarded by Serkowski and Chojnacki (1969) to be "not incontestable". Serkowski and Chojnacki measured a number of CP2 stars and found no evidence for a linear polarization variability at the 0.1% level. In particular, the variability reported earlier for HD 71866 (0.1 to 0.3%, Thiessen 1961) was not confirmed.

For the star HD 215441, Kemp and Wolstencroft (1973) detected large changes, up to 2%, in the linear polarization of the Hß center and wings. Continuum values were relatively constant. For the star 53 Cam, the detected small variations (less than 0.1%) in both Hß and the broad-band continuum could be correlated with the star's magnetic phases.

Polarimetric studies of both the Hß profile and the broad-band continuum in magnetic CP2 stars appear very promising. A number of bright candidates are available.

In the previous sections we could list only some astrophysical areas for which linear polarimetry with small telescopes appears promising. The high accuracy of modern polarimeters permits us to examine bright stars to study new problems. It is regrettable that the highly polarized polars (AM Her stars) had to be omitted from our list of proposed programs for small telescopes because of the combination of their relative faintness and rapid polarization variability. (We note that AM Her itself, with an intermediate V magnitude of $12^{\mathrm{m}}.9$ could be studied with a one-meter telescope.)

REFERENCES

Bastien, P. 1983, Rev. Mex. Astron. Astrof., **7**, 198
Behr, A. 1959, Nachr. Akad. Wiss. Göttingen, Math-Phys., **K1**, 7185
Breger, M. 1979, Ap.J., **233**, 97
Breger, M. 1982, Ap.J., **263**, 199
Demkina, L.B., and Obridko, V.N. 1973, Soviet Astron., A.J., **16**, 853
Dyck, H.M., 1968, A.J., **73**, 688
Hayes, D.P. 1981, P.A.S.P., **93**, 752
Hsu, J.C., and Breger, M. 1982, Ap.J., **262**, 732
Kemp, J.C., and Wolstencroft, R.D. 1973, Ap.J.(Letters), **179**, L33
Meyers, K.A., Snow, T., Federman, S.R., and Breger, M. 1985, Ap.J., **288**, 148
Odell, A.P. 1979, P.A.S.P., **91**, 326
Odell, A.P. 1981, Ap.J.(Letters), **246**, L77
Poeckert, R., Bastien, P., and Landstreet, J.D. 1979, A.J., **84**, 812
Schwarz, H.E., and Clarke, D. 1984, Astr. Ap., **132**, 370
Schwarzschild, M. 1975, Ap.J., **195**, 137
Serkowski, K. 1974, in Planets, Stars and Nebulae Studied with Photopolarimetry, ed. T. Gehrels (Tucson: Univ. of Arizona), p.135
Serkowski, K., and Chojnacki, W. 1969, Astr. Ap., **1**, 442
Serkowski, K., Mathewson, D.S., and Ford, V. 1975, Ap.J., **196**,261 (SMF)
Stamford, P.A., and Watson, R.D. 1980, Acta Astr., **30**, 193
Thiessen, G. 1961, Astron. Abh. Sternw. Hamburg-Bergedorf, **5**, 273
Underhill, A., and Doazan, V. 1982, NASA SP, **456**, 346
White, R.E., Shawl, S.J., and Coyne, G.V. 1984, A.J., **89**, 480
Wilking, B.A., Lebofsky, M.J., and Rieke, G.H. 1982, A.J., **87**, 695
Wolstencroft, R.D., and Smith, R.J. 1984, Mon.Not.R.A.S., **208**, 461

DISCUSSION

Burki: Is the polarization technique more sensitive than the study of line profile variations for the determination of non-radial modes?

Breger: No, but the study of line profile variations requires large telescopes and high dispersion spectrographs. In order to understand non-radial pulsations we need to use as many techniques as possible : line profiles, photometry for variability and relationships between light and colour variations, radial velocity and polarisation measurements.

Finkenzeller: I fear that the main problem when doing polarimetry of pre-main sequence objects is not that of getting the vector, but its astrophysical interpretation. Almost all objects are surrounded by reflection nebulosity which introduces strong aperture-size effects. Thus, one would wish to have both high data quality and high spatial resolution. Together, with typical magnitudes of T Tauri stars fainter than 11, one is likely to rely on intermediate and large size telescopes.

Breger: I believe that important work can still be done with small telescopes. An example might be the star W90 in NGC 2264. It appears to lie below the main sequence and moves parallel to the main sequence. Its polarization is about 2%. Are the polarization and abnormal position in the HR diagram caused by intracluster dust or shells?

Rucinski: I think that eclipsing binaries and close binaries in general should not be forgotten. The polarization methods can be used to obtain additional information on the geometry and, indeed, can be used to discover pole-on close binaries which should show a constant but rotating vector of polarization.

Warner: The statistics of limiting magnitudes for linear polarization is similar to that for circular polarization. As we now know, many AM Her stars which have circular polarizations in excess of 10 percent, accuracies of only ~ 1 percent are adequate for the study of time variability. With broad-band polarimetry, magnitude 15½ can be reached with time resolution of ~ minutes even with telescopes of 1 metre aperture or less. Because of the great variability of the AM Her stars, on timescales ranging from minutes to years, a much greater body of data needs to be accumulated for analysis and this can only be done with the large amounts of time available on small telescopes.

Breger: Before the discovery of the high polarization of polars (AM Her stars), the amount of known circular polarization was too small to be measured with small telescopes. The interesting circular polarization of AM Her stars can be observed with small telescopes.

Warner: The problem is that we don't have standards with 50% circular polarization and we therefore have to extrapolate from objects with a few tenths of a percent polarization.

Schober: Very few polarimeters are working at visiting observatory sites. ESO has tried for 15 years to set up a polarimeter. It requires an experienced resident observer. Also the alignment of optical elements is critical for instrumental calibration and with frequent instrument changes on telescopes this can be difficult.

Breger: Firstly, my suggestion is to take your own polarimeter, which you can probably trust, to the other observatories. The instrumental correction (which is usually small) can be checked by observing unpolarised standards (see Serkowski 1975, and Hsu and Breger 1982) and at least one or two of these should be observed each night.

Schober: It is very expensive to have one's own instrument!

A'Hearn: Yet another area in which useful polarimetry can be done from small telescopes is the polarimetry of cometary dust. Thus far, polarimetry of comets has not been very fruitful both because the interpretation has been done usually with Mie theory rather than with laboratory measurements of irregular particles and because, with a few exceptions such as the recent work by Myers and Nordsieck at Wisconsin, the data have not been obtained systematically. These measurements can be made with a very small telescope for bright comets such as Halley and only relatively low precision is needed, say 1%.

Breger: Since we are mentioning the many promising applications of polarimetry that had to be omitted from the talk, I will mention an omission myself. The polarimetric study of asteroids to determine their rotation and surface variation appears to be very important.

CONTACT BINARIES

S.M. Ruciński
David Dunlap Observatory
University of Toronto
P.O. Box 360, Richmond Hill,
Ont. L4C 4Y6, Canada

ASBTRACT. The most promising mechanism for the formation of contact binaries involves the orbital angular momentum loss (AML) and the resulting orbital decay of detached but close synchronized binaries. The efficiency of magnetic wind braking should abruptly decrease upon formation of a contact binary because of the transformation into a system of earlier spectral type and (possibly) of longer orbital period. The new primary of the contact system should have convective zone thinner than indicated by the surface temperature of the common envelope. The decrease in the coronal (X-rays and radio) activity of contact binaries, which is indeed observed, is used as an agrument that the AML efficiency in contact is relatively low and that the contact stage is considerably prolonged relative to adjacent stages. This small modification to the AML models is capable of explaining why many *different* contact binaries are observed in old systems like NGC188. The AML evolution is not the only mechanism leading to formation of contact binaries; some of them may have originated *via* Algol-like evolution. Thus, the observed contact binaries are probably a mixture of systems formed in different ways.

1. INTRODUCTION

The W UMa-type binaries need no introduction to participants at this meeting. Their short orbital periods make them prime targets for small telescopes, particularly as "easy" objects for photometric studies. This opinion about W UMa systems is not always profitable for their understanding because many papers have been prepared by first-time observers or students who were given "easy" term projects. This has resulted in too many incidental papers which obscure rather than enlighten the view. In addition, the pleasure of obtaining a really *curved* light curve with relatively little effort frequently eclipses the need of using a calibrated photometric system: Still quite common in papers are expressions such as: "...our light curve was obtained through the B and V filters which closely match..." (which usually means that everything was left in the instrumental system!). The spectroscopic observations of W UMa systems are more difficult and the situation here is a

J. B. Hearnshaw and P. L. Cottrell (eds.), Instrumentation and Research Programmes for Small Telescopes, 159–172.

bit better but still frequent are orbital solutions based on less than 20 observations and with sine fits which neglect the *large* rotational effects of contact binaries.

However, my intention is not to complain about improper ways of obtaining and handling observations of W UMa systems but rather to address the questions of why we should study these systems and what do we know about their origin and evolution. Quite a few reviews have appeared relatively recently: The evolutionary state and activity of W UMa systems were described in the same volume by Mochnacki (1985) and by Ruciński (1985a); other reviews are by Smith (1984), Duerbeck (1984) and Ruciński (1985b). In this situation of the relative abundance of reviews, I decided to discuss here only some problems related to the formation and evolution of W UMa systems, assuming that the reader needs no detailed introduction to the field.

2. CONTACT NATURE OF W UMA SYSTEMS

It can hardly be argued that the W UMa systems are *not* in contact. This assertion results not only from solutions of light and radial-velocity curves but primarily from the mere fact that their components have identical temperatures and hence luminosities scaling as $L \propto R^2 (\propto M$ for Roche geometry) in spite of differing masses. We know that the W UMa systems avoid having equal components ($q = M_2/M_1 \neq 1$) and may, in fact, prefer small mass-ratios which are expected to be discriminated against by observational selection effects (Van't Veer 1978). Since the cores produce energy as $L \propto M^4$ (or steeper if the cores are evolved), it is an inevitable conclusion that very large amounts of energy must be transported through the optically thick "neck" between components. This picture must certainly be true for all systems which show EW-type light curves (i.e. with equal minima), irrespectively of orbital period. However, as noticed already by Lucy (1976), for $P > 0.4$ days, the EB-type systems (light curves with unequal minima) start to appear. Some of them may be semi-detached but some are in genuine contact but with strange surface-brightness distributions indicating the presence of bright spots in the neck area which can be modelled by unphysical values of the reflection albedo (Lucy and Wilson 1979; Kalužny 1983 and in press). The lesson given to us by the early-type system SV Cen may be instructive here. This system was once considered to be in unusually deep contact (Wilson and Starr 1976; Ruciński 1976); we know now that a strong and very hot light source, which is located somewhere between the components and results from the mass-transfer flow (Drechsel et al. 1982), mimics large distortion of components. We cannot exclude the possibility that at least in some "normal" EB-type binaries the mass-transfer and mass-accretion phenomena also produce systems which, perhaps similarly to SV Cen, *appear* to be in good physical but poor thermal contact.

The significance of the division at about 0.4 days is still not clear. It is interesting to note that contact binaries with the A-type light curves which, in fact, best conform to the contact model, also start appearing at roughly $P > 0.35 - 0.4$ days and that the range $0.22 < P < 0.35$ is almost exclusively populated by systems with W-type light curves.

3. MODELS OF THE INTERNAL STRUCTURE

The two best-known models are both modifications of the "classic" model of Lucy (1968); we shall summarize their properties in turn.[1] The TRO (thermal-relaxation-oscillation) model was proposed by Flannery (1976) and by Lucy (1976) and was later more broadly explored by Robertson and Eggleton (1977). It is a direct extension of the original models of Lucy (1968) but now the matter - and not only the energy - is also permitted to be exchanged between components. When the total mass and angular momentum are preserved, such models reveal relaxation-type oscillations with alternate long-lasting contact phases and relatively short semi-detached phases. During the contact phase the mass-ratio decreases and the separation increases until disruption of contact; in the semi-detached phase the matter flows back until contact is re-established. The relative duration of both phases depends on the ratio of thermal time scales of both components. Since these time-scales depend on the masses roughly as $t \propto M^{-3}$, the relative durations of contact and semi-detached phases should scale as $t_{cont}/t_{sd} \propto q^{-3}$. Thus, for small mass-ratios, the semi-detached phase may be relatively brief.

The AML (angular-momentum-loss) model (Mochnacki 1981; Rahunen and Vilhu 1982) was proposed to explain the non-existence of semi-detached configurations of the TRO models, at least among short-period W UMa binaries. It also derives from the known strong magnetic activity of W UMa systems (to which we shall return later) and the presumed importance of the loss of angular momentum *via* the magnetized wind.[2]

---

[1] We will not discuss the contact-discontinuity model (DSC) which is in fact an alternative to the original Lucy models (Shu et al. 1976; for a full list of later references, see Shu et al. 1980). Its main idea is that one can hide the cooler component inside the mantle extended over it by the hotter component. This model presented *no* explicit predictions as to the testable properties of contact binaries (except the by-product concerning the non-existence of the gravity darkening: Anderson and Shu 1977) but generated a hot debate which, from the present perspective, seems to have played an important rôle in visualizing the extreme complexity of the energy transfer between components. This hydrodynamic Pandora's Box is relatively little explored, except for a few pioneering efforts such as those by Robertson (1980) and Hazlehurst (1985). It should be noted that the DSC models may correctly describe contact binaries at *some* stages of evolution such as immediately after establishing contact in the TRO models. The concept of the DSC models also helped to recognise the over-constrained nature of the original models of Lucy (1968).

[2] Importance of the AML, especially at the very early stages of binary-star evolution, was pointed out much earlier by Okamoto and Sato (1970).

In effect, this loss is used here to gently squeeze the systems and thus to prolong the contact phase of the TRO model to such a degree that the semi-detached phase never appears (or appears very rarely). The system drifts to smaller and smaller mass-ratios along contact branches of the TRO cycles. It is not clear what regulates the AML in such a way that it is strong *enough* to eliminate the semi-detached configuration yet not *too* strong to lead to a premature merger. It is an observed fact, however, that models constructed according to the precepts of the self-regulated AML explain the period - color relation very well. The expected time-scales of evolution *via* this mechanism are of the order of 5×10^8 years (Rahunen 1981) or may be even longer (Mochnacki 1981).

The TRO and AML models have explained many observational facts: the preference for marginal contact, the evolution towards small mass-ratios, and the shape of the period - color diagram. They have also given many leads where to look for possible observational tests. Particularly easily testable seemed to be the occurrence of the broken-contact phase and the correlation of period changes with the branch of the oscillation cycle. Only later on it was realized that the period changes are dominated by magnetic effects which have more in common with those observed in RS CVn systems (Kreiner 1977; Herczeg 1979; Hall and Kreiner 1980; Van Buren and Young 1985) than with real mass-transfer effects. As we already said, systems which appear to be in the semi-detached phase (or at least with poor thermal contact) do seem to exist: They start at about P = 0.35 days; there are few of them for $0.35 < P <$ 0.45 days but they become more frequent at longer periods. In none of the better studied cases (Kalužny 1983; Hilditch et al. 1984; King and Hilditch 1984; Mochnacki et al. 1985; further as yet unpublished studies by Kalužny and Hilditch et al.) can we be sure that these are really broken-contact exemplifications of the TRO models; they look more like evolved Algols which started with much longer periods and through mass-exchange, mass-ratio reversal and loss of angular momentum followed by further evolution of former secondaries approached configurations just bordering on contact (see Budding 1984, or the scenario suggested by Tapia and Whelan (1975) to explain the properties of ε Cr A). We should not, however, dismiss the possibility that these objects *are* related to the contact systems. In fact, if we consider what must happen to them, we come to the conclusion that these semi-detached objects may very well form contact binaries in the not too distant future. As is well known from the experience of calculating contact models (Lucy 1968, 1976; Moss and Whelan 1970; Hazlehurst 1970; Moss 1971), it is much easier to obtain stable contact binaries using *evolved* models, preferebly with quite differently evolved cores. Indeed, the further we depart from homology relations for both components, the easier it is to form a contact binary. But contact binaries consisting of *unevolved* Main Sequence components certainly also exist, as documented by their large mean densities and locations on the H-R diagrams of clusters (Mochnacki 1981, 1985). How to explain those systems? It seems that to solve that, we really must consider contact binaries in the broader perspective of the evolution of close but detached binaries under the influence of the angular momentum loss. This loss *must* operate in all cool, rapidly-rotating stars and may produce dramatic effects in close synchronized binaries. Before we do that in Sec. 5, we will have a look at the magnetic activity of W UMa systems.

4. ACTIVITY OF W UMA SYSTEMS

The W UMa systems are very active stars, fully confirming the well known correlation between activity on one hand and the rotational period and spectral type on the other hand. A single quantity, the inverse Rossby number τ/P (Noyes et al. 1984), combines the spectral type and period in one convenient parameter[3] which can be used to compare activity levels in single stars or even to predict rotation periods from the observed activity levels (Soderblom 1985). Apparently, it can also be used for synchronized binary stars, including W UMa systems (Vilhu 1984). But there are some important differences between contact binaries and all other stars and we will describe them shortly.

One of the indications of elevated activity, photometric spots, are know to exist on surfaces of W UMa systems. Such spots *may* exist in ordinary systems such as W UMa itself (Eaton et al. 1980) but they *must* exist in unusual ones such as TZ Boo (Hoffmann 1978; Ruciński 1985a), where there is no other way to explain weird changes in light curves on time scales of years (conversions from A- to W-type of light curve, etc.). It is difficult to quantize the spot coverage; it is much easier to do that for various non-thermal emissions of these stars. In terms of the chromospheric (Riciński 1985c) and transition-region (Vilhu and Ruciński 1983; Eaton 1983) emissions, the W UMa systems seem to be identical to those few single stars and detached-binary components which have equally short rotation periods and hence equally large τ/P (Vilhu 1984). In fact, strengths of these emissions increase very slowly with the increase of τ/P for these very active stars; this can be explained as resulting from the total saturation of stellar surfaces by small-scale active regions. But interesting things start to happen when we consider the coronal exissions: Both in X-rays (Cruddace and Dupree 1984; Vilhu 1984) and in the radio (Hughes and McLean 1984, 1985) the W UMa systems are definitely *less active* than are similar single stars or components of non-contact binaries. Apparently, *the W UMa systems are unable to generate and maintain large magnetic-loop structures which could retain coronal plasma.* What is the reason for that?

We can hypothesize an explanation for the lowered coronal activity of contact binaries by making an observation that sizes of magnetic loops tell us something about dimensions of interior regions where these loops are generated and then temporarily anchored. The largest coronal structures should have sizes of the order of the vertical extent of the whole convective zone. The important distinction for contact binaries

---

[3] The quantity τ (which is a function of the spectral type) gives the convective turnover time in the deep layers of the convective envelope where the time scales for magnetic bouyancy losses are long; the rotational period P describes, to first order, the rotational velocity field which can amplify the magnetic field.

momentum. The surface areas for the emission of the wind of the new contact system and of the original detached system are comparable so that the relative efficiencies of the wind braking should depend only on the rotational periods and spectral types. Using the inverse Rossby number as a guide, we estimate that the drop in coronal activity should be ~ 20 times. To arrive at this very approximate number we note that the change in τ/P for the transformation from G 2 to F 2 is by a factor ~ 50 but the activity is not linear in τ/P so the drop should be smaller; however, one star is lost from the AML budget compensating for this effect. We think that the AML efficiency should drop by a comparable factor, possibly retaining the same dependence on Ω as before establishment of contact. Thus, only the constant of proportionality in $dJ/dt \propto -\Omega^{1.5}$ would have to change. The subsequent evolution of the contact binary towards smaller mass-ratio will result in some *increase* of the orbital period (decrease of Ω); the spectral type will also become earlier. If we again use τ/P as a guide, we see that both effects should lead to a further decrease of $|dJ/dt|$; this is in accordance with the models of Rahunen (1981) which indicate that progressively less of the AML is needed to keep a system in marginal contact. This type of braking can extend up to relatively early spectral types, perhaps as early as ~A 7 where the convective dynamo still seems to operate (Schmitt et al. 1985).

It is not clear how the mechanism described in Figure 1 regulates itself so precisely that the W UMa systems always stay close to marginal contact. Therefore, much remains to be done to obtain a coherent picture of the whole process from general tendencies we have only enumerated here.

6. CONCLUSIONS

As a result of much observational and theoretical work, a picture of the contact binary formation and evolution starts to emerge. Apparently, these systems can be formed from detached binaries by a variety of processes with the common denominator being the loss of angular momentum. Thus, some of them may have been formed by the Algol-type evolution with a mass-ratio reversal; some of them may have evolved directly into contact from detached Main-Sequence binaries *via* the magnetic-wind braking. Therefore, the contact stage seems to be the *purgatory* for all close binaries which precedes the ultimate merging of components and disappearance in the sea of single stars. We think that the contact stage is relatively long-lived because the magnetic braking is less efficient at that time than in stages which precede and follow the contact configuration. This drop in efficiency results from two effects: at first, there is a jump to earlier spectral types and (probably) longer periods; later, there is the evolution in contact to smaller mass-ratios and even earlier spectral types. Both effects should lead to a drop in activity and to a less efficient wind braking. We see this drop only in the coronal radiation; the much cooler common convective envelope generates quite strong surface activity but this activity has no influence on the large-scale magnetic structures which survive intact.

Possibly related to these considerations of the binary survival and contact binary formation are reasons why we do not observe cool contact binaries with spectral type later than ~K 5 and periods shorter than 0.22 days (Mochnacki 1985). This is not the only subject which has been left undiscussed here: One of the still remaining problems is the explanation of the W-type light curves (Ruciński 1985) but - almost certainly - this phenomenon is also related to the magnetic activity which is generally so important for contact binaries.

I would like to thank Stefan Mochnacki for many stimulating discussions and for comments concerning this paper.

This work was supported by operating grants from the Natural Sciences and Engineering Research Council of Canada to C.T. Bolton, S.W. Mochnacki, J.D. Fernie and J.R. Percy.

REFERENCES

Anderson, L. and Shu, F.H.,: 1977, Astrophys. J., 214, 798.
Baliunas, S.L. and Guinan, E.F.: 1985, Astrophys. J., 294, 207.
Batten, A.H. and Wenxian Lu: 1985, preprint.
Bopp, B.W. and Ruciński, S.M.: 1981, in *Fundamental Problems in the Theory of Stellar Evolution*, eds. D. Sugimoto et al. (Reidel Publ. Co.), p. 177.
Bopp, B.W. and Stencel, R.E.; 1981, Astrophys. J. (Lett.), 247, L 131.
Budding, E.: 1984, Astron. Astrophys., 130, 324.
Campbell, C.G. and Papaloizou, J.: 1983, Mon. Not. Roy. astr. Soc., 204, 433.
Cruddace, R.G. and Dupree, A.K.: 1984, Astrophys. J., 277, 263.
Drechsel, H., Rahe, J., Wargau, W. and Wolf, B.: 1982, Astron. Astroph., 110, 246.
Duerbeck, H.W.: 1984, Mitt. Astr. Gesell., 62, 87.
Eaton, J.A.: 1983, Astrophys. J., 268, 800.
Eaton, J.A., Wu, C.-C. and Rucinski, S.M.: 1980, Astrophys. J., 239, 919.
Flannery, B.P.: 1976, Astrophys. J., 205, 217.
Hall, D.S. and Kreiner, J.M.: 1980, Acta Astron., 30, 387.
Hazlehurst, J.: 1970, Mon. Not. Roy. astr. Soc., 149, 129.
Hazlehurst, J.: 1985, Astron. Astrophys., 145, 25.
Herczeg, T.: 1979, Bull. Amer. Astr. Soc., 11, 438.
Hilditch, R.W., King, D.J., Hill, G. and Peockert, R.: 1984, Mon. Not. Roy. astr. Soc., 208, 135.
Hoffmann, M.: 1978, Inf. Bull. Var. Stars., No. 1478.
Hughes, V.A. and McLean, B.J.: 1984, Astrophys. J., 278, 716.
Hughes, V.A. and McLean, B.J.: 1985, in *Radio Stars*, eds. R.M. Hjellming and D.M. Gibson (Reidel Publ. Co.), p. 271.
Irwin, M.J. and Trimble, V.: 1984, Astron. J., 89, 83.
Kalużny, J.: 1983, Acta Astr., 33, 345.
King, D.J. and Hilditch, R.W.: 1984, Mon. Not. Roy. astr. Soc., 209, 645.
Kreiner, J.M.: 1977, in *Interaction of Variable Stars with their Environment*, eds. R. Kippenhahn et al. (= Veröff. Bamberg, 11, No. 121), p. 393.

Lucy, L.B.: 1968, Astrophys. J., 151, 1123.
Lucy, L.B.: 1976, Astrophys. J., 205, 208.
Lucy, L.B.: and Wilson, R.E.: 1979, Astrophys. J., 231, 502.
Mochnacki, S.W.: 1981, Astrophys. J., 245, 650.
Mochnacki, S.W.: 1985, in *Interacting Binaries*, eds. P.P. Eggleton and J.E. Pringle (Reidel Publ. Co.), p. 51.
Mochnacki, S.W., Fernie, J.D., Lyons, R., Schmidtt, F.H. and Gray, R.O.: 1985, preprint.
Moss, D.L.: 1971, Mon. Not. Roy. astr. Soc., 153, 41.
Moss, D.: 1985, Astron. Astrophys., 150, 343.
Moss, D.L. and Whelan, J.A.J.: 1970, Mon. Not. Roy. astr. Soc., 149, 147.
Noyes, R.W., Hartmann, L.W., Baliunas, S.L., Duncan, D.K. and Vaughan, A.H.: 1984, Astrophys. J., 279, 763.
Okamoto, I. and Sato, K.: 1970, Publ. Astr. Soc. Japan, 22, 317.
Popova, E.I., Tutukov, A.V. and Yungelson, L.R.: 1982a, Pisma v Astr. Zh., 8, 297.
Popova, E.I., Tutukov, A.V. and Yungelson, L.R.: 1982b, Astrophys. Space Sci., 88, 55.
Rahunen, T.: 1981, Astron. Astrophys., 102, 81.
Rahunen, T. and Vilhu, O.: 1982, in *Binary and Multiple Stars as Tracers of Stellar Evolution*, eds. Z. Kopal and J. Rahe (Reidel Publ. Co.), p. 289.
Robertson, J.A.: 1980, Mon. Not. Roy. astr. Soc., 192, 263.
Robertson, J.A. and Eggleton, P.P.: 1977, Mon. Not. Roy. astr. Soc., 179, 359.
Ruciński, S.M.: 1976, Publ. Astr. Soc. Pacific, 88, 244.
Ruciński, S.M.: 1982, Astron. Astrophys., 112, 273.
Ruciński, S.M.: 1983, Astron. Astrophys., 127, 84.
Ruciński, S.M.: 1984, The Observatory, 103, 280.
Ruciński, S.M.: 1985a, in *Interacting Binaries*, eds. P.P. Eggleton and J.E. Pringle (Reidel Publ. Co.), p. 13.
Ruciński, S.M.: 1985b, in *Interacting Binary Stars*, eds. J.E. Pringle and R.A. Wade (Cambridge Univ. Press), pp. 85 and 113.
Ruciński, S.M.: 1985c, Mon. Not. Roy. astr. Soc., 215, 615.
Ruciński, S.M. and Kałużny, J.: 1981, Acta Astr., 31, 409.
Ruciński, S.M. and Kałużny, J.: 1982, Astrophys. Space Sci., 88, 433.
Scharlemann, E.T.: 1981a, Astrophys. J., 246, 292.
Scharlemann, E.T.: 1981b, Astrophys. J., 246, 298.
Schmitt, J.H.M.M., Golub, L., Harnden Jr., F.R., Maxson, C.W., Rosner, R. and Vaiana, G.S.: 1985, Astrophys. J., 290, 307.
Shu, F.H., Lubow, S.H. and Anderson, L.: 1976, Astrophys. J., 209, 536.
Shu, F.H., Lubow, S.H. and Anderson, L.: 1980, Astrophys. J., 239, 937.
Skumanich, A.: 1972, Astrophys. J., 171, 565.
Smith, R.C.: 1984, Quart. Journ. Roy. astr. Soc., 25, 405.
Soderblom, D.R.: 1985, Astron. J., 90, 2103.
Tapia, S. and Whelan, J.: 1975, Astrophys. J., 200, 98.
Van Buren, D. and Young, A.: 1985, Astrophys. J. (Lett.), 295, L 39.
VandenBerg, D.A.: 1985, Astrophys. J. Suppl., 58, 711.
Van't Veer, F.: 1978, Astron. Astrophys., 70, 91.

Van't Veer, F.: 1982, in *Binary and Multiple Stars as Tracers of Stellar Evolution*, eds. Z. Kopal and J. Rahe (Reidel Publ. Co.), p. 279.
Van't Veer, F.: 1984, Astron. Astrophys., 139, 477.
Vilhu, O.: 1982, Astron. Astrophys., 109, 17.
Vilhu, O.: 1984, Astron. Astrophys., 133, 117.
Vilhu, O. and Ruciński, S.M.: 1983, Astron. Astrophys., 127, 5.
Webbink, R.F.: 1976, Astrophys. J., 209, 829.
Williams, P.S. and Roxburgh, I.W.: 1976, Mon. Not. Roy. astr. Soc., 176, 81.
Wilson, R.E. and Starr, T.C.: 1976, Mon. Not. Roy. astr. Soc., 176, 625.
Zahn, J.-P.: 1977, Astron. Astrophys., 57, 383.

DISCUSSION

Breger: What important observations would best be done with small telescopes?

Rucinski: Period changes and light curves, but be careful with instrumental effects.

Breger: How large are the period changes?

Rucinski: Quite large changes over small time intervals. One needs to be able to make intensive observations, when everything is happening, for a few weeks every 10 years. The rest of the time one needs to just monitor the object.

Peery: Is the ultraviolet activity (e.g. strong CIV) of these systems observed to be phase dependent?

Rucinski: No, it's phase independent. That is, the stars show active regions distributed evenly over the surface.

Peery: It's intriguing to know if this sort of activity in late-type stars always implies a companion?

Rucinski: These stars are normal as far as their chromospheric and transition regions are concerned. What is abnormal is their coronal and X-ray activity which is lower than normal.

Cottrell: Do you have any comment on the recent paper by Bruce Campbell about the coalescence of contact binaries producing the globular cluster CN, Na/Aℓ anomalies?

Rucinski: Violent events occur over one or two cycles when stars go from detached to contact systems. One would expect even more disturbance when stars go from contact to single stars.

Cameron: A useful intermediate step towards answering that question might be to look for such abundance anomalies in the FK Com stars, which may be the post-coalescence descendants of contact systems.

Barwig: In the case of differential rotation, would the Rossby relationship still work?

Rucinski: This is another way of explaining the underactivity of contact binaries. Contact systems can be considered to be rotating more like solid bodies than other stars. If you disturb the star, by contact with another, you create strong turbulence, and the convective elements break into smaller cells which leads to a more solid body rotator which has less differential rotation.

Plavec: What is the present opinion about the subsurface discontinuity proposed by Shu and Lubow?

Rucinski: In their model they have one hotter and one cooler star. The hotter star envelops the cooler one. This is potentially a stable configuration. However the problem is that there is a temperature gradient going both ways - increasing inwards and outwards which is an unstable situation. This short phase of contact may equate with the rapid period changes which I alluded to earlier.

Mochnacki: With the lower rate of braking needed for the systems to survive more than a few times 10^8 years, the braking is insufficient to maintain the Vilhu-Rahunen mechanism for suppressing the thermal relaxation cycles.

INTERACTING BINARIES: AN EXCELLENT PROJECT FOR SMALL TELESCOPES

Mirek J. Plavec
Department of Astronomy
University of California at Los Angeles
Los Angeles, CA 90024
U.S.A.

ABSTRACT. Discussed are meritorious projects for small and moderate telescopes in the field of interacting binaries with non-degenerate component stars. These interacting binaries are undergoing a mass transfer process, as a consequence of which the mass-accreting star may be partly or completely hidden in an accretion disk, and the system may be shrouded in dense clouds of circumstellar matter. This makes the observation, interpretation and modeling difficult; but it is important to study these "bizarre" binaries since they tell us a lot about stellar evolution in binary systems. Needed are various observations: Timing of eclipses; observation and re-observation of light curves in several colors (in uvby rather than in UBV; and in the red and infrared); radial velocity studies; spectrophotometry of crucial regions of the spectrum. As examples for these needs, the following systems are discussed in some detail, and their problems revealed: RX Cassiopeiae, W Serpentis, and W Crucis as examples of the strongly interacting systems (W Serpentids); υ Sagittarii as a helium-rich binary in a rare evolutionary stage, which will be better understood if we decide whether the star eclipses or not; and KX Andromedae as a representative of non-eclipsing interacting binaries and of Be and shell stars, which may or may not be binaries.

1. INTRODUCTION

My favorite celestial body is the Moon. Not that I care much about it; already in my amateur years, I somehow felt that everything within about 20 parsecs would sooner or later become geology rather than astronomy. No; I love the Moon because when it shines, the sky background is so bright that extragalactic astronomers must leave the big telescopes to stellar astrophysicists. We occasionally hear that everything about stars has already been discovered and they are no longer interesting; yet the Moon has the last word, and is therefore the stellar astrophysicist's best friend. Recently, however, we seem to have found another powerful ally: all the smart instrumentalists who have developed the Reticons and CCD's and all the other gadgets to such a perfection that even with rather small telescopes, one can nowadays perform photometric and spectrographic studies on quite faint objects. This is why I believe that this Symposium is extremely important. I have come here to lobby

J. B. Hearnshaw and P. L. Cottrell (eds.), Instrumentation and Research Programmes for Small Telescopes, 173–186.

among the users of small and moderate-size telescopes to point them to interacting binary stars. Occasional as well as regular photometric observations in several colors, spectrophotometric studies and radial velocity measurements are highly desirable. For the photometric work, the most sophisticated gadgets are not needed; a good photoelectric photometer will do a splendid job, too, and today such a photometer is not rare even among amateur observers.

2. MEET THE INTERACTING BINARIES

Interacting binaries are those binary systems which transfer mass between the components on such a large scale that the structure and future evolution of both stars are significantly altered compared to single stars. Astronomically speaking, mass transfer proceeds on a relatively short time scale. Thus when you observe interacting binaries, you are more directly in touch with stellar evolution than in many other observations. Naturally a supernova explosion beats everything else, but if you had been waiting for a bright galactic supernova on every night since 1604, you would still be waiting. Betelgeuse and Antares may blow up pretty soon now, but "now" means anything between tonight and about 100,000 years hence.

The interacting binary stars come in two varieties, a violent and a non-violent one. The violent kind, the cataclysmics, includes novae and dwarf novae, certainly attractive because of their outbursts. However, these outbursts may be mere showmanship, just showing off, and no real stuff. Holland Ford (1978) argued that a classical nova can undergo as many as 10,000 outbursts and still remain the same white degenerate dwarf reacting irritably to a slow inflow of cool hydrogen from a red dwarf. I am pleading here for the non-violent type of interacting binaries, in which both components are non-degenerate stars. A very crude but very nice model of such an interacting binary is in Fig. 1. The component that was initially more massive has evolved faster and eventually filled all its available volume (the critical Roche lobe). It then has the shape of a droplet. Fig. 1 is not too suitable to show its shape, so please look at Fig. 2 which represents a model I do not dare to call nicer, but it is hopefully closer to reality and in any case mathematically more tractable. The conical point of the "droplet", called the First Lagrangian Point, plays the same role as a hole in the inner tube of a bicycle tire: gas pressure from inside pushes gas molecules through this orifice into a stream towards the other component. Depending on the relative size of that "gainer" with respect to system dimensions, the stream either impinges on the gainer or forms an accretion disk around it. Although most of the gas will spiral inward and eventually fall on the gainer, conditions in both cases are sufficiently chaotic so that circumstellar matter surrounds the stars and may become so dense that it will contaminate the spectra, the flux from the system, and even obscure the stars. Certainly much depends on the available volume and on the rate of mass transfer. Every eclipsing binary with an eclipse deeper than 0.75 mag is an interacting binary, for only binary star evolution with mass transfer produces a loser which is cooler and less massive but larger than the gainer. A large majority of eclipsing binaries with deep

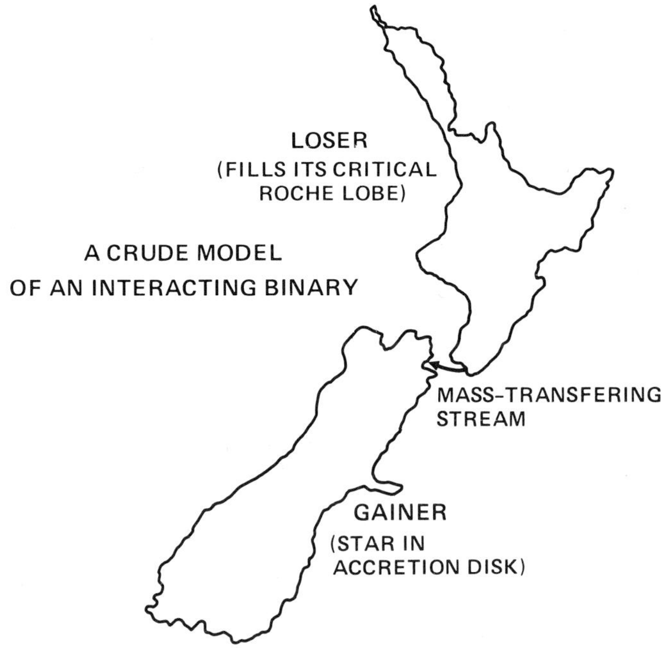

Fig. 1. - An extremely crude but nice model of an interacting binary.

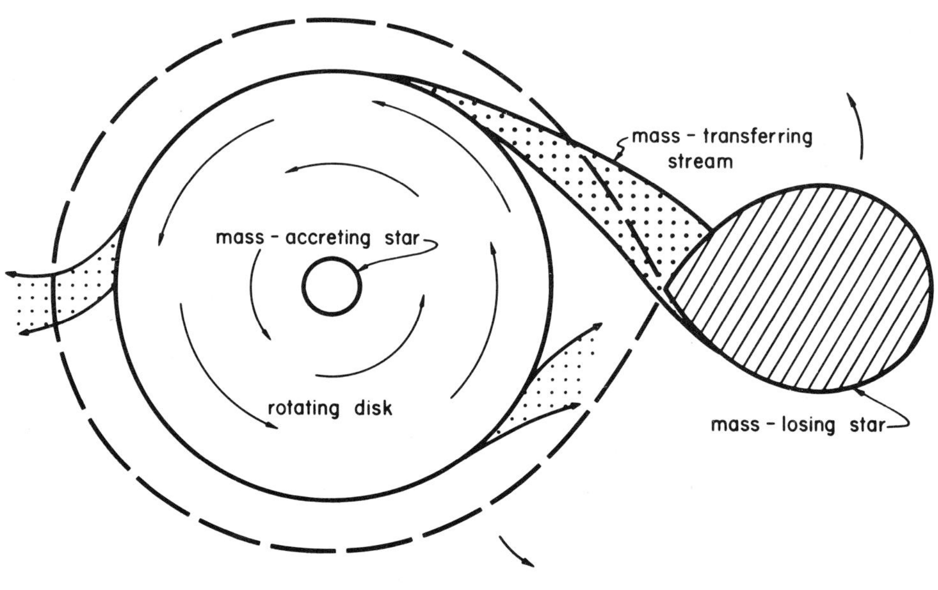

Fig. 2. - A (hopefully) more realistic model of an interacting binary. (after S. Kříž and P. Harmanec).

eclipses are Algols, a well-behaving variety near the end of the mass-transfer stage, in which the mass transfer rate is so small that it is easy to see the stars but difficult to detect the circumstellar matter. Only an emission at Hα and distortions of absorption line profiles may indicate that the system is not perfectly clean. So why try hard to see the circumsttelar matter in Algols? The weak structures and processes we observe in them offer clues to the much rarer and much more difficult cases of a high mass transfer rate.

The circumstellar matter in Algols is best observed if and when the hotter component (the gainer) is totally eclipsed. Fig. 3 shows that at that time, the out-of-eclipse spectrum, dominated by the hotter component (of spectral type B7 in this system) is replaced by a K0 III spectrum of the loser. But note that there exists excess radiation at and shortward of the Balmer limit at 3650 Å, due to optically thin radiation of circumstellar hydrogen. More exciting is the ultraviolet spectrum shown (for U Cephei) in Fig. 4, and obtained with the International Ultraviolet Explorer satellite (IUE). We note emission lines of C IV, Si IV, N V, and similar species, quite conspicuous in totality but invisible outside it because of the strong flux from the accreting star.

3. AN IMPORTANT TASK: TIMING THE ECLIPSES

And here is where the need for cooperation with photometrists enters. The IUE observations in totality must be extremely well timed. What happens when they are not, shows the ultraviolet spectrum of RW Tauri in Fig. 5. Here I was late by only about 15 minutes. The strongest emission lines remain just barely visible against the quickly increasing flux from the hotter, B8 star, just emerging from behind the K0 III subgiant. The entire totality in RW Tauri lasts about 80 minutes, but 60 minutes are needed for a good exposure. Thus very accurate timing is needed, but the interacting binaries do not keep constant periods. This is natural, since mass and angular momentum are on the move in them. The change, expressed as dP/P, is usually quite small, but its effects accumulate in such a way that eventually the difference O − C, i.e. observed minus calculated time of eclipse, may become quite large.

A good example is RX Cassiopeiae, an eclipsing binary with a period of 32.3 days. For years, Martynov of Moscow exhorted observers world-wide to watch this rather bizarre system. At the end of his article published in 1975 (Martynov 1975), he remarked to the effect that his older linear ephemeris, based on scattered observations during the first half of the century, was still good. In fact, by that time, the (O-C) deviations were already growing rapidly, as Fig. 6 shows, and by now, the delay with respect to the original ephemeris is already more than four full days. Now please realize the plight of an IUE observer who wants to study the eclipse spectrum of RX Cas. The competition for observing time is tremendous, and if he is lucky, he will be allocated three 8-hours shifts per year. He must apply for a specific shift several months in advance. The fact about the large (O-C) deviation in RX Cas would be unknown were it not for one or two photometrists. So our IUE observer would obtain an ordinary out-of-eclipse spectrum and the star would laugh behind his back and gracefully go into eclipse four days later.

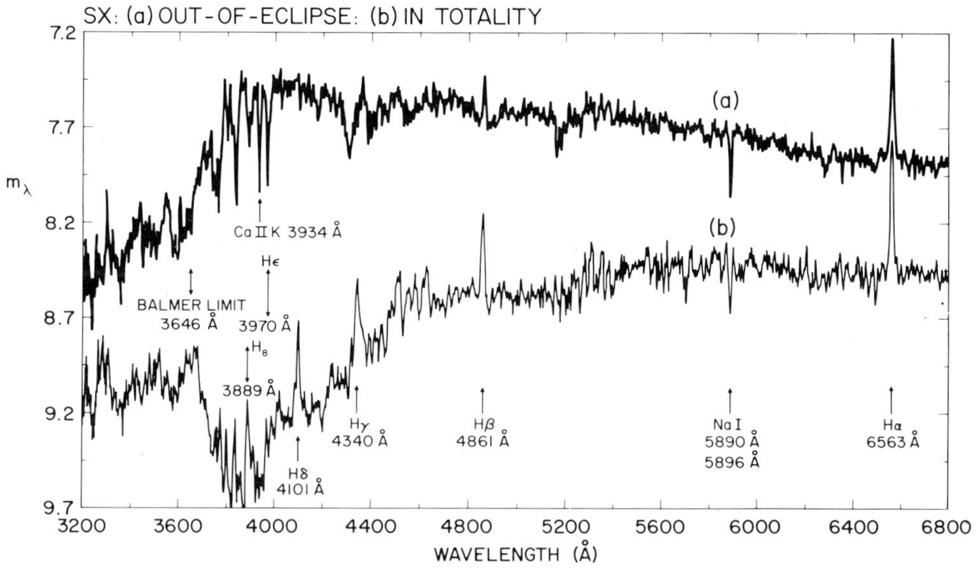

Fig. 3. - Optical spectrum of SX Cassiopeiae, outside eclipse and in total eclipse. Observations with ITS scanner, Lick Obs.

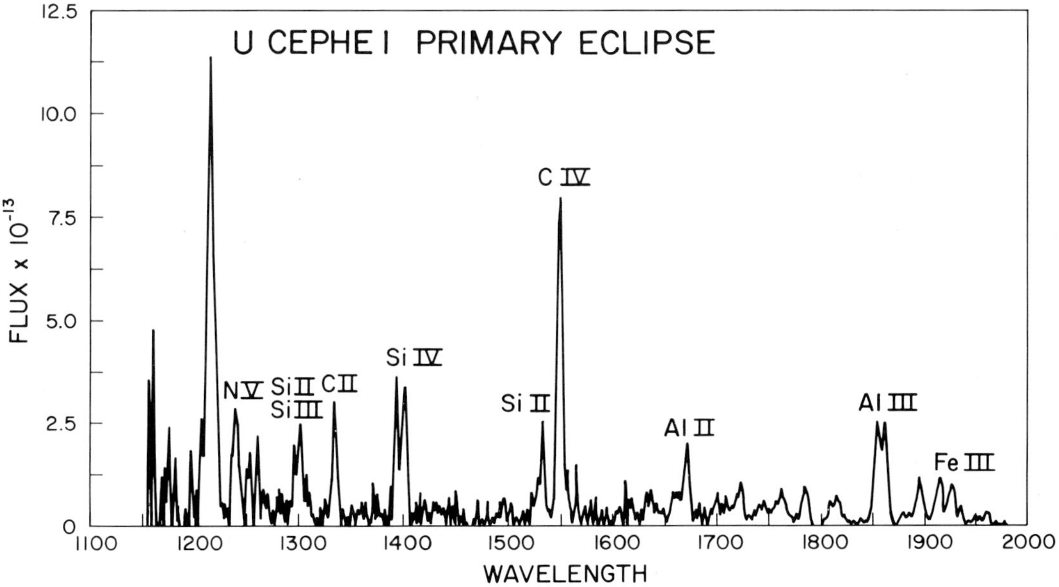

Fig. 4. - Ultraviolet (IUE) spectrum of U Cephei in total eclipse.

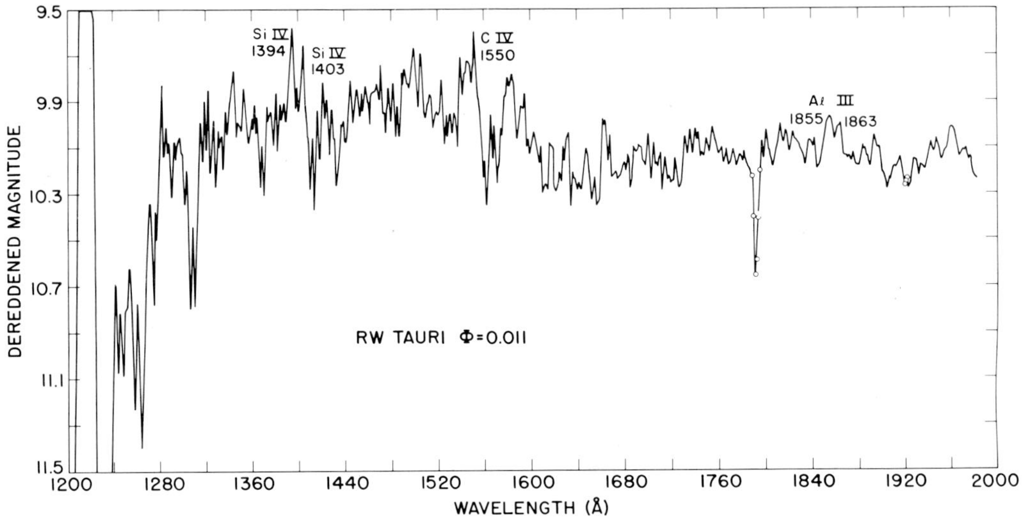

Fig. 5. - Ultraviolet spectrum of RW Tauri, observed at end of totality.

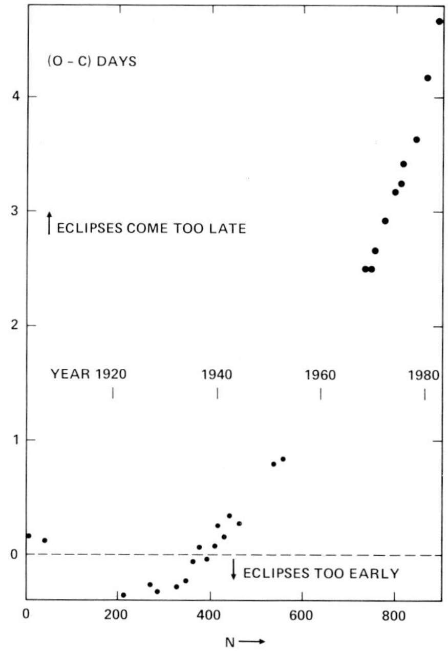

Fig. 6. - RX Cassiopeiae: Deviations of eclipse times from linear formula.

I had exactly this worry recently with W Crucis, whose period is 198.5 days. A very brief total eclipse is suspected, probably no longer than one day. But with so long a period, a very small fractional change dP/P may shift the eclipse easily by several days with respect to an ephemeris derived thirty years ago. Thus I appealed to southern observers for help (Plavec 1984); and it indeed worked: Messrs. Marino and Walker from Auckland, New Zealand, and several South African observers determined the time of eclipse and enabled me to time my IUE observations correctly. The period changes in eclipsing binaries do have a positive aspect, since they are telling us something about the process of mass transfer. For a process conserving mass and orbital angular momentum of the system, the formula for the period change reads:

$$dP/dt = 3\ P\ (dM_1/dt)\ (M_1 - M_g)\ /\ (M_1 \times M_g) \qquad (1)$$

Here M_1 and M_g are the masses of the loser and gainer, respectively. We see that if the period is increasing, this means that gas is flowing from the less massive star to the more massive one. From a parabolic fit to the deviations in Fig. 6, we conclude that the rate of mass transfer in RX Cas is somewhere between 10^{-5} and 10^{-7} solar masses per year. This figure is both large and very uncertain, and the two aspects of it are closely related. The system of RX Cas is so "clouded over" by circumstellar material that it is nearly impossible to get a clear picture about the masses and spectral types of the components. RX Cas is not a classical Algol; it is one of the strongly interacting systems which I called the W Serpentis stars (Plavec 1980). In these binaries, the circumstellar material plays a dominant role in the photometric and spectroscopic appearance of the system. Since this implies a higher rate of mass transfer, the W Serpentis stars promise to tell us more about it, but their messages are so garbled that deciphering requires very concentrated effort. Here comes the need for complete light curves.

4. LIGHT CURVES NEEDED, IN PARTICULAR FOR THE MORE BIZARRE SYSTEMS

No eclipsing binary displays an ideally simple light curve; there is always some light fluctuation, although often only at about the level of observational errors. But the light curve of RX Cas shows an enormous scatter, which is shown half-schematically in Fig. 7, based on a selection of the observations accumulated by Kalv (1979). The Figure conveys the impression strongly advocated by Kalv, namely that the system undergoes another periodic variation on a time scale of about 516 days, during which it gradually shifts its out-of-eclipse light level from "MAX" to "MIN", in Kalv's terminology. More recent photoelectric observations by the German amateur astronomer Mario Fernandes (1985) cast some doubt on the strict periodicity of this long wave, but not on its existence.

The shape of the light curve of RX Cas indicates that both the loser and the gainer are non-spherical objects. The large light fluctuations cease during the short total primary eclipse, so that their source is associated with the gainer, which is eclipsed at that time. We have concluded (Weiland and Plavec 1986) that the gainer is a star surrounded

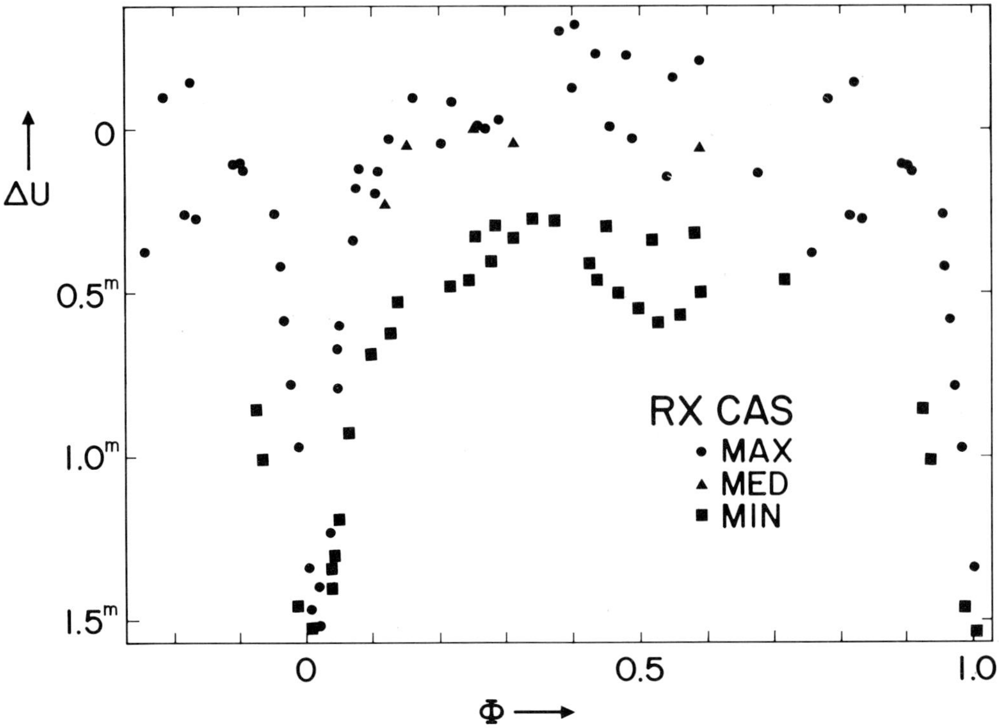

Fig. 7. - Light curve of RX Cassiopeiae after Kalv (1979).

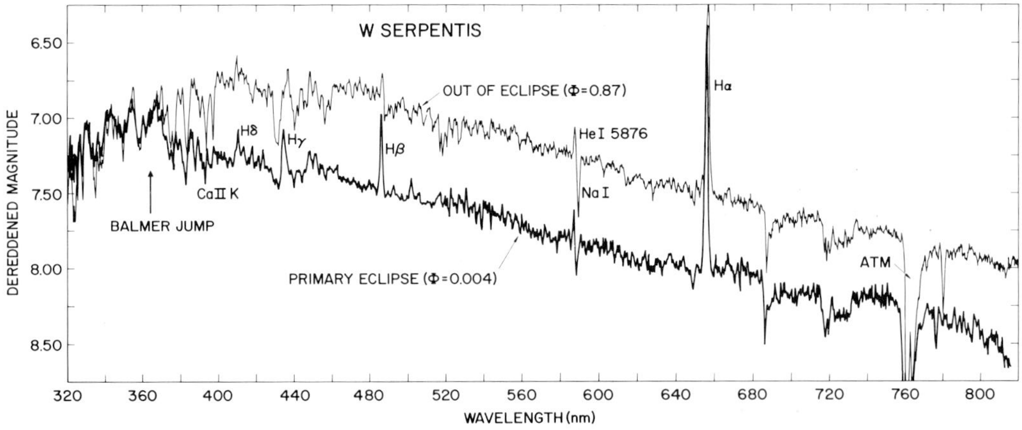

Fig. 8. - Optical spectrum of W Serpentis in eclipse and outside eclipse.

by a thick accretion disk. The disk varies in size, and is largest but coolest at the "MIN" stage, when it is about equal in radius and in surface temperature to the loser, which is a K1 III giant. At the "MAX" stage, the disk shines like an F8 - G2 star. Thus we have two cool objects in RX Cas, and UV observations will not tell us much about them. Visual and infrared photometry is needed, first to cover the light curve well and then to follow its changes. One day the circumstellar clouds may clear up a little and perhaps grant us a look at the accreting star.

5. PLEASE SEARCH FOR THE COMPANION IN W SERPENTIS!

Poorly visible even from California is the prototype of the strongly interacting binaries, W Serpentis. At a declination of -15°, this eclipsing binary with 14.16 days period is waiting for a dedicated southern observer. Its light curve is at least as badly distorted as that of RX Cas. The secondary eclipse, which is variable, distorted, and unusually broad but present in RX Cas, is missing in the light curve of W Serpentis. Yet the star is a binary system, and the star causing the primary eclipse is unlikely to be completely non-luminous, and must be quite large. So what is it?

The trouble is that we are not even sure about the nature of the visible star. Its many strong, sharp absorption lines suggest a spectral class of F5 II. However, Hack (1963) suggested that this spectrum probably originates in a circumstellar shell, less dense and cooler than the photosphere of the underlying star. From the appearance of the Mg II line at 4481 Å, it appears that the star may be an early A type. My own observations, if anything, tend to compound the puzzle. The flux in the optical region, studied with the IDS scanner at the 3-m Shane telescope of Lick Observatory, seems to confirm the F5 classification. But I know, from our work on SX Cas (Plavec, Weiland and Koch 1982) that flux contribution from the secondary component and from circumstellar matter may change the slope of the Paschen continuum seriously and cause it to simulate a rather different temperature. Moreover, as my IUE observations show, the flux continues to rise across the Balmer jump and reaches levels and shape in the far ultraviolet reminiscent of about a B7 star. So one could imagine a B7 star surrounded by a cooler disk in such a way that only a part of the star's surface is visible. Or we can envisage another model in which the star is completely invisible and the ultraviolet flux comes from the transition region between the star and its accretion disk. The theory of this transition layer (Pringle 1977), applied to a non-degenerate star (Plavec 1980) makes this model viable. Janet Weiland thinks that the UV continuum is a "warm spot" on a disk. During the primary eclipse, the far ultraviolet fluxes almost do not change; in fact, as Fig. 8 shows, the eclipse is perceptible only just longward of the Balmer jump. This seems to indicate that the eclipsed star (or disk) may indeed be F5 or something not much earlier in spectral type, and that the "warmer" radiation in the ultraviolet comes from an extended circumstellar source.

Even more puzzling is the other star in the system. I believe that some of its absorption lines may be observable in the red and infrared spectral regions, either at all phases or at least at primary eclipse, when the light from the primary star is dimmed by about one magnitude.

6. W CRUCIS -- ANOTHER CHALLENGE: SEARCH FOR THE SECONDARY STAR!

If W Serpentis is a challenge mainly to the southerners, W Crucis is of course entirely within their domain. Very few facts are known for sure about this eclipsing binary with a long period (198.5 days) (Woolf 1961): The only visible spectrum looks like a G1 supergiant, except for some circumstellar Balmer line components and Hα in emission. Its radial velocity curve appears reasonably reliable, but leads to a terrible dilemma. If the two component stars are equally massive, then each has a mass of about 24 M_o. This is perhaps acceptable for a supergiant, but where is the other supergiant? It leaves no trace in the spectrum. If we assume that it is less massive than the observed star, both masses become unacceptably large: for a mass ratio of 1:2, they would be 55 M_o for the unknown star and 110 M_o for the observed supergiant! Clearly, if we want to stay within the realm of common sense, we must assume that the star we do not see is more massive than the one we do see. For example, should the mysterious ghost be twice as massive, then the mass of the visible supergiant would be only 7 M_o, but then we must ask: Where is the light from the star with 14 M_o?

This situation is not quite unheard of, and in β Lyrae we are almost sure that the "invisible" star is more massive than the B8 II giant which dominates the optical spectrum. But in β Lyrae, the companion is not really invisible, only we are unable to detect its spectral lines. It does generate measurable continuous flux in the optical region, and in the IUE ultraviolet, its flux is equal to that of the B8 II star. The companion is a flattened, optically thick disk. It is natural to make the same assumption about W Crucis. But there is no trace of the companion in the ultraviolet: I have found only radiation from a huge, rarefied circumstellar plasma at a surprisingly high electron temperature, near 100,000 °K. The companion of W Crucis is not entirely invisible, since its eclipse seems to cause a measurable minimum in the light curve (unless the minimum is due only to the ellipticity of the G1 supergiant). Perhaps farther in the red and infrared, its signature will be better legible. The system calls for a good photoelectric light curve and a thoughtful solution for elements. A thorough examination of the optical spectrum with a high signal-to-noise ratio is also desirable.

7. THE NEED FOR LIGHT CURVES IN SEVERAL COLORS

It is rather obvious that a reliable determination of the nature of the component stars in many binaries prerequires good photoelectric light curves in more than one color. I would like to point out that the UBV system is not the best system for this purpose, because of the frequent presence of radiation from a circumstellar hydrogen shell or cloud. The U filter of the UBV system is sensitive in the spectral range of about 3,100 - 4,100 Å, with its peak sensitivity going right across the Balmer jump. This is an outright disaster for studies of interacting binaries, in which an optically thin hydrogen cloud may contribute strong

but often very variable continuous flux shortward of the Balmer jump, but a much weaker flux longward. Many light curves in the U color are practically useless for determining the stellar characteristics and dimensions, but then we are left only with the (B-V) color index, which is often insufficient.
The Strömgren four-color photometry is much better, since the u and v filters peak at opposite sides of the Balmer jump.

8. UPSILON SAGITTARII: DOES IT ECLIPSE OR NOT?

The Roche lobe overflow can carry away so much mass from the loser that exposed now may be the layers that originally were deep inside the star and underwent CNO processing during which carbon was mostly converted into nitrogen, and helium has become four times more abundant than hydrogen (Lambert 1982, Plavec 1986). This helium abundance, however, is nothing compared to what we see in two peculiar binaries, υ Sagittarii and KS Persei, in whose atmospheres there may be some 10,000 atoms of helium per one atom of hydrogen (Drilling and Schönberner 1982). This extreme He/H ratio can be explained by a complicated evolution with two separate stages of mass loss from the same star, but this type of evolution can occur only for stars within a fairly narrowly defined initial, intermediate, and present masses (Plavec 1973; Schönberner and Drilling 1983; Plavec 1986). Therefore, a good determination of the component masses is important for verifying the model.
υ Sagittarii has a good radial velocity curve for the star that dominates the optical spectrum, and there is hope that the other radial velocity curve can be obtained from far ultraviolet observations of the companion. But a really good mass determination also requires the knowledge of orbital inclination. In the past, shallow eclipses were reported (Eggen, Kron and Greenstein 1950), but they are unconfirmed by other observers. If they do occur, the inclination will be much better known. It takes a very skilled photometrist and a very good climate to detect eclipses or light fluctuations of less than 0.1 mag. It is also possible that the eclipse depths will be greater at other wavelengths.

9. THE NEED FOR IMPROVED RADIAL VELOCITY CURVES

Radial velocity curves remain our most important way of determining the masses and absolute dimensions of the component stars, without which it is impossible to model the interacting systems and compare the observations with theoretical calculations. It was in fact the work of Otto Struve and his collaborators in the 1940's that led to the discovery of interacting binaries. But that pioneering work was done on spectrograms whose dispersion and quality are far below those required for reliable mass determination (Popper 1980). Since the masses vary with the third power of the radial velocity range, tremendous errors may be incurred if old crude data are still used today.
RX Cassiopeiae may serve as a case in point. Struve (1944) classified the component stars as gG3 + gA5e and gave their radial velocity ranges as 34 km s^{-1} and 36 km s^{-1}, respectively. Combined with P = 32.3 days,

this gives us a system of two dwarf stars, 0.59 M_O and 0.56 M_O, separated by only 5 R_O, and crawling around the center of gravity of this minuscule system once in a month. This led to various speculations about collapsed objects in the system, but in fact the dispersion used by Struve failed to reveal serious blending, the spectral types are actually different (Weiland and Plavec 1986), and both velocity ranges must be rejected. A new value for the loser, obtained by Johannes Andersen with CORAVEL, is 97.7 km s^{-1}; although correct masses are still not available, all reasonable values of mass ratio lead to entirely normal masses. I hope that this example shows that in many important interacting binaries, new radial velocity curves are badly needed.

10. KX ANDROMEDAE: A CASE FOR OBSERVING Be AND SHELL STARS

A large majority of interacting binaries are not seen as eclipsing from the Earth. Some of them may be known as Be stars or shell stars (Kříž and Harmanec 1975). They gave KX Andromedae (HD 218393) as a good example, since its radial velocities show an almost-periodic variation with a period of 38.9 days. And, indeed, the secondary component was subsequently discovered in the near infrared (Polidan 1976) and tentatively classified as a K1 III star. This is the loser in this interacting system.

The gainer presents a more complicated problem and may indeed be a complicated object. From its absorption lines and Balmer jump, the star has been variously classified as everything between A5 and B0. This is no small difference, since often we must estimate masses on the basis of the spectral types of the stars, but in this case the possible range is between 2 and 20 solar masses! An extension of the spectral coverage into the ultraviolet did not bring any definite answer either. In 1978, I obtained the first IUE spectra which bear close resemblance to the strongly interacting star SX Cas: A strong ultraviolet continuum suggests a star of about 11,000 K, while the Paschen continuum indicates only about 9,000 K (Plavec 1979; Goraya, Singh, and Chaubey 1984). It could be an optically thick disk as in β Lyrae. Or it could be a B star partly obscured by a shell. Our optical scanner observations show that the transparency of the shell in the continuum is quite variable: so, one day we may see the accreting star almost without a veil. But there is another approach possible and promising: To examine the He I absorption lines with an apparatus yielding high signal-to-noise ratio, and derive the spectral type from the line profiles of He I 4471 Å etc.

KX Andromedae is by no means a unique case. Many Be stars show light and spectrum variations which may help decide if they are interacting binaries or if some other model (stellar wind? non-radial oscillations?) applies. An international campaign is currently being conducted (see Harmanec, Horn, and Koubsky 1982; Barker 1982).

11. CONCLUSION: MT. EVEREST, HURRICANES, AND INTERACTING BINARIES

I have attempted to present suggestions as to what could be studied in interacting binaries, using small and moderate telescopes. I discussed individual cases. They have one thing in common: All of them are rather bizarre systems, difficult to understand. The question may be asked: Why study extreme cases when there are many well-behaving systems? I am all in favor of observing simple systems for the sake of determining fundamental characteristics of stars. However, the complicated cases should not be ignored. One reason is the classical Mt. Everest argument: We must study them because they are there. I would like to present a more powerful argument: It may be comfortable for the meteorologists to study only gentle sea breezes; but one probably learns much more about the dynamics of atmospheric circulation if one tries hard to understand tornadoes and hurricanes.

And one last opinion: I hope that it transpires from all the cases that the proper way to success is cooperation -- between photometrists, spectroscopists, and theoreticians; between optical and infrared and ultraviolet people; between the northern and southern hemispheres, and so on.

REFERENCES

Barker, P.K. 1982, in Be Stars, ed. M. Jaschek and H.-G. Groth, (Dordrecht: Reidel), 269.

Drilling, J.S. and Schönberner, D.: 1982, Astr. Ap. 113, L22.

Eggen, O.J., Kron, G.E., and Greenstein, J.L.: 1950, P. A. S. P. 62, 171

Fernandes, M. 1985, BAV Rundbrief 34, 49.

Ford, H.C. 1978, Ap. J. 219, 595.

Goraya, P.S., Singh M., and Chaubey, U.S. 1984, Inf. Bull. Var. St. 2519.

Hack, M. 1963, Mem. Soc. Astr. Ital. 34, 3.

Harmanec, P., Horn, J. and Koubsky, P. 1982, in Be Stars (see above), 275

Kalv, P. 1979, Tartu Astr. Obs. Teated 60, 3.

Kříž, S. and Harmanec, P. 1975, Bull. Astr. Czech. 26, 65.

Lambert, D. L. 1982, in Advances in UV Astronomy, NASA Conf. P. 2238, 114

Martynov, D. Ya. 1975, Astr. Circ. USSR 891, 4.

Plavec, M.J.: 1973, in Extended Atmospheres etc., ed. A. H. Batten, (Dordrecht: Reidel) 216.

Plavec, M.J. 1979, Bull. Amer. Astron. Soc. 11, 648.

Plavec, M.J.: 1980, in Close Binary Stars: Observation and Interpretation ed. M.J. Plavec, D.M. Popper, and R.K. Ulrich (Dordrecht: Reidel), 251

Plavec, M.J. 1984, Inf. Bull. Var. Stars 2524.

Plavec, M.J., Weiland, J.L., and Koch, R.H. 1982, Ap. J. 256, 206.

Plavec, M.J. 1986, IAU Coll. 87 on Hydrogen Deficient Stars etc.

Polidan, R.S. 1976, in Be and Shell Stars, ed. A. Slettebak (Dordrecht: Reidel), 401.

Popper, D.M. 1980, Ann. Rev. Astr. Aph. 18, 115.

Pringle, J.E. 1977, Mon. Not. R.A.S. 178, 195.

Schönberner, D. and Drilling, J.S.: 1983, Ap. J. 268, 225.

Struve, O. 1944, Ap. J. 99, 295.

Weiland, J.L. and Plavec, M.J. 1986, submitted for publication.

Woolf, N.J. 1961, Mon. Not. R.A.S. 123, 399.

DISCUSSION

Warner: I agree with your attack on cataclysmic variables. If you want to observe white dwarfs the worst place to look is in a cataclysmic variable since it is immersed in the accretion disk. The real reason to look at cataclysmics is to investigate the disks.

Plavec: I agree with Dr Warner that the cataclysmics are very important for studying the properties of accretion disks. But the results about the α-disks in the cataclysmics cannot be transferred to non-degenerate systems. As I showed for U Cephei (Ap. J. 1983), the latter systems have geometrically much thicker circumstellar structures. I think we should agree that all types of interacting binaries should be ardently observed.

OCCULTATIONS, PAST AND FUTURE

David S. Evans
Department of Astronomy
University of Texas
Austin, Texas 78712
U.S.A.

ABSTRACT. The techniques, advantages and limitations of photoelectric observations of occultations of stars by the Moon are reviewed. The results for timings, lunar limb slopes, double and multiple stars, and the determination of angular diameters of stars are considered. Possible effects of lunar limb irregularities are discussed. Observers are strongly recommended to pursue the upcoming series of occultations of Antares and other important stars.

1. INTRODUCTION

Modern observations of occultations of stars by the Moon date from the work of Whitford (1939) who first demonstrated photoelectrically the phenomenon of diffraction of starlight at the lunar limb. His paper was accompanied by an analysis by Williams (1939) of the effects produced by stars of different angular diameters (Figure 1). The fringes had previously been detected photographically by Arnulf (1936) who used them to determine the angular diameter of Regulus.

Following a suggestion of David Thackeray occultation observations of Antares were made at Pretoria (Evans et al. 1953), Johannesburg (Evans, 1955) and Cape Town (Cousins and Guelke, 1953) of the Antares series of the early nineteen fifties and some other stars. My interpretation of these results was severely criticised with such effect that other observers were deterred for many years. I shall return to this topic later in my talk. I believe that a renewed and quite widespread interest in photoelectric observation of occultations can be attributed to the initiative of R. E. Nather and myself at the McDonald Observatory of the University of Texas starting in 1968.

The data derived from photoelectric observations of occultations relates to the following fields which I shall discuss in turn. (i) Timings, leading to information on the position of the Moon in relation to the star background. (ii) Information on the character of the lunar limb. (iii) The discovery and in some cases, investigation of double and multiple stars. (iv) The determination of angular

J. B. Hearnshaw and P. L. Cottrell (eds.), Instrumentation and Research Programmes for Small Telescopes, 187–198.

diameters of stars with numerous astrophysical applications of the results. (v) Some stars of special interest.

2. TIMINGS AND THE LUNAR LIMB

Timings of occultations of stars by visual means have a long history and are surprisingly accurate. The rate of movement of the lead point of the lunar limb against the star background only surpasses 0.5 arc seconds per second of time when the Moon is near the horizon, often not even then. Even in such adverse circumstances a visual timing by a skilled observer good to one fifth of a second specifies the position of a point on the lunar limb to about 0.1 arc seconds or some 200 metres linearly.

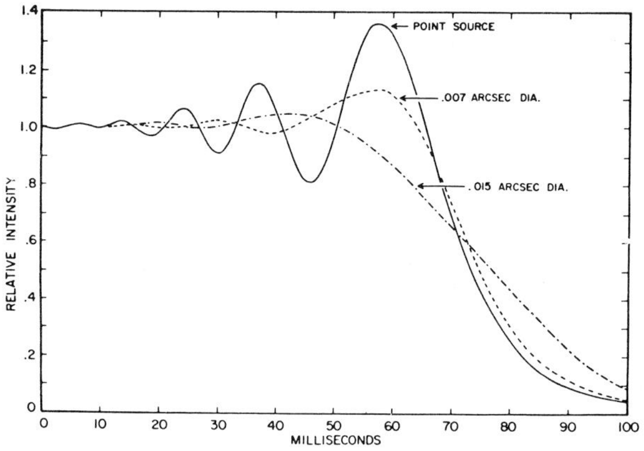

Figure 1: Typical theoretical traces for various angular diameter stars (McGraw et al. 1973).

In Figure 1 we see an ideal photoelectric trace for a point source in which the brightest diffraction fringe attains a peak level of 1.37 if the unobscured light of the star is counted as 1.00. The moment of geometrical occultation occurs at level 0.25. In practical cases the ideal trace is modified by the bandwidth of the filter used, the colour of the star, the photocell response, by seeing noise and in rather rare cases by the fact that the star may have a sensibly non-zero angular diameter.

The source of seeing noise is largely the scattered light from the illuminated part of the Moon. In the case of a faint star the intensity of this may exceed that of the star by a hundred times or more, so that the fringes are obliterated completely and all that is seen is a sudden drop or rise in the light level as the star disappears or reappears. The classical photoelectric technique is thus almost entirely restricted to events at the dark limb of the Moon, though we have on rare occasions for bright stars using specially selected narrow interference filters made observations at the bright limb.

I should mention here the technique of Dommanget (1980) who uses a TV camera to record the field on tape. On playback he can at his leisure determine the fraction of a picture scan at which the star disappears or reappears, even at the bright limb and so obtains timings, though he cannot observe the diffraction fringes or any subtle details of the fading or reappearance. He does however provide data on the position of the lunar limb at phases not ordinarily accessible to the standard photoelectric method.

Even for the noisiest photoelectric trace a proper computer reduction program which takes account of star colour, system response and the rest, can produce a timing good to a few milleseconds corresponding to fixing a point on the lunar limb with an accuracy of a metre or even less.

There are elevations and depressions on the lunar limb which are presented differently at different conditions of libration (Watts 1963). These may exceed some tenths of an arc second up to an arc second and it is clear that their effects can swamp the intrinsic accuracy of photoelectric timings when used to improve the lunar ephemeris. Indeed Morrison (1971) while recognising that photoelectric occultation timings should be free of systematic and accidental errors considered that they should not be given much more weight than visual ones.

Considering that lunar laser ranging to reflectors placed on the surface by American and Russian missions began over ten years ago and can produce range figures accurate to a few centimetres it might be supposed that occultation timings are no longer useful for ephemeris studies. However I consulted Dr. Raynor L. Duncombe formerly Director of the U. S. Nautical Almanac Office, who hoped the program would continue and remarked that these observations were the best available up to the beginning of the laser ranging program. Dr. Peter J. Shelus who directs the lunar laser ranging program at Texas points out that occultation observations serve to locate the Moon with respect to the standard star positions of the FK4 catalogue, whereas the lunar laser ranging program does not do this directly. The occultation results were used by Mulholland (1981) to locate the centre of figure of the Moon with respect to its centre of mass.

There thus seems to be some expert opinion in favour of the continuation of a program of timings. At Texas we have made nearly 7000 observations starting in 1968 and so have not quite emulated the example of Edmond Halley who, starting at age 63, observed the position of the Moon through an entire rotation of the nodes of its orbit, which takes 18.61 years, but we have come fairly near it. After this interval the same zodiacal stars tend to be occulted again. This shows up in the entries on my card catalogue of all photoelectric observations. Our program is no longer supported by the U. S. National Science Foundation and I intend to retire in 1986. I hope other observers will continue the work and perhaps someone will take over the catalogue and even prepare it for publication.

If an occultation trace shows good fringes we can compare their rate of passage with that predicted for a level limb, and can attribute the differences to the presence of a limb slope which applies, I think, to some 50 metres of its length. The deduced value is usually ambiguous since all we are determining is the angle between the limb and the line of relative travel of the star. One value is usually small, almost never exceeding 30°, whereas the other is often very steep or overhanging and can almost always be rejected with confidence. The Moon models of James Nasmyth with their exaggeration of vertical scale led to a general belief in the existence of steep slopes on the Moon, but even a cursory study of visible shadows on the lunar disk shows that steep slopes are rare, while at the lunar limb, the fact that double disappearances of stars only occur at grazes, supports the same conclusion.

Our convention is that a slope which spreads the fringes counts negative, one that compresses them counts positive. Only negative slopes can occur for occultations at nearly normal incidence and values with large uncertainties since it requires a large slope to make a significant change in rate. At large angles of incidence most events take place on forward slopes and errors tend to be small. The situation is illustrated in Figure 2. (Evans, 1971)

Figure 2: Statistics of limb slope determinations as a function of contact angle.

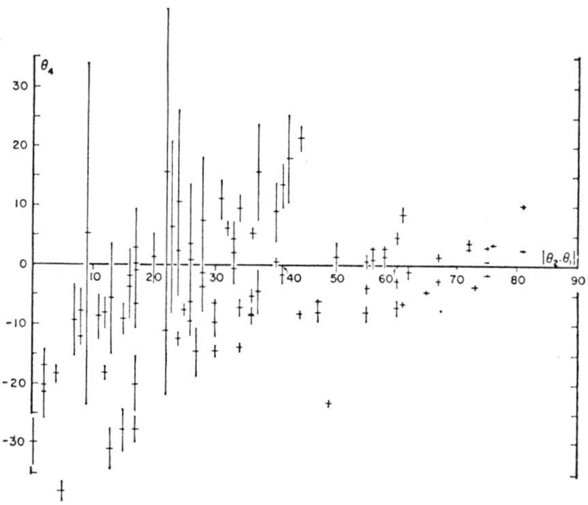

3. DOUBLE AND MULTIPLE STARS

The occultation method has proved valuable for the detection and in some cases analysis of double and multiple stars. In the simplest case of a double star the disappearance or reappearance is in two distinct steps (Figure 3) (Africano, et al. 1978). The separation

Figure 3: Disappearance of 115 Tau A. Vector separation 0.0987 arc seconds.

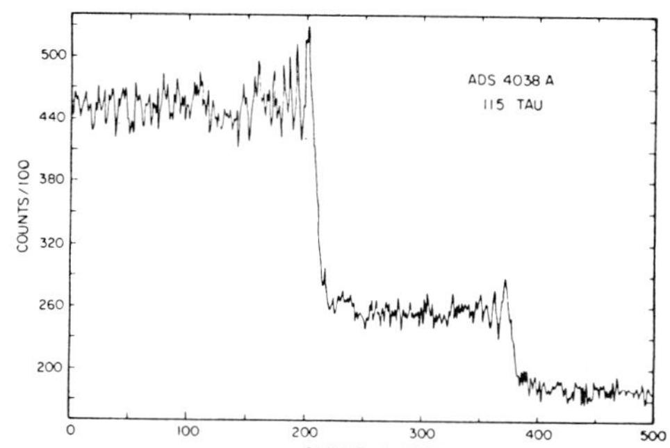

between these in time, multiplied by the lunar rate adjusted for the local lunar limb slope defines the "vector separation" between the pair that is their true separation projected on to a line perpendicular to the lunar limb, taking slope into account, at the point of occultation.

If the separation is small the fringes due to the two components may overlie each other producing what appears to be a distorted trace (Figure 4) (McGraw et al. 1974). If the trace is very noisy it may be

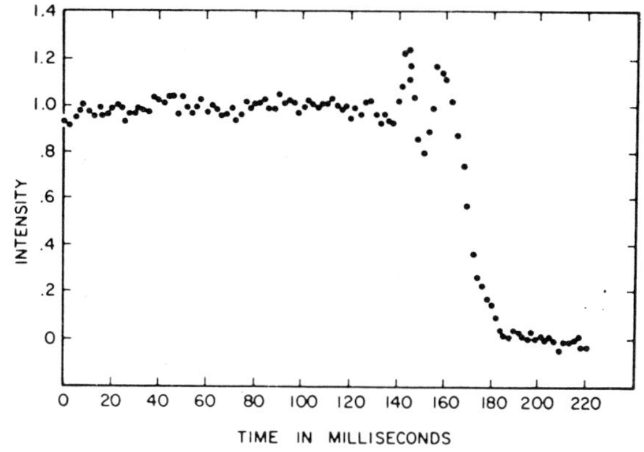

Figure 4: Disappearance of the Pleiades double star, Atlas.

difficult to be certain of possible duplicity. In this case a device introduced by Bartholdi is of great help. This is to plot the integral of the deviation of the measures from the mean value for the whole trace. This plot is little affected by seeing noise and consists of two straight line segments for a single star, three for a double and four for a triple star (Figure 5) (Dunham et al. 1973). We have adopted the practice of assigning a quality parameter 3 for a certain

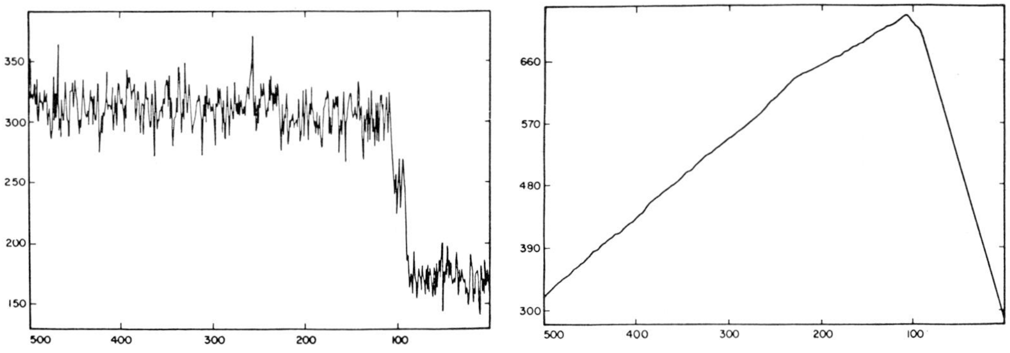

Figure 5: Original trace and integral plot for disappearance of of triple star, SAO 118786.

double, 2 for a probable double, 1 for a possible double, and zero for no duplicity. In the best cases vector separations down to a few arc milliseconds can be determined, together with magnitude difference between the components, and, if a dual channel system is used, differences of colour.

For the brighter A, B and O stars some 20-30 percent are found to be double with smaller values for later types (Evans, 1983) and a few years ago a catalogue was produced. Because of severe selection effects, these numbers must gravely underestimate the real incidence of duplicity, which, for these stars seems to increase the closer the pairs under consideration. At the present time this is one of the most powerful methods of angular resolution available, having the advantage over speckle interferometry that it is not subject to the Dawes limit and is not ambiguous as to the order of the components. It has, of course, the severe limitation of applying only to the stars of the zodiacal band. The statistics have been of interest to those concerned with the guidance systems of the Space Telescope and Hipparcos missions.

A single observation of a double star tells that the secondary lies on a certain position line parallel to the effective lunar limb. Simultaneous observations of the same event from two different observatories, or in the case of no orbital motion, from the same or different observatories on different occasions, can yield measures of true separation and position angle from the intersection of two or more position lines. Some judgement must be exercised, since if the difference in position angle is too small, observational errors will make the position lines nearly parallel and a false result will be obtained. Best results are usually obtained for position angles on the lunar limb differing by at least 30 degrees.

Many zodiacal double stars have been discussed so I shall select only two for special mention. The first is the quintuple star Beta Scorpii (Evans et al. 1978) in which New Zealand observers played an important part. Beta Scorpii is overtly a pair separated by 13.6 arc seconds. Star A, B0V with V=2.63, is itself a close spectroscopic binary with a fainter companion Star B about half an arc second away. Star C is B2V with V=4.92, with a close faint companion. This star system was of special interest because it was occulted by Jupiter in 1971 (Hubbard et al. 1972) and the analysis of the observations made then required a knowledge of the star's multiplicity, not yet known in detail. Occultation observations were made in 1975 at Dunedin and Auckland and at various American observatories in 1976. These served to fix the separation and position angles of the pairs AB and AC as shown in Figure 6. I have no time to go into detail but this figure shows how well the close pair could be analysed, to give a separation of 0.463±0.017 arc seconds in P.A 116°.5±2°.6. The analysis of the wider pair is not so satisfactory, demonstrating that to make this method work both components of a pair must be close enough to be occulted on essentially the same piece of lunar limb.

My second example is Beta Capricorni (Evans and Fekel, 1979). This is a triple star of which the primary is a K0 II-III in orbit, with a very long period for a spectroscopic binary, of 1374 days, with a B8V star, itself a spectroscopic binary with a period of 8.68 days. Fekel made an intensive spectroscopic study of the system and obtained all the usual parameters with their usual limitations. It will be realised that if a visual orbit were also available for the primary pair then complete data for the system could be obtained and this with a certain amount of contriving could be got from 10 occultation observations made in various places between 1975 and 1977. There was also

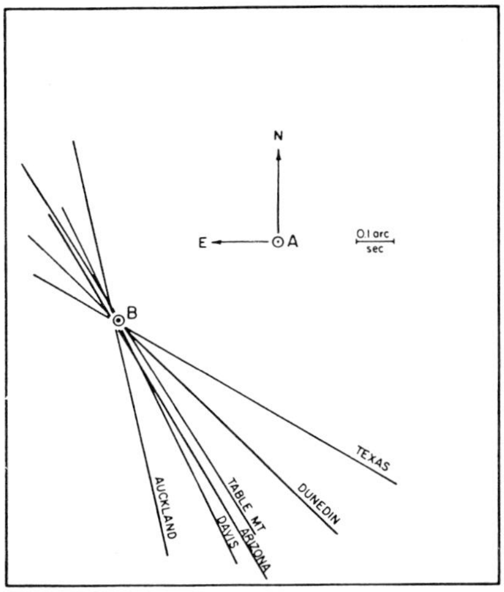

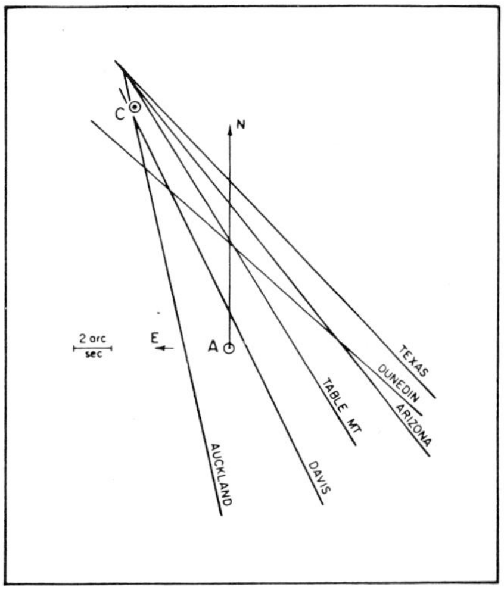

Figure 6: Concurrences of position lines for occultation observations of components of Beta Scorpii.

an early speckle observation (Blazit et al. 1977) which did not fit well and we now think ought to have been rejected, which distorted our original conclusions to a minor degree (Figure 7). The point is that

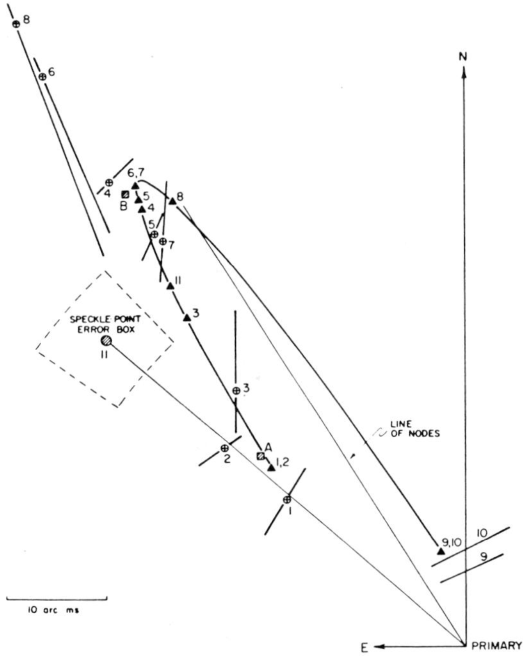

Figure 7: Visual orbit of the B-type secondary of Beta Capricorni as deduced from occultation observations.

through occultation observations this system became one of the best known in the sky with relative magnitudes (Figure 8), masses of the primary and secondary pair, parallax, linear diameter of the primary, and age all determined.

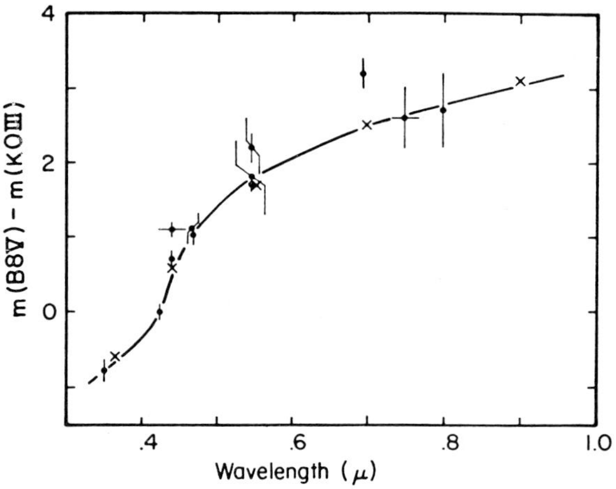

Figure 8: Relative magnitudes of B-type and K-type components of Beta Capricorni from occultation observations.

4. ANGULAR DIAMETERS

If a star has a significant angular diameter each strip of its surface parallel to the lunar limb produces its own set of diffraction fringes, which, suitably displaced, fall on top of each other. The consequence is that the fringes become less and less prominent for larger stars (Figure 1). In particular the first fringe becomes more and more depressed and some early workers have used the amount of the depression to determine the angular diameter. This short cut is risky because of the influence of seeing noise and the traces should be analysed using a suitable computer program. A recent compilation of angular diameter measures by Feierman (1985) lists more than 300 measures of 113 stars ranging from 67 arc milliseconds for R Leonis in the near infrared to just under 2 arc milliseconds. Some infrared sources have given much larger values and this is also true of R Leonis in the far infrared. Leaving out these exceptional objects the lower limit for such determination seems to lie between 1 and 2 arc milliseconds for traces of very high quality where the seeing effects on the fringes are minimal. The contribution to knowledge of angular diameters from occultation observations is thus considerable though new interferometry techniques are promised capable of measuring much smaller stars following the successful lead of Hanbury Brown et al. (1970). In spite of these new developments I do not think we need pronounce the funeral oration of the occultation method just yet.

Measurements of angular diameters of stars have had a wide impact on stellar astrophysics. It is easy to prove that

$$F_V = \log T_e + 0.1C = 4.2207 - 0.1V_o - 0.5 \log \phi'$$

where T_e is the effective temperature, C the bolometric correction,

$V_\odot$ the apparent V magnitude adjusted for interstellar extinction and $\phi'$the angular diameter in arc milliseconds of a star. The constant is derived from the solar calibration. F_V is a visual surface brightness parameter which Barnes et al. (1978) showed, based on the assemblage of known angular diameters, could be deduced from photometric measures in the V (Johnson) and R(Johnson) bands. Thus angular diameters of any sort of star can be determined. Applications have been made to Cepheids, RR Lyraes, red giants, white dwarfs, flare stars and even to portions of star surfaces such as starspots on red dwarf stars and RS CVn stars. The addition of new angular diameter measures will help to improve the definition of this calibration curve.

5. ANTARES AND ALDEBARAN

These stars are each occulted in series separated by about 19 years with their series separated by about 9 years. Figure 9 shows the trace obtained at Pretoria in 1952 (Evans et al. 1953). A close examination of this shows that the trace begins to rise far sooner than expected and this must be taken seriously because a few seconds previously the trace obtained for the companion was hardly larger than this deviation. Several authors have tried to attribute this to irregularities on the lunar limb (e.g. McCants and Nather, 1971) but their argument is faulty. The effect can however be produced by a sort of

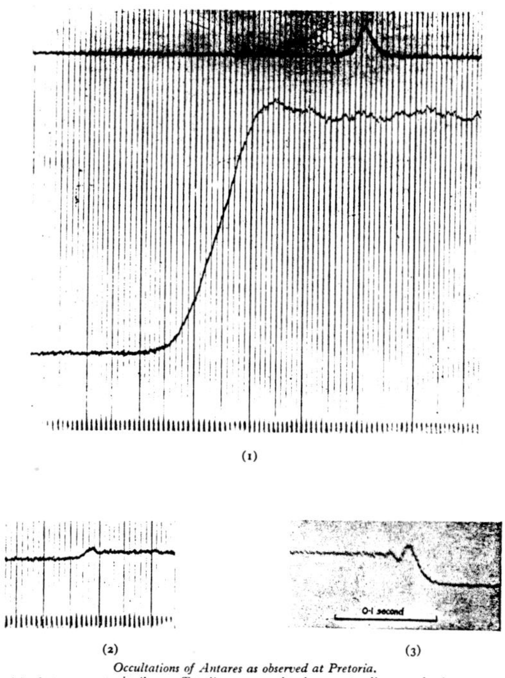

Figure 9

Occultations of Antares as observed at Pretoria.
(1) *Antares*, 1952 *April* 13. *Top line*, 1 *second pulses; centre line, occultation trace; bottom line*, 100 *c/sec timing trace.*
(2) *Antares B*, 1952 *April* 13. *Same scale as* (1).
(3) *Antares B*, 1953 *March* 8.

dome on the limb carefully placed with respect to the star disk. Many critics thought that lumps on the lunar limb would commonly invalidate occultation results. In principle they can even produce bogus double stars (Evans 1970) but the effects are very sensitive to the wavelength band used, which should help in rejecting dubious cases. One certain result is that if the first fringe exceeds the canonical level of 1.37 this must be due to limb irregularities. The first and I think the only times we have encountered this was on Runs Nos 1218, 2754 and 4160 (Evans, 1971; Africano et al. 1975; Africano et al. 1977) and we believe such incidents are very rare. Antares was observed on the same occasion (Figure 10) at Cape Town (Evans 1957) and produced essentially the same result. Again in 1953 Antares was observed at Johannesburg and Cape Town with matching results which differed from the previous ones. It therefore seems very improbable that lunar limb

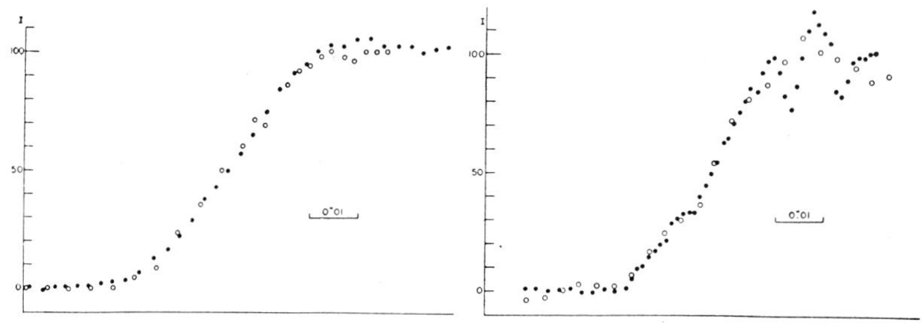

Figure 10: Duplicate observations of Antares occultations. Left: 1952 April 13, Pretoria and Cape Town. Right: 1953 March 8, Cape Town and Johannesburg.

irregularities could have been responsible. What is more Feierman's compilation lists 45 observations of Aldebaran during the last series for that star, with closely accordant results and apparently no necessity to appeal to lunar limb irregularities.

My analysis of the Antares results suggested that they were not consistent with a star model which was circular in outline and centrally symmetrical. At the time this was regarded as heresy but now with suggestions that stars such as Betelgeuse may have large blotches or starspots this does not seem so far-fetched.

I mention Antares now since a new series of occultations starts in 1986 - and there will be other interesting stars such as the Pleiades and Sigma Scorpii coming up soon - and I would strongly encourage observers to go after Antares and other interesting objects in the expectation that there are some important astrophysical results waiting to be found.

REFERENCES

Africano, J. L., Cobb, C. L., Dunham, D. W., Evans, D. S., Fekel, F. C. and Vogt, S. S. (1975), Astron. J. 80, 689.
Africano, J. L., Evans, D. S., Fekel, F. C. and Montemayor, T. (1977), Astron. J. 82, 631.
Africano, J. L., Evans, D. S., Fekel, F. C., Smith, B. W. and Morgan, C. A. (1978), Astron. J. 83, 1100.
Arnulf, M. A. (1936). Compt. Rend. 202, 115.
Barnes, T. G., Evans, D. S. and Moffett, T. J. (1978), Monthly Notices Roy. Astron. Soc. 185, 285.
Blazit, A., Bonneau, D., Koechlin, L. and Labeyrie, A. (1977) Astrophys. J. Lett. 214, L79.
Cousins, A.W.J. and Guelke, R., (1953), Monthly Notices, Roy. Astron. Soc., 113, 776.
Dommanget, J. (1980), Bull. Ast. Obs. Belgique, 9, 154.
Dunham, D. W., Evans, D. S., McGraw, J. T., Sandmann, W. H. and Wells, D. C., (1973), Astron. J. 78, 482.
Evans, D. S. (1955), Monthly Notices Roy. Astron. Soc., 115, 468.
Evans, D. S. (1957), Astron. J. 62, 83.
Evans, D. S. (1970), Astron. J. 75, 589.
Evans, D. S. (1971), Astron. J. 76, 1107.
Evans, D. S. (1983), Lowell Obs. Bull., No 167, pp 63,73.
Evans, D. S. and Fekel, F. C., (1979), Astrophys. J. 228, 497.
Evans, D. S., Heydenrych, J.C.R. and van Wyk, J.D.N. (1953), Monthly Notices Roy. Astron. Soc., 113, 781.
Evans, D. S., Elliot, J. L. and Peterson, D. M. (1978), Astron. J. 83, 438.
Feierman, B. H. (1985) Personal communication.
Hanbury Brown, R., Davis, J., Herbison-Evans, D., and Allen, L. R. (1970), Monthly Notices Roy. Astron. Soc., 148, 103.
Hubbard, W. B., Nather, R. E., Evans, D. S., Tull, R. G., Wells, D. C., Van Citters, G. W., Warner, B., and Vanden Bout, P. (1972), Astron. J. 77, 41.
McCants, M. M. and Nather, R. E. (1971), Highlights Astron. 2, 668.
McGraw, J. T., Wells, D. C. and Wiant, J. R. (1973), Rev. Sci. Inst. 44, 748.
McGraw, J. T., Dunham, D. W., Evans, D. S. and Moffett, T. J. (1974), Astron. J. 79, 1299.
Morrison, L. V. (1971), Highlights Astron. 2, 589.
Mulholland, J. D. (1981), Astron. J. 86, 800.
Watts. C. B. (1963), Astron. Papers Wash., XVII.
Whitford, A. E. (1939), Astrophys. J. 89, 472.
Williams, J. D. (1939), Astrophys. J. 89, 467.

Acknowledgements: This is a suitable moment to acknowledge with thanks the support given this program over the years by the U.S. National Science Foundation, and the University Research Institute of the University of Texas at Austin. Such success as we have enjoyed would never have been possible without the generous help and skills of colleagues especially those whose names appear in the list of references.

DISCUSSION

Rowe: With β Sco, spikes were observed during occultation by Jupiter, usually in pairs. An interpretation has been made in some (obscure) papers in terms of a resolved, close-spaced pair - 10 or 20 microarcseconds - a record in both optical and VLB radio astronomy and explicable in terms of the number of halfwaves of light refracted towards the common observing telescope by a high Jovian atmospheric refracting layer or layers.

Evans: Thank you for your remarks; of course this was a planetary not a lunar occultation. Evidently this could indicate a detection of the secondary component found in 1976.

Graham: Could you outline the basic equipment needed for occultation observations?

Evans: A photometer with recording equipment capable of say two or four millisecond data-rate and good tie in to an absolute time system. One should use a circulating buffer to avoid being swamped by the quantity of data. (The total amount of useful telescope time we have used in undertaking 7000 observations is ~ 3 minutes.)

Orchiston: Would you care to comment of the value of visual observations of grazing occultations?

Evans: A great deal of effort has been expanded on such observations and I would like to see a synthesis of the results.

Kurtz: Could you estimate what percentage of B stars are double?

Evans: Among HR stars we found 29 per cent. As there are several selection factors working against detection by the occultation method, the real percentage must be higher. Independent evidence plus occultation data for the Pleiades would suggest two or three times this percentage.

OCCULTATION STUDIES WITH SMALL TELESCOPES

Robert L. Millis
Lowell Observatory
Mars Hill Road, 1400 West
Flagstaff, Arizona 86001
U.S.A.

ABSTRACT. Various types of occultations and eclipses offer individuals with access to small telescopes the opportunity to contribute significantly to modern solar system research. In fact, often it is only through the cooperation of several such observers that the desired scientific objectives can be realized. This review discusses three classes of occultations which are of high current interest and which are particularly suitable for investigation with small telescopes.

1. INTRODUCTION

When one considers the types of solar system research that a ground-based observer can productively undertake with a modest-aperture telescope, investigations based on observation of occultations immediately spring to mind. Within the last decade and a half, observations of occultations (and eclipses) have resulted in several important discoveries, in precise measurement of size and shape for a variety of objects, in accurate determinations of the temperature and density profiles of planetary ionospheres, and in marked refinement of the orbits of the Galilean and certain Saturnian satellites.

The high potential of occultations for solar system research results from a variety of factors. Not the least of these is the fact that occultations have not been observed for very long or very widely; hence the chance for unexpected discovery is relatively high. Additionally, in an occultation of a star, for example, we observe the solar system object in transmitted, refracted, or diffracted light rather than in scattered or reflected light, as is more usually the case. Since different physical processes are involved, in an occultation it is often possible to measure physical parameters that are extremely difficult or impossible to measure with more conventional Earthbased techniques. Occultations are also ideally suited for measuring dimensions and positions of solar system bodies primarily because the problem is converted from one of angular resolution to a simple matter of timing. Resolution in an occultation is limited by the Fresnel scale rather than the seeing, with the result that order-

J. B. Hearnshaw and P. L. Cottrell (eds.), Instrumentation and Research Programmes for Small Telescopes, 199–211.

of-magnitude improvements in accuracy over conventional techniques can be achieved.

It is interesting to note that most occultation observations have been made with telescopes of 1-meter aperture or less. There are several reasons: First of all, certain types of occultations must be observed with a distributed network of telescopes if the maximum scientific return is to be achieved. Usually, it is only through the use of small, sometimes portable telescopes that an appropriate network can be put together. Secondly, the objects involved in occultations are often bright, giving an adequate signal-to-noise ratio with a very modest telescope. Finally, for many occultations completely satisfactory observations can be obtained with simple, even rudimentary photometric equipment at observing sites that are less than optimum. Moreover, a large time commitment is not required to prepare for and conduct the observations. In short, occultations offer the busy professor or student at a small college an ideal way to contribute meaningfully to astronomical research. Occultations by their very nature require wide cooperation and collaboration; many individuals isolated at out-of-the-way institutions have found this type of work to be an avenue to active involvement with colleagues at major observatories.

In this paper no effort will be made to review the entire spectrum of occultation studies. Rather, a few classes of occultations that seem to the author to be of greatest current interest and that are most amenable to observation with small telescopes will be highlighted. The intent here is to explain the scientific objectives to be achieved by observing these events and to outline the instrumental requirements. Readers seeking a broader treatment of the subject are directed to reviews by Elliot (1979) and Millis (1983).

2. ASTEROID OCCULTATIONS

When one considers the thousands of asteroids moving against the backdrop of distant stars, it is evident that occultations of stars by minor planets will occur with great frequency. Millis and Elliot (1979) have estimated that thirteen of the larger asteroids would alone produce more than 160 potentially observable occultations per year. In spite of the near-hourly occurrence of these events, all but the tiniest fraction of a percent pass unobserved. Most are not predicted; of those that are, many are rejected because of one unfavorable characteristic or another--they are observable only over the ocean, the asteroid is too close to the Sun, the star is too faint, the asteroid's angular diameter is too small to permit accurate prediction of the region of visibility, etc. Yet these events can tell us much about the asteroids that we can learn in no other way from Earth.

During an occultation the shadow cast by the asteroid in the light of the occulted star sweeps across the Earth, describing a ground track such as the one illustrated in Figure 1. Observers within the ground track will see the asteroid approach the star and momentarily block it from view, while those near the track will observe only a close appulse. Because the asteroid's shadow, when projected onto a plane perpendicular to a line connecting the star and asteroid, has the same size and shape as the asteroid's profile on the sky and because the velocity of the shadow is known, an observer has only to record the times of the star's disappearance and reappearance to determine the length of one chord across the minor planet. By combining observations from sites at various distances from the centerline of the track, the complete profile of the asteroid can be mapped. In principle the occultation technique can yield asteroid dimensions significantly more accurately than other presently existing techniques. A typical occultation lasts 10 to 20 seconds, though some fall well outside this range. If immersion and emersion are timed at a given site to within 0.1 second--a level of accuracy easily achievable with simple photoelectric equipment--then the length of the corresponding chord is known to about one percent. By combining an adequate number of equally accurate chords, the dimensions of the face of the asteroid seen at the time of occultation can be determined to a percent or two. The occultation method has the added virtue that it is direct, with no need for simplifying assumptions about limb darkening or shape.

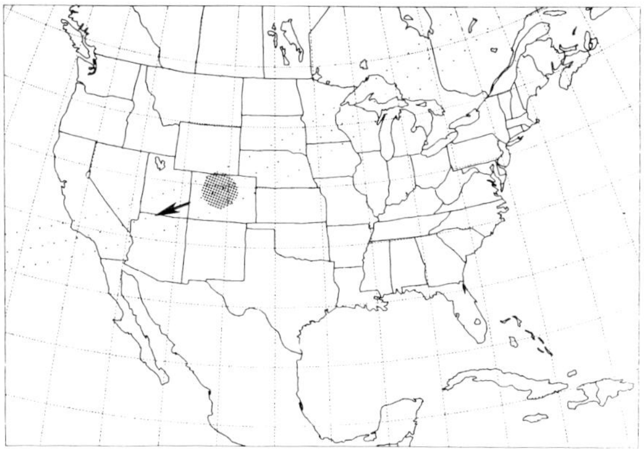

Figure 1. Schematic illustration of an asteroid shadow sweeping across the Earth during an occultation.

Why do we want to know precise sizes and shapes of asteroids? There are many reasons. In the cases of Ceres, Pallas, and Vesta, whose masses are known accurately (Schubart and Matson 1979), well-determined diameters permit calculation of the mean density, providing valuable clues to gross composition. This objective has been accomplished for Pallas (Wasserman **et al.** 1979; Millis and Elliot 1979) and Ceres (Millis **et al.** 1985a), but the density of Vesta remains to be learned.

Accurate diameters of asteroids spanning a broad range in size and taxonomic type are needed to permit the best possible calibration and refinement of the radiometric method of size determination (e.g., Morrison and Lebofsky 1979). This indirect, model-dependent technique has been applied widely not only to asteroids, but to other solar system objects ranging from cometary nuclei to planetary satellites. Significant improvement of the radiometric method has been accomplished already on the basis of occultation results (Brown **et al.** 1982; Lebofsky **et al.** 1986). Figure 2 shows a comparison of occultation and radiometric diameters. The agreement between the two methods is much improved compared with the situation five years ago, but it is important to extend the points in the diagram to cover smaller asteroids and to fill in more points at intermediate sizes. When developed to its full potential, infrared radiometry will provide a tool with which to measure with confidence size and albedo for large numbers of small solar sýstem bodies and, perhaps, to identify those whose surfaces have unusual thermal properties.

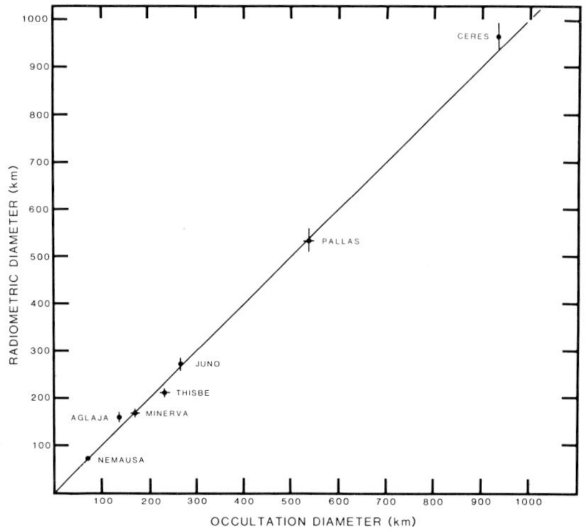

Figure 2. Radiometric diameters of asteroids plotted against their occultation diameters. The solid line corresponds to perfect agreement between the two methods.

Just as occultation observations complement infrared observations, they similarly enhance the potential of radar studies of minor planets. Radar has the unique capablity to probe the porosity, roughness (on the scale of centimeters to meters), and metal concentration of asteroidal regoliths. However, the radar observations can be interpreted fully only when size, shape, and spin vector of the target asteroid are known (Ostro 1985). Occultation observations can provide the first two parameters, while knowledge of the spin vector comes from rotational light curves recorded at many different aspects.

Knowledge of the shapes of asteroids, in addition to aiding interpretation of radar observations, is of interest for its own sake. Our understanding of the collisional history of the asteroid belt and the clear evidence provided by the complex and often large-amplitude rotational light curves of asteroids lead us to expect that many, if not most, small minor planets are irregular in shape. Asteroid occultations provide us with the only presently available means of accurately measuring the figure of an asteroid, short of a space mission. Observation of a single occultation gives us the shape of the asteroid's limb profile at one particular viewing aspect. For most well-observed occultations by minor planets of 200 km-diameter or larger, it has been possible to represent the limb reasonably well with an ellipse or a circle (e.g., Wasserman **et al.** 1979; Millis **et al.** 1981; Millis **et al.** 1984; Millis **et al.** 1985b; Dunham **et al.** 1984), but in all cases real limb irregularities having vertical scales of a few kilometers have been detected. A measure of the three-dimensional figure of an asteroid can be obtained by observing two or more occultations by the same minor planet occurring at different rotational phases and/or viewing aspects. To date this feat has been accomplished only for Pallas (Wasserman **et al.** 1979; Dunham **et al.** 1983), whose figure was found to be well represented by a triaxial ellipsoid (see Figure 3). When dealing with the larger asteroids such as Pallas and Ceres, departures from sphericity place useful constraints on the strength of the material of which these bodies are composed and on their internal structure (e.g., Dermott 1979).

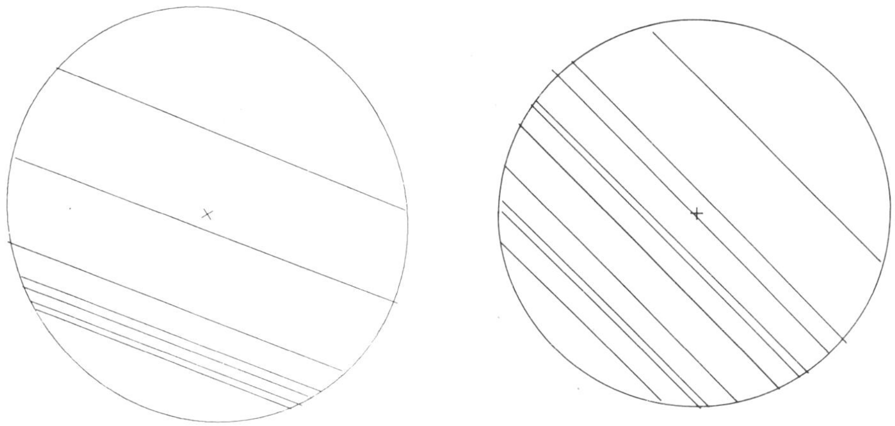

Figure 3. The figure of Pallas seen at two different aspects. The profile on the left is from a 1978 occultation (Wasserman **et al.** 1979); the one on the right is from the 1983 occultation of 1 Vulpeculae (Dunham **et al.** 1983).

An aspect of asteroid research that has been greatly stimulated by occultation observations concerns the possible existence of minor planet satellites. On occasion observers well outside of the primary asteroid occultation ground track have reported brief obscurations of the target star (Williamon 1980; Arlot **et al.** 1985). These reports have been interpreted in terms of minor planet satellites (e.g., Van Flandern **et al.** 1979), as have the binary-like light curves of certain asteroids and the higher-than-expected frequency of double craters on the Earth, Moon, and other cratered bodies (Hut and Weissman 1985). Some have found the occultation evidence for the existence of minor planet satellites to be less than convincing (Reitsema 1979). Certainly, reports of secondary occultation events, as the presumed satellite detections are called, have been less frequent in the last few years. Moreover, deep, coronographic CCD exposures covering the immediate vicinity of a few asteroids have failed to detect satellites (Gradie **et al.** 1985; Terrile and Smith 1985). These negative results at first glance seem surprising, since satellites have been claimed to be stable within a zone extending out to about 100 times the diameter of the asteroid (Van Flandern **et al.** 1979). Weidenschilling (1985), however, has recently shown that the only configurations likely to be stable are contact binaries whose mass ratio is near one or small satellites at large distances from the primary. It is of interest in this regard that a speckle search for duplicity among a sample of about 30 asteroids failed to find any close binaries (Franz 1985). If CCD searches continue to find no small satellites, we will be forced to conclude that minor planet satellites are rare, if they exist at all. In the meantime, it is possible that occultation observations will unequivocally turn up a satellite, but any such case must be extremely well documented.

Most observations of asteroid occultations have been attempted with the aim of surveying the occulting asteroid. However, these events, in principle, offer the opportunity to measure the angular diameter of the occulted star as well by recording the diffraction pattern which is present in the immersion and emersion light curves. All other things being equal, the limiting angular resolution achievable goes as $1/\sqrt{D}$, where D is the distance between the observer and the occulting body (Ridgway 1977). The asteroid belt is roughly 800 times farther from Earth than is the Moon, so one might expect by observing asteroid occultations to measure the angular diameters of stars which are 20 to 30 times smaller than the limit for lunar occultation measurements. This advantage is often largely offset by the fact that the diffraction pattern in an asteroid occultation typically sweeps across the telescope 10 to 20 times faster than in a lunar occultation. To date, efforts to derive stellar angular diameters from asteroid occultation data have not been successful (e.g., Reitsema **et al.** 1981), primarily because of inadequate signal-to-noise. The Fresnel diffraction pattern was clearly seen in observations of the 11 April 1985 occultation of AG+20°1138 by Antigone (see Figure 4), but unfortunately only a strip-chart tracing of the occultation was obtained. In any case, future observers should bear in mind the potential of asteroid occultations for measuring stellar angular diameters.

Additional help is badly needed in observing asteroid occultations. Table I lists the occultations that have been observed for which the author could find reasonably reliable reports in the scientific literature. Note that many of the events were observed only at one or two sites or were observed primarily by visual observers. In such cases, a diameter determination of useful accuracy is usually not possible. If more photoelectrically equipped observers were available, there is no question that the success rate in observing asteroid occultations could be markedly improved. Predictions are available on

TABLE I. SUMMARY OF OBSERVED OCCULTATIONS OF STARS BY ASTEROIDS

Date	Asteroid	No. of Chords*	Reference
19 Feb 58	3 Juno	1 v	J. Brit. Astron. Assoc. 72, 212, 1962.
2 Oct 61	2 Pallas	1 p	J. Brit. Astron. Assoc. 72, 212, 1962.
6 Feb 73	2 Pallas	2 p	J. Roy. Astron. Soc. Canada 67, 198, 1973.
24 Jan 75	433 Eros	8 v	Icarus 28, 133, 1976.
5 Mar 77	6 Hebe	2 v	Icarus 34, 89, 1978.
29 May 78	2 Pallas	7 p	Astron. J. 84, 259, 1979.
7 Jun 78	532 Herculina	1 p/2 v	Bull. Amer. Astron. Soc. 10, 594, 1978.
19 Jul 78	3 Juno	1 v	Occultation Newsletter 2, 12, 1979.
11 Dec 78	18 Melpomene	6 p/2 v	Astron. J. 85, 174, 1980.
17 Aug 79	51 Nemausa	2 p	Astron. Astrophys. Suppl. 44, 375, 1981.
17 Oct 79	65 Cybele	1 p/2 v	J. Brit. Astron. Assoc. 92, 1, 1981.
11 Dec 79	9 Metis	2 v	IAU Circular No. 3437.
11 Dec 79	3 Juno	15 p	Astron. J. 86, 306, 1981.
4 Sep 80	78 Diana	10 v	Occultation Newsletter, March 1981.
10 Oct 80	216 Kleopatra	9 v	Occultation Newsletter, March 1981.
24 Nov 80	134 Sophrosyne	5 v	Occultation Newsletter, March 1981.
8 Jan 81	44 Nysa	1 v	Occultation Newsletter, March 1981.
19 Mar 81	48 Doris	3 v	Occultation Newsletter, March 1982.
7 Aug 81	18 Melpomene	2 p	Sky and Telescope, January 1982.
5 Oct 81	105 Artemis	1 p	Sky and Telescope, January 1982.
7 Oct 81	88 Thisbe	3 p/9 v	Astron. J. 88, 229, 1983.
23 Mar 82	386 Siegena	1 p/1 v	Sky and Telescope, January 1983.
30 Mar 82	15 Eunomia	1 v	Sky and Telescope, January 1983.
18 Apr 82	146 Lucina	1 p/1 v	Icarus 61, 224, 1985.
17 Sep 82	19 Fortuna	1 p	Sky and Telescope, January 1983.
14 Nov 82	690 Wratislavia	2 v	IAU Circular No. 3747.
15 Nov 82	375 Ursula	2 p/3 v	Astron. J. 89, 592, 1984.
22 Nov 82	93 Minerva	5 p/5 v	Icarus 61, 124, 1985.
19 Jan 83	106 Dione	10 v	Astron. Nachr. 305, 207, 1984.
29 May 83	2 Pallas	~100 p&v	Bull. Amer. Astron. Soc. 15, 822, 1983.
11 Sep 83	51 Nemausa	6 p	Astron. J. 89, 1755, 1984.
16 Sep 84	47 Aglaja	3 p/7 v	Bull. Amer. Astron. Soc. 16, 1027, 1984.
13 Nov 84	1 Ceres	13 p	Bull. Amer. Astron. Soc. 17, 729, 1985.
11 Apr 85	129 Antigone	3 p/2 v	Occultation Newsletter, August 1985.
15 Apr 85	275 Sapientia	1 p/1 v	Occultation Newsletter, August 1985.

*p = photoelectric or video observations; v = visual observations.

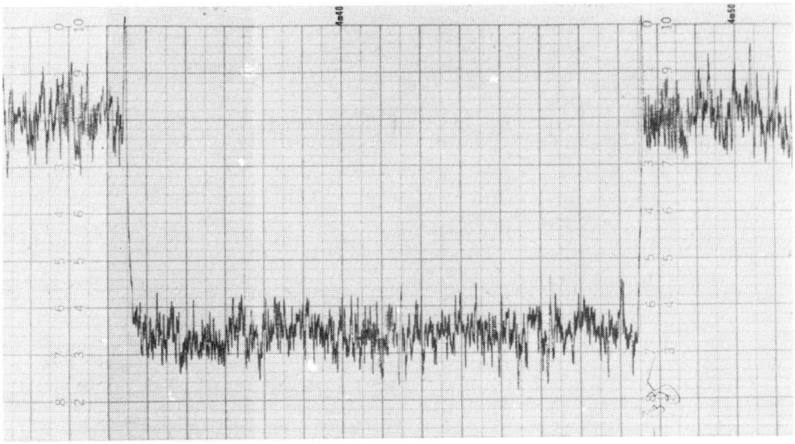

Figure 4. A light curve of the 11 April 1985 occultation of AG+20°1138 by Antigone. Note the diffraction pattern at immersion and emersion. Observations were made by O. G. Franz and R. Oliver of Lowell Observatory using a 0.35-meter portable telescope.

a worldwide basis (e.g., Wasserman **et al.** 1985), and the techniques of accurately refining the location of the ground track for any individual occultation are well developed. The only instrumental requirements for this work are a small telescope (instruments as small as 8-inches aperture have been used successfully), a simple photometer, and a data system capable of recording the signal from the photometer at a timing resolution of 0.1 second.

3. THE PLUTO-CHARON MUTUAL EVENTS

In 1978 Christy and Harrington (1978) discovered that Pluto possesses a relatively large satellite which is now commonly called Charon. Soon after the discovery, Andersson (1978) pointed out that the orientation of Charon's orbit was such that Pluto's orbital motion would soon cause the satellite's orbital plane to sweep across the Earth and Sun. Andersson noted that this process would extend over five or six years, during which time alternate transits of the satellite across Pluto and occultations of Charon by Pluto would occur.

Photometric observations of the Pluto-Charon mutual events can tell us much about the Pluto-Charon system. If a large number of the events are observed, it will be possible to measure the size of each object; their relative magnitudes, albedos, and colors; the precise parameters of Charon's orbit; and hence, the total mass and bulk density of the system. Low-resolution albedo maps of one face of each object may also be derived.

The first of this series of Pluto-Charon mutual events was

detected in early 1985 (Binzel **et al.** 1985). Since the next series will not begin until about the year 2109, it is important that the best job possible be done this time. Tholen (1985) has published predictions for the events occurring during the 1985/1986 apparition. Eighty-one events will occur, each lasting between four and five hours, a time span which near the ends of the apparition exceeds the available observing interval at any given observatory. Presumably similar numbers of occultations and transits will occur during subsequent years of the present series.

Clearly, worldwide cooperation is needed to insure adequate observation of the Pluto-Charon events. Given the amount of telescope time that will be required, it seems certain that much of the observing must be done with meter-class or smaller telescopes. Pluto will be near 14^m in V so photometry accurate to 1% in V or B with a time resolution on the order of 5 minutes should be possible with a 1-meter telescope. For reasons of uniformity, observations in V or B are preferred, but individuals using telescopes of significantly less than 1-meter aperture may have to observe without a filter in order to achieve the maximum signal-to-noise ratio in their data.

A worldwide campaign is being organized to observe the Pluto-Charon mutual events. Individuals interested in participating should contact Dr. Edward Tedesco at the Jet Propulsion Laboratory, Pasadena, California 91109.

4. OCCULTATIONS BY COMETS

Because of their relatively rapid motion across the sky, comets, like asteroids, have frequent close appulses to moderately bright stars. The small size of cometary nuclei and the difficulty of doing very high-precision astrometry of comets cause the probability of observing an occultation of a star by the nucleus of a comet to be vanishingly small. However, valuable information about the solid grains in the inner coma of a comet can be learned from events where the starlight has passed within a few hundred kilometers of the nucleus (Bowell **et al.** 1984). At greater distances the coma is likely to be too tenuous to produce detectable diminution of the starlight.

The reflectance, R, of the coma at a given point is given by

$$R = N \sigma A , \qquad (1)$$

where N is the total number of grains in a column of unit cross-section extending completely through the coma, σ is the cross-section of a single grain, and A is the albedo of the grains as defined by A'Hearn **et al.** (1984). The optical thickness, τ, of the coma at that point is related to the filling factor, Nσ, by

$$N\sigma = (1-e^{-\tau}) . \qquad (2)$$

Combining Eqs. (1) and (2), we get

$$R = (1-e^{-\tau})\ A. \tag{3}$$

Furthermore, the optical thickness of the coma is by definition

$$I/I_0 = e^{-\tau}, \tag{4}$$

where I is the apparent brightness of a star seen through the coma at the point in question, and I_0 is the unobscured brightness of the star. Hence, photometric observations of a close appulse of a star and a cometary nucleus will yield by Eq. (2) the distribution of grains in the coma along the apparent path of the star. When combined with measurements of the surface brightness of the coma along that path, the grain albedo can be found from Eq. (3).

Although predictions of comet occultations have been published during the past few years (e.g., Bowell and Wasserman 1985), very few events of this type have been observed, and those few have been recorded serendipitously rather than as a result of predictions. The best-documented observations are those by Larson and A'Hearn (1984), who observed the passage of the light from a 15th-magnitude star to within about 540 km of the nucleus of Comet Bowell. These investigators reported a maximum extinction of the star of 3% ± 1%. Obviously, more definitive results could be obtained with events involving dustier comets, closer appulses, and brighter stars; but events of that type are unlikely to be observed accidentally. Perhaps the best approach would be for large numbers of observers to attempt observations of all predicted comet occultations in their geographical region. While success cannot be guaranteed for any particular observer, over time the cooperative effort should produce good results. Here again, as with the two types of events discussed earlier, only small-telescope observers can achieve the necessary density and distribution of observing sites.

5. CONCLUDING REMARKS

This review has emphasized three particular classes of occultations: occultations of stars by asteroids, mutual phenomena of the Pluto-Charon system, and occultations of stars by comets. Other types of occultations sometimes also can be productively observed with small telescopes. The Uranian rings were discovered with telescopes of less than 1-meter aperture during an occultation of the 9th-magnitude star SAO 158687 (Elliot **et al.** 1977; Millis **et al.** 1977). Subsequent observations of similar events have made possible remarkably detailed kinematic models of the nine rings (e.g., French **et al.** 1985). Recently, observations of occultations by Neptune have shown exciting evidence for a fragmented ring around that planet (Brahic **et al.** 1985). The contributions of occultation observers to studies of the thermal and density structure of planetary atmospheres, to measurements of the oblateness of distant planets, and to the search for atmospheres of Pluto and outer planet satellites likewise should be mentioned. Clearly, continued observations of these kinds will be valuable. They have not been treated in detail in this review, primarily because few such events involving bright stars have been found in searches extending through the end of this decade (e.g., Mink and Klemola 1985; Killian and Dalton 1985). It appears that for the next several years stellar occultations involving the outer planets, their rings, and their satellites of necessity will be observed with large telescopes. Eventually, however, events within the grasp of small telescopes will occur, and observers should be alert for these opportunities.

Many predictions of occultations are published well in advance in widely circulated scientific journals. Most early predictions, however, are only approximate, requiring refinement a few weeks or days prior to the occultation. The refined predictions are disseminated through formal and informal networks of observers. Since it is usually only these last-minute predictions which are sufficiently accurate to serve as the basis for actual observing plans, individuals interested in observing occultations should seek to become a part of a network. The Working Group on Occultations of IAU Commission 20 operates one such network. Observers interested in receiving predictions from the Working Group are urged to contact the author.

ACKNOWLEDGMENTS

This review paper was prepared with support from NASA Grant NSG-7603. Figures 1 and 3 are based on figures originally published in the **Astronomical Journal** and are reprinted here with permission of that journal.

References

A'Hearn, M. F., Schleicher, D. G., Feldman, P. D., Millis, R. L., and Thompson, D. T. (1984). **Astron. J.** 89, 579-591.

Andersson, L. E. (1978). **Bull. Amer. Astron. Soc.** 10, 586.

Arlot, J. E., Lecachaux, J., Richardson, C. H., and Thuillot, W. (1985). **Icarus** 61, 224-231.

Binzel, R. P., Tholen, D. J., Tedesco, E. F., Buratti, B. J., and Nelson, R. M. (1985). **Science** 228, 1193-1195.

Bowell, E., and Wasserman, L. H. (1985). **International Halley Watch Newsletter** No. 7, 30-37.

Bowell, E., Wasserman, L. H., Baum, W. A., Millis, R. L., and Lumme, K. (1984). In **Cometary Astrometry** (D. K. Yeomans, R. M. West, R. S. Harrington, and B. G. Marsden, Eds.), pp. 105-122. Jet Propulsion Laboratory, Pasadena.

Brahic, A., Hubbard, W. B., Bouchet, P., Eliser, L.-R., Hafner, R., Manfroid, J., Roques, F., Sicardy, B., and Vilas, F. (1985). **Nature,** in press.

Brown, R. H., Morrison, D., Telesco, C. M., and Brunk, W. E. (1982). **Icarus** 52, 188-195.

Christy, J. W., and Harrington, R. S. (1978). **Astron. J.** 83, 1005-1008.

Dermott, S. F. (1979). **Icarus** 37, 575-586.
Dunham, D. W., Van Flandern, T. C., Millis, R. L., Chapman, C. R., Maley, P. D., and Povenmire, H. (1983). **Bull. Amer. Astron. Soc.** 15, 822.
Dunham, E. W., Baron, R. L., Conner, S., Dunham, D. W., Dunham, J. B., Schneider, G., Cohen, H. L., Helms, V. T., Croom, M., and Safko, J. (1984). **Astron. J.** 89, 1755-1758.
Elliot, J. L. (1979). In **Annual Reviews of Astronomy and Astrophysics** (G. Burbidge, D. Layzer, and J. G. Phillips, Eds.), Vol. 17, pp. 445-476. Annual Reviews, Inc., Palo Alto.
Elliot, J. L., Dunham, T., and Mink, D. (1977). **Nature** 267, 328-330.
Franz, O. G. (1985). Presented at the June 1985 meeting of the Astronomical Society of the Pacific.
French, R. G., Kangas, J. A., and Elliot, J. L. (1985). **Bull. Amer. Astron. Soc.** 17, 718.
Gradie, J., Hammel, H., and Pilcher, C. (1985). **Bull. Amer. Astron. Soc.** 17, 729.
Hut, P., and Weissman, P. R. (1985). **Bull. Amer. Astron. Soc.** 17, 690.
Killian,A. M., and Dalton, A. S. (1985). **Astron. J.** 90, 2372-2376.
Larson, S. M., and A'Hearn, M. F. (1984). **Icarus** 58, 446-450.
Lebofsky, L. A., Sykes, M. V., Veeder, G. J., Tedesco, E. F., Matson, D. L., Brown, R. H., Gradie, J. C., Feierberg, M. A., and Rudy, R. J. (1986). Submitted to **Icarus.**
Millis, R. L. (1983). In **Solar System Photometry Handbook** (R. M. Genet, Ed.), pp. 7-1 to 7-18. Willmann-Bell, Richmond.
Millis, R. L., and Elliot, J. L. (1979). In **Asteroids** (T. Gehrels, Ed.), pp. 98-118. University of Arizona Press, Tucson.
Millis, R. L., Wasserman, L. H., and Birch, P. V. (1977). **Nature** 267, 330-331.
Millis, R. L., Wasserman, L. H., Bowell, E., Franz, O. G., and Klemola, A. (1984). **Astron. J.** 89, 592-596.
Millis, R. L., Wasserman, L. H., Bowell, E., Franz, O. G., Nye, R., Osborn, W., and Klemola, A. (1985b). **Icarus** 61, 124-131.
Millis, R. L., and 37 other authors. (1981). **Astron. J.** 86, 306-313.
Millis, R., Wasserman, L., Franz, O., Nye, R., Oliver, R., Jones, S., Kreidl, T., A'Hearn, M., Schnurr, R., Osborn, W., and Klemola, A. (1985a). **Publ. Astron. Soc. Pacific,** in press.
Mink, D. J., and Klemola, A. (1985) **Astron. J.** 90, 1894-1899.
Morrison, D., and Lebofsky, L. A. (1979). In **Asteroids** (T. Gehrels, Ed.), pp. 184-205. University of Arizona Press, Tucson.
Ostro, S. J. (1985). **Publ. Astron. Soc. Pacific,** in press.
Reitsema, H. J. (1979). **Science** 205, 185-186.
Reitsema, H. J., Hubbard, W. B., Zellner, B. H., and Lebofsky, L. A. (1981). **Astron. J.** 86, 121-126.
Ridgway, S. T. (1977). **Astron. J.** 82, 511-515.
Schubart, J., and Matson, D. L. (1979). In **Asteroids** (T. Gehrels, Ed.), pp. 84-97. University of Arizona Press, Tucson.
Terrile, R. J., and Smith, B. A. (1985). **Bull. Amer. Astron. Soc.,** in press.
Tholen, D. J. (1985). **Astron. J.,** in press.
Van Flandern, T. C., Tedesco, E. F., and Binzel, R. P. (1979). In

Asteroids (T. Gehrels, Ed.), pp. 443-465. University of Arizona Press, Tucson.
Wasserman, L. H., Bowell, E., and Millis, R. L. (1985). **Astron. J.** 90, 2124-2129
Wasserman, L. H., and 17 other authors. (1979). **Astron. J.** 84, 259-268.
Weidenschilling, S. J. (1985). **Bull. Amer. Astron. Soc.** 17, 731.
Williamon, R. L. (1980). **Astron. J.** 85, 174-176.

DISCUSSION

Evans: I don't want to discourage anybody, but trying to do minor planet occultations from fixed observatories is very much of a gamble, since one must be within a shadow track only about as large as an asteroid. This may be fine for the larger ones but it is difficult for the smaller ones.

Millis: One needs to make a concerted effort using portable instruments. Fixed telescopes are a little more difficult, but many good observations have come from them.

Penhallow: The first international expedition to observe an asteroid occultation was successful and produced the first photograph of such an event. This involved Metis in South America in 1979. The photograph was reproduced in Sky & Telescope. The prediction was based on last minute astrometry from Lick, Lowell, R.G.O. and the Quonochontang Observatory.

Evans: Are timings of close approach worthwhile observations?

Millis: They have astrometric value, and any observation of a near miss makes it more unlikely that there are minor planet satellites. Speckle interferometry and deep CCD images have also failed to find any satellites.

Woodward: To compliment the occultation observations we have been imaging planets to try and detect rings and sattelites. We have recently resolved the rings of Uranus with the University of Rochester infrared array camera. Our 2.2μm broad-band image has appeared in Sky and Telescope (in the October 1985 issue). We are writing up our results currently, and will soon submit the manuscript for publication in the astronomical literature.

Millis: We see the ring structure all together. We are still doing some image processing to look for any finer structure in the rings.

ASTEROIDS

H.J. Schober
Institut für Astronomie
Universitätsplatz 5
A-8010 Graz
Austria

ABSTRACT. Using a number of different conventional observing methods of astronomy like spectrophotometry, polarimetry, IR-radiometry and UBV-photometry, physical properties of asteroids can be derived, such as diameters, spectral reflectivity and albedo. Based on observable parameters a classification in terms of taxonomic types can be made and the mineralogy of the surface can be determined and compared with meteoritic analogues. UBV-photometry at different phase angles can reveal something about surface properties and diameters, whereas from accurate photometry during different phases of the rotational cycles we are able to derive rotation periods, geometric irregularities of the asteroid-body and topographic features on the surface. If observations of lightcurves are obtained during several oppositions at different ecliptic positions, the orientation of the spin axis in space and sometimes the sense of rotation can be derived. Observations of asteroids are especially well suited to small and medium-sized telescopes, as most of the programmes need long telescope runs. In addition there is a definite need for a large number of observations of asteroids in the magnitude range 12-15 mag.

1. INTRODUCTION

At the present stage of solar system space research we must state that we have had different space missions to nearly all the objects in the planetary system, which include the investigations of the moon, the sun, the planets and their satellites and of course interplanetary space. With the reappearance of Comet Halley a space mission to such an object has been undertaken. We must wonder why there has so far been no fly-by of a specific selected object in the asteroid belt. It would be logical to explore asteroids directly at a fly-by by means of in situ investigations of asteroids whose orbits are more or less between the planets Mars and Jupiter.

Projects for such flights were presented and proposed to ESA in missions like ASTEREX in 1979 or AGORA in 1983. A fly-by of an asteroid is planned by NASA in the CRAF mission between 1990 and 2000. Looking

J. B. Hearnshaw and P. L. Cottrell (eds.), Instrumentation and Research Programmes for Small Telescopes, 213–221.

into the future it seems at least possible to talk about possible mining and exploiting such relatively small rock-like objects. Of special interest are nearby asteroids with earth-crossing orbits and it would be of special interest to know the mineralogy of a small asteroid (with a diameter of about a kilometer). Also the property of the surface would be of interest and could tell us something about the impact history.

The interesting physical parameters of asteroids are especially their masses, volumes and densities, as well as the geometry of the body, its rotation and surface properties. As far as masses and densities are concerned, it is extremely difficult to obtain precise values. Schubart and Matson (1979) have collected and presented results for the asteroids 1 Ceres, 2 Pallas and 4 Vesta and get for example, 2.3 ± 1.1 g/cm^3 for 1 Ceres and 3.3 ± 1.5 g/cm^3 for 4 Vesta, which give a rough indication that there probably are density differences for different types of asteroids - but the precision is only about ± 50%. Masses are usually derived from perturbation computations, but it also should be mentioned that it is difficult to obtain accurate diameters. The best diameters are based on occultation measurements - until now obtained only for a few asteroids with larger diameters, and they represent only diameters at certain rotational phases. Asteroids up until now have been resolved directly, the largest asteroid, 1 Ceres with a listed diameter of 1025 km has only 0.7 arcsec apparent diameter at most, and many asteroids are much smaller, or farther out. From earth there is no chance of getting direct images of the surface, though speckle interferometry is used to get elliptical images of asteroids.

2. COLLISIONAL EVOLUTION OF ASTEROIDS

Today we believe that the study of asteroids contributes a lot to our understanding of the origin and evolution of the solar system. There are other objects, too, but we think that asteroids are small enough that they have not undergone the complicated chemical evolution like the larger planets, and on the other hand they are large enough, when compared to comets, that their orbits have not been perturbed too much. In addition, masses should be higher than for comets, since asteroid orbits are not too close to the sun, so that little disintegration has taken place. We believe that in general asteroids are still found in those regions where they were formed.

Wilkening (1979) has given a schematic diagram of the possible formation and evolution of asteroids. It is assumed that out of the primitive solar nebula in the asteroid zone a number of primordial asteroids were formed, roughly with diameters of a few hundred kilometers, by condensation or by accretion of material. If no chemical evolution takes place, such an asteroid should go only through the process of impact cratering, produced by smaller objects. It should end up as a typical C-type asteroid with surface structures due to impacts. As indicated by Paolicchi et al. (1981) there could occur a kind of catastrophic impact which would produce a number of smaller objects, depending on the size, on the kinetic energy involved and on the solid-state forces. The smaller parts could, under the influence of self-gravitation, form a kind of loose pile of rubble, or binary and multiple asteroids, or the asteroid could be disintegrated and the resulting product could probably be a more physically defined asteroid family than those defined only by orbital characteristics.

If the temperatures are higher an asteroid would go through melting processes and would be heated up and cooled down. Through the repeated process of fractional differentiation, layers of different densities would be formed around the central core. If again we have impacts or catastrophic collisions, deep cratering or fragmentation would occur. The asteroid, e.g. an S-type asteroid with a dense core, would exhibit deeper layers with different properties, and if observed from earth in reflected sunlight - that is the only information we get - it would show variations on the surface during its rotation.

3. SPECTROPHOTOMETRY

The number of asteroids with known orbital characteristics (in order to identify and recover any given object) has increased during the last few years. In the Ephemerides of Minor Planets for 1979 we find 2050 numbered asteroids. In 1985 we have 2958, which means an increase of roughly 900 asteroids in six years, and for the year 1986 we find 3143 asteroids listed which is 200 more than last year. Though most of the newly discovered asteroids are fainter objects, we would estimate that in the magnitude range V = 8 to 15 there are about 600 to 700 observable asteroids, depending on the distances between the sun,the earth and the asteroid. These are within easy reach of any telescope of 0.5 m up to 1.5 m aperture. Doing simple photometry even V =17 can be reached. Modern detectors can also enhance the finding and identification of asteroids.

Bobrovnikoff (1929) stated that 2 Pallas was bluer than other asteroids - that was a kind of qualitative spectrophotometry and was a first indication of possible surface composition differences. Only a few years ago detailed studies of the brighter asteroids were made by Gaffey and McCord (1979), using a 25-filter spectrophotometer, covering the optical wavelength range from 0.3 to 1.1 μm. From a comparison with reflectance spectra of meteorites and artificial mixed powders of olivine, pyroxene and nickel-iron containing material, analogues for asteroids were derived. Quantitative spectral curves were measured for about 300 brighter asteroids by Chapman and Gaffey (1979) and average spectra were found for the major asteroid compositional type. C-types, which show low reflectivity and a flat spectrum, are silicates and opaque carbons, similar to carbonaceaous meteorites. S-type asteroids have higher reflectivity, typically 15-20%, and consist of metallic silicates. Their spectral curve is deep and not flat and exhibits an absorption near 0.95 μm in the olivine-pyroxene band. The third and major class is named M (medium, metallic) with high reflectivity. They generally rotate faster and might have higher densities. Chapman and Gaffey (1979) tried to group the reflectance spectra due to the composition of the material with C : S ranging from 1 : 6 to 6 : 1, but the resulting classes were far too complicated and had only laboratory importance. Among all asteroids 4 Vesta turned out to be a unique case, classified as U (unknown) and showing a deep olivine-pyroxene absorption band.

It should be noted, that this type of specrophotometry has not been continued. Zellner et al (1985) used an eight-colour asteroid survey system for nearly 600 asteroids, and derived a new classification system. Recently there are reports of using IDS systems for continuous spectral-scanning with about 1 to 2 nm resolution over the optical wavelength range. There is a need to continue this type of work and to extend it to about magnitude 15, which will require a lot of observing time.

4. POLARIMETRY AND IR-RADIOMETRY

In order to study the surface texture and the broadband reflectivity, Dollfus and Zellner (1979) measured the scattered partly polarized light of asteroids. The polarization curve is characterized by the minimum negative polarization at small phase angles and the width of the negative part up to the inversion angle, where polarization becomes positive. These are also the parameters which describe the surfaces in terms of roughness and porosity. Additionally, the linear slope of the polarization curve, if polarization is measured over a longer phase angle range, is linked to the albedo (in the visual spectral range p_v) and in consequence with the diameter. The albedo of about 300 asteroids has been measured and an interesting bimodality due to surface reflectivity was found for the C-type asteroids with p_v from 0.065 down to a peak between 0.035 and 0.04, and for the S-type objects with p_v between roughly 0.065 to 0.23 and a peak at about 0.14 (Zellner and Gradie (1976) and Morrison (1977)).

In addition Morrison and Lebovsky (1979) derived radiometric albedos and diameters for about 200 asteroids, measuring the thermal emission of asteroids at infrared wavelengths (10 and 20 μm), combined with simultaneous B and V measurements. They assumed a standard thermal model of high thermal inertia,caused by an insulating and poor conducting regolith, which was responsible for the fact that both the illuminated and non-illuminated sides of an asteroid have no temperature difference. In general, the diameters and albedos, derived from infrared methods do not differ much from those derived from polarimetry, and they usually agree within 10% which is within the accuracy limits of diameters derived from modern occultation measurements. It must be noted, that this type of work has not been extended to a larger sample of fainter asteroids to enhance the statistics.

5. UBV-PHOTOMETRY

The most complete list of UBV data for asteroids was created (and is still maintained) by Bowell et al. (1979). This list contains not only colours, but types, diameters and magnitudes. This type of work is still going on and UBV photometry has the advantage that measurements can be made quickly and that a large number of asteroids, including fainter ones down to 17-18 mag, can be analyzed. For the taxonomy of asteroids the albedo, polarization, spectroscopic parameters and colour indices B-V and U-B were used. For a total of 750 asteroids with reported UBV measurements the following results were obtained: about 75% are dark C-type asteroids, 15% bright S-asteroids, the rest, 10% are of type M (metallic), E (enstatic achondrites), R (reddish) or U (unknown). Asteroid classification will become more complicated, since Gradie and Tedesco (1982) announced new categories based on the new eight-filter system. Using this system Zellner et al. (1985) find an interesting clustering of asteroids within a given class if analysed for various distances from the sun.

The mean brightness of an asteroid depends on the geometry and on the scattering properties of the asteroid surface. If the magnitude is measured over a larger phase angle range, reduced for a defined absolute distance, we obtain a phase curve, called the magnitude-phase relation. It consists of a (nearly) linear part for phase angles larger than 10 degrees (which gives the phase coefficient) and the non-linear part at smaller phase angles, called the opposition effect, where asteroids become brighter than they should. Scaltriti and Zappalá (1980) showed that the opposition effect is similar for all asteroids, i.e. it is not affected by type or albedo.

The opposition effect in the phase curve is mainly controlled by the volume density or porosity whereas the linear part in the phase curve is determined by the roughness of the surface. From more than 1500 UBV observations and based on a multiple scattering theory of the surface of atmosphereless bodies, Bowell and Lumme (1979) found that asteroid surfaces are similar in texture, moderately porous, purely as a consequence of the similarity of the opposition effect. From different slopes of the magnitude-phase relation they concluded that asteroids are moderately rough and that the surface roughness varies with types. It also should be noted that just recently at the IAU General Assembly at New Delhi this theory was adopted by Commission 20 of IAU for the future prediction of the magnitudes of asteroids, which will finally provide a system which is based upon photometric observations.

6. PHOTOMETRY AND SURFACE VARIEGATION OBSERVED DURING THE ROTATION

A large research field during the last few years - and also in the future - is the study of asteroid rotation rates. All asteroids rotate somehow, with spinrates from only hours up to days. Most of the asteroids have rotation rates of between 8 and 15 hours, but very long rotation rates of up to 50 days have been found. Light variations of 0.01 to 1.5 mag. are measured and expressed in terms of the variation of the cross-section. Light curves turn out to be generally double-wave and very symmetric. Therefore with respect to surface geometry properties this would give a direct determination of the geometry of an asteroid. That is also the easiest method to get an idea about the shape of an asteroid. Nearly all observed asteroids do show such variations during rotation, but for some it might be at the limit of detectability.

In order to get information about details on the surface itself, e.g. the possible variation in composition on different sides of an asteroid, one needs to carry out spectrophotometry during one rotation which is time consuming and not applicable for a large number of asteroids at smaller telescopes. But to find any kind of colour-patches or albedo spots, a mixture of E-type and S-type material, on the surface, or even topographic features like mountains or the shadowing effects of such features, the best method to be applied would be polarization and/or colour measurements during rotation. Lists presented by Degewij et al. (1979) and by Schober and Schroll (1981) could show that there is evidence for about 20% of investigated asteroids to exhibit such variations - but only to the order of a few hundredths of a magnitude. Among them 3 Juno and 4 Vesta seem to be especially good candidates.

Simple photometry is still one of the best means to investigate the surface and the rotational properties of asteroids but lightcurves of asteroids sometimes turned out to be more complicated. Frequently rotation rates have to be revised after repeated observations, mostly by a factor two and often there are features in the lightcurves which are caused by surface properties and result in triple or even quadruple lightcurves with triple extrema. Interpretation is difficult and sometimes repeated observations could give completely different results. Also there is the possible existence of satellites or binary asteroids (with lightcurves quite similar to eclipsing binary stars) which must be considered. Evidence for such binary lightcurve objects may be found, from: triple or quadruple extrema, colour variations during rotation, different apparitions which show different lightcurves, very slowly spinning asteroids showing tidal effects and nonperiodic irregular lightcurve features within the lightcurves of an asteroid. It must be mentioned that there is still a lot of speculation and that until now no real proof has been found for any binary object or satellite.

7. POLE DETERMINATION OF ASTEROIDS

One of the most important parameters of asteroid physics is certainly the knowledge of the orientation of the spin axis - or pole coordinates with respect to the plane of the ecliptic. For the study of the collisional evolution of asteroids the distribution of spin axis orientation is important. Also for future studies, including possible space research projects, knowledge of the pole orientation and if available the sense of rotation is important. During the last years an increasing effort has been made to improve our knowledge of pole positions, mainly due to the effort of Zappalà et al. (1983) and Magnusson (1983).

In principle, the lightcurve of an asteroid depends on the aspect of the asteroid, which means that an asteroid, if observed at different earth-sun-asteroid geometry, shows changing lightcurve amplitudes, changing timings of the extrema, and changing appearance of the lightcurve. That is, it depends on the line of sight of the observer towards the surface of the asteroid. The most extreme possibilities are pole-on and equator-on views with the respective result that there is either no light variation or maximum light variation.

If we do not consider the most direct method of using speckle interferometry to derive pole orientations, there are in principle four different methods to determine pole positions: graphic pole determination (using amplitudes and ecliptic longitudes) as given by Vesely (1971), semianalytic methods (using amplitudes, ecliptic longitudes and the V magnitude at lightcurve maximum) as given by Zappalà (1981), astrometric photometry as introduced by Gehrels and reviewed by Taylor (1979) and combined methods by Magnusson (1983). Until now only about 30 asteroids have been successfully observed and analysed, with good accuracy for pole orientations - a program which definitely merits future attention and telescope time. But again it must be noted that it is time consuming as every two or three years, depending on the asteroid, observations are necessary including observations for a large phase angle range. On the other hand this project is very well suited for smaller telescopes, as many of the brighter asteroids from V = 8 to 13 are still to be observed.

7. CONCLUSION

Although I know that this review is far from covering all the different aspects of asteroid research (I completely omitted the still necessary astrometric work for improving orbits and for positions and discoveries, and I did not even mention the modern CCD-imaging and automatic discovering of moving targets), I hope that a few topics were mentioned, which makes it clear that observing asteroids is a useful project. I also should mention that there are a number of groups working in asteroid physics, and that there is a number of publications, which must and should be mentioned. The last official asteroid meeting was held in 1979 at Tucson and a book "Asteroid" edited by T. Gehrels (1979) came out of that meeting. That book is unique and it will be hard to make a better one in the future. Every potential observer is certainly recommended to look at that book. The next asteroid conference will again be held at Tucson in 1987. There are two other Asteroid-related books available, both the result of conferences: "Asteroids, Comets, Meteors" I and II, both edited by Lagerkvist and Rickman (1983) and part II is just in print, in 1985 or 1986. They contain the current programs in a more detailed form. I have tried to concentrate on modern observing techniques, mostly based on photometry or related methods, but I should not forget to mention the "Minor Planet Bulletin" now edited by Binzel and appearing several times per year. I would also like to make reference to Binzel's (1983) article on the Photometry of Asteroids in the Solar System Photometry Handbook, edited by Genet.

ACKNOWLEDGEMENTS

A great part of the current project on the physics of asteroids is supported financially by the Austrian "Fonds zur Förderung der Wissenschaftlichen Forschung" under Project No. P4852. Also I would like to thank the Directors of Cerro Tololo Interamerican Observatory, CTIO, and of the

European Southern Observatory, ESO, both in Chile, for the generous allotment of time on telescopes with apertures between 0.5 and 1.5 m.

REFERENCES

Books:
Asteroids, ed. T. Gehrels, Univ. of Arizona Press, Tucson, Arizona,1979.
Asteroids, Comets, Meteors, eds. C.I. Lagerkvist and H. Rickman, Uppsala universitet, Uppsala, 1983.
Asteroids, Comets, Meteors II, eds. C. I. Lagerkvist and H. Rickman, Uppsala universitet, Uppsala, 1986.
Solar System Photometry Handbook, ed. R. M. Genet, Willmann-Bell Inc., Richmond, 1983.
Articles:
Binzel, R.P.: 1983, *Solar Photometry Handbook,* Chapter 1.
Bobrovnikoff, N.T.: 1929, *Lick Obs. Bull.*,14, 18.
Bowell, E., Gehrels, T., Zellner, B.: 1979, in *Asteroids,* p.1108.
Bowell, E., Lumme, K.: 1979, in *Asteroids,* p.132.
Degewij, J., Tedesco, E., Zellner, B.: 1979, *Icarus,* 40, 346.
Gaffey, M.J., McCord, T.B.: 1979, in *Asteroids,* p.688.
Gradie, J., Tedesco, E.F.: 1982, *Science,* 216, 1405.
Morrison, D.: 1977, in *Comets, Asteroids, Meteorites: Interrelation, Evolution Origin,* ed. Delsemme, A.H., Univ.of Toledo Press, Toledo, p.177.
Morrison, D., Lebovsky, L.A.: 1979, in *Asteroids,* p.184.
Magnusson, P.: 1983, in *Asteroids, Comets, Meteors,* p.75.
Paolicchi, P., Farinella, P., Zappalà, V.: 1982, in *Sun and Planetary System,* eds. W. Fricke, G. Teleki, Reidel, Dordrecht, p.295.
Scaltriti, F., Zappalà, V.: 1980, *Astron. Astrophys.,* 83, 249.
Schober, H.J., Schroll, A.: 1982, in *Sun and Planetary System,* eds. W. Fricke, G. Teleki, Reidel, Dordrecht, p.285.
Schubart, J., Matson, D.L.: 1979, in *Asteroids,* p.84.
Taylor, R.C.: 1979, in *Asteroids,* p.480.
Vesely, C.D.: 1971, in *Physical Studies of Minor Planets,* ed.T.Gehrels, NASA SP-267, Washington, p.133.
Wilkening, L.L.: 1979, in *Asteroids,* p.61.
Zappalà, V.: 1981, *Moon and Planets,* 24, 319.
Zappalà, V., M. di Martino, Farinella, P., Paolicchi, P.: 1983, in *Asteroids, Comets, Meteors,* p.73.
Zellner, B., Gradie, J.: 1976, *Astron. Journ.,* 81, 262.
Zellner, B., Tholen, D.J., Tedesco, E.F.: 1985, *Icarus,* 61, 443.

DISCUSSION

Millis: The diameters quoted by Schubart and Matson (1979) for Ceres and Vesta are not occultation diameters. We still don't know the diameter of Vesta, but the diameter and density of Ceres are now known and they are significantly different from the values given by Schubart and Matson.

Graham: What type of asteroid observations, using small telescopes, are required at present?

Schober: Probably polarisation measurements.

Rowe: In your bimodal distribution, which asteroids are on the inner parts of the solar system?

Schober: The darker ones (C-type) should be more on the inside.

COORDINATED PROGRAMS FOR COMET HALLEY

Michael F. A'Hearn
Astronomy Program
University of Maryland
College Park, MD 20742
USA

ABSTRACT. Because comets are so difficult to observe, it is essential to coordinate observations from a variety of telescopes in order to fully understand their behavior. The apparition of Comet Halley represents a unique opportunity for coordinated programs because this is the only predictable comet which is routinely bright enough for the many observational programs desired. The International Halley Watch is playing an advocacy and coordinating role in the observations of Halley. Coordinated photographic observations will provide a record of the development of jets in the coma and of the motion of features in the tail. Coordinated photometry will provide both the heliocentric variation in vaporization and also the details of the outbursts thought to occur in most comets. Coordination with larger optical telescopes and with radio and infrared telescopes will allow a more complete understanding of the physical and chemical processes occurring in comets. Examples, of course, must still be drawn from observations of previous comets but they illustrate the results expected from Halley.

1. INTRODUCTION

There are a number of reasons for studying comets and many of these reason require coordinated programs using small telescopes. Perhaps the most obvious reason for studying comets is that they are the most dramatic appearances in the sky other than a total solar eclipse. The most dramatic aspect of comets is the existence of tails and these provide fascinating phenomena which must be studied with small telescopes in order to have a large enough field of view. A second reason for studying comets is that they allow us to probe the solar wind, both the variation in the solar wind's properties with direction, heliographic latitude and longitude, and the way in which the plasma of the solar wind interacts with the primarily neutral gas in the cometary coma. These effects also are often most obvious in the tails of comets and again often best studied with small telescopes in order to get large fields of view. Many astronomers would argue, however, that the most important reason for studying the comets is that they may have preserved

J. B. Hearnshaw and P. L. Cottrell (eds.), Instrumentation and Research Programmes for Small Telescopes, 223–235.

material from the formation of the solar system - probably the nebular condensates and possibly even interstellar grains that were incorporated directly. Even in this area of understanding the chemical nature of comets there are a number of programs, primarily photometry and CCD imaging but also to some extent spectroscopy and spectrophotometry, which can readily be carried out on small telescopes. In order to illustrate the importance of coordination, I will discuss some examples of past comets for which coordinated observations were extremely valuable.

2. EXAMPLES FROM THE PAST

Comets are among the most difficult of astronomical objects to study because they are often large in angular size, low in surface brightness, moving rapidly, and located close to the sun when they are bright. Furthermore, they are known to vary dramatically on relatively short timescales, i.e. within hours. The simplest form of coordination is required by the fact that, unlike all other objects in the solar system, cometary orbits have high inclinations to the ecliptic. Even short-period comets have orbits which are, on average, inclined more than those of asteroids but the long-period and parabolic comets have randomly oriented orbits. Many comets, therefore move from high declinations in the northern hemisphere to high declinations in the southern hemisphere or vice versa. This is, of course, one of the features which makes comets valuable as probes of the solar wind - they probe the solar wind far from the ecliptic. Nevertheless, it is also a problem because it requires coordination between northern and southern observatories to properly study such comets. Such coordination is, in fact, quite common in astrometry and not uncommon in photometry although it is not so common in other forms of cometary observations. The need for astrometric observations over the entire observable orbit is sufficiently obvious that it need not even be stated but the need, e.g., for photometric observations over the entire observable portion of the orbit is not so intuitively obvious. The general reason is that the variation in production of gas with heliocentric distance is important in understanding the physics and chemistry of the nucleus. At the simplest level, observations over a sufficiently wide range of heliocentric distances enable us to learn the volatility of whatever species controls the vaporization of the nucleus. Other fundamental questions about the cometary nucleus can also be studied by using photometry from both pre-and post-perihelion phases.

A very recent example is shown in Figure 1. Comet Encke is usually visible only from the northern hemisphere before perihelion and only from the southern hemisphere after perihelion. Several years ago, Whipple and Sekanina (1979) developed a model for the precession of the rotation axis of Encke's nucleus by assuming that the visual light curve, which is highly asymmetric about perihelion, represented the total outflow of gas. Since publication of that model, the actual production of gas has been measured. The data are based primarily on narrow-band photometry of Encke's comet obtained at Lowell Observatory before perihelion and at both Mt. John University Observatory and Perth

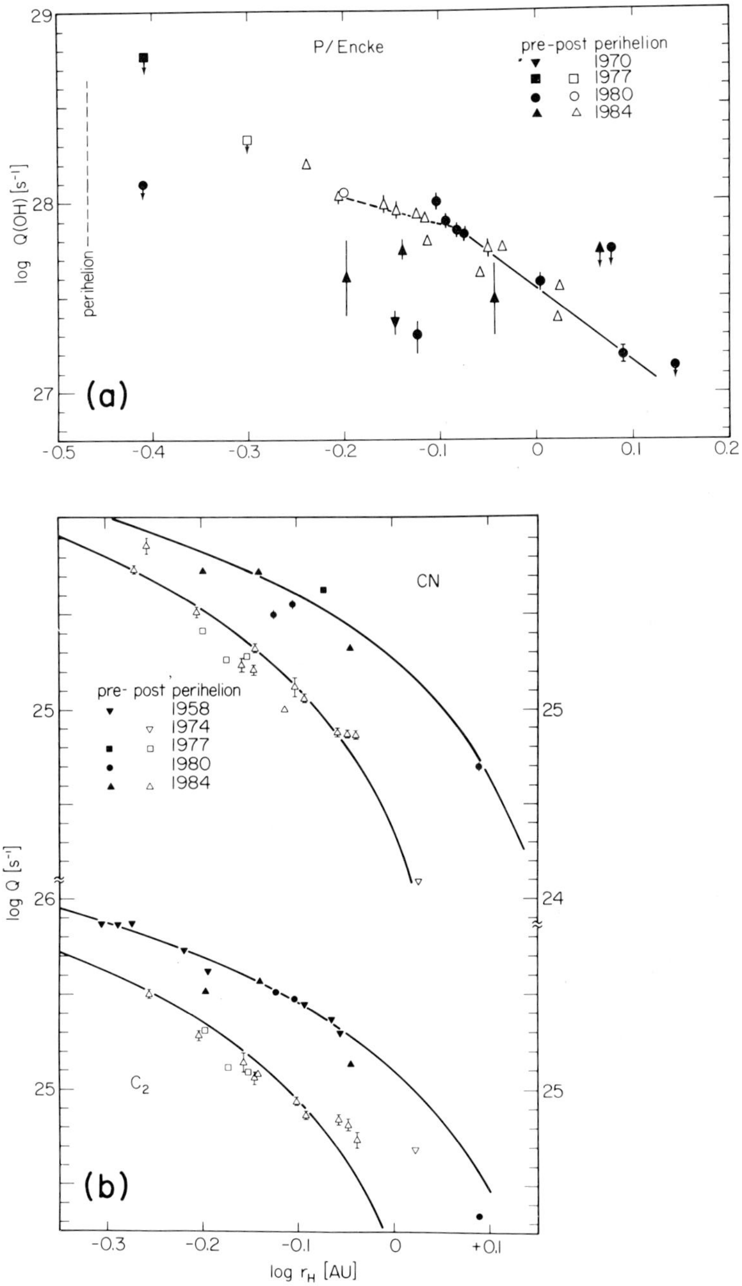

Figure 1. Production of OH, CN and C_2 in Comet Encke. The solid lines in the lower panel represent the visual light curve. Data are from northern hemisphere (pre-perihelion) and southern hemisphere (post-perihelion) observatories as well as IUE. From A'Hearn et al. 1985.

Observatory after perihelion (A'Hearn et al. 1985). Nearly all the data were obtained with small telescopes. It is clear that the production rate of trace gases C_2 and CN does mimic the visual light curve as you would expect. Most of the gas, however, is from H_2O and Figure 1 shows measurements of OH taken from Lowell before perihelion and from Mt. John after perhelion. Some additional points were also obtained from the IUE satellite (illustrating yet another type of coordination). Obviously the production of OH does not mimic the visual light curve at all. Stimulated by this result, Sekanina (1986) has looked further into his model for the precession and found a quite different physical effect which produces an asymmetry in the non-gravitational forces even without an asymmetry in the production rate of gas; by looking at images of sunward fans in the coma of Encke's comet, he has shown that the lag angle is large before perihelion but very small after perihelion. He now finds that the actual precessional path of Encke's rotation axis is similar to that in the earlier model but we get a quite different picture of the physical processes in the nucleus which produce this precession. The surface varies not in the fraction of ice cover but in the composition of the ice and the thermal inertia of the surface layer. The lack of an asymmetry in the production of OH coupled with the asymmetry in other species was detected only because of coordination among different observatories. The original data also show the need for more than one observatory in each hemisphere. In particular, the post-perihelion phases are well described only because two southern observatories (Mt. John and Perth) with different weather patterns were active during the 1984 apparition of Encke's comet. For comets which vary greatly with time and which must be observed now or not until many years from now, it is also important to have this latter redundancy which circumvents the weather patterns.

A much more critical type of coordination is required for studies of the short term variability of comets. It is well known that outbursts in the comae of comets take place on a time scale of order a day. Because of their proximity to the sun, bright comets are rarely observable for more than an hour or two from a given longitude. In order to understand these outbursts, which appear photometrically as brief increases in brightness but which appear even more dramatically in images as jets and haloes, it is essential to have data from several observatories at different longitudes. A prime example of this is shown in Figure 2 which is taken from work by Larson and Sekanina (1985) using images of Halley's comet from 1910. All images were taken with telescopes less than one meter in size and located at Vienna (V), Lowell (L), Helwan (H) and Lick (C). The images were processed using a shift-difference algorithm to enhance the low-contrast features. There was little coordination in taking these images and it is sometimes difficult to identify features in successive images. These images do illustrate what could be learned from a properly coordinated effort. They show dramatically the changes over a short interval of time, changes that could not be observed from a single site. It is this work that has allowed Larson and Sekanina to deduce both an expansion velocity for the coma of 1.4 km/sec and a rotation period of roughly 40 hours for the nucleus of Halley's comet. Without the observations at different longi-

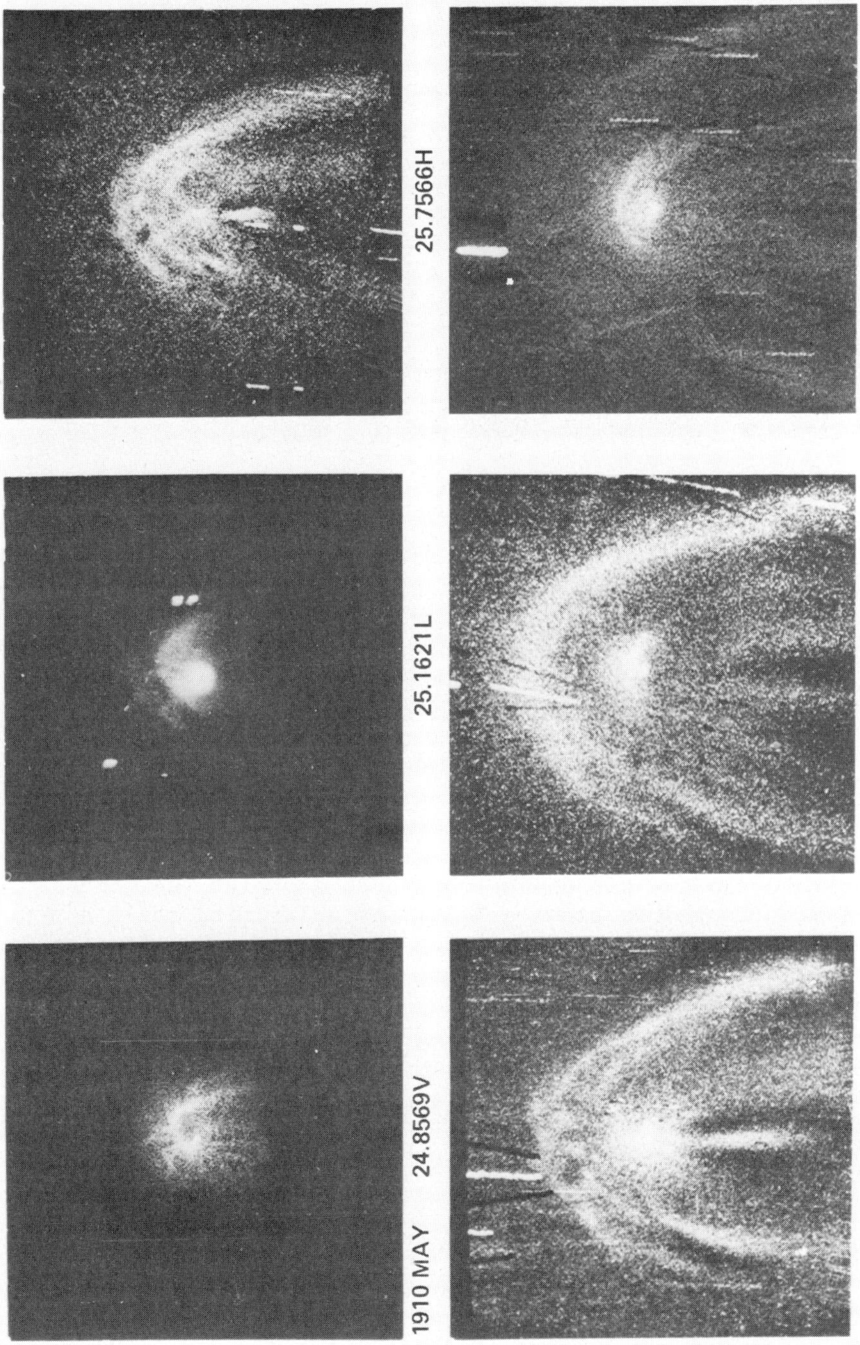

Figure 2. Highly processed images of Comet Halley in May 1910. Images labelled with UT data and observatory: V = Vienna, L = Lowell, H = Helwan, C = Lick. From Larson and Sekanina, 1985. Photo compliments of S. Larson.

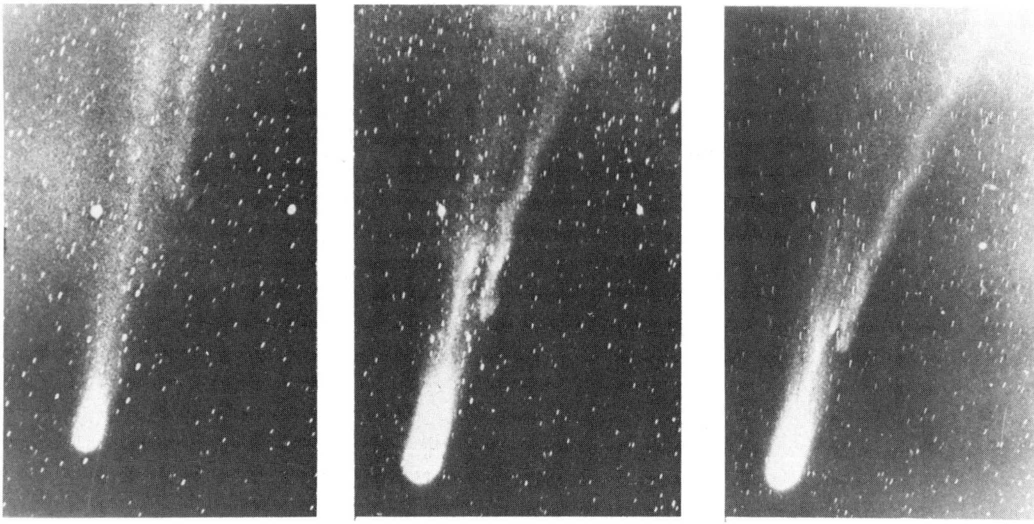

Figure 3. A disconnection event in the tail of Comet Halley, 1910 June 6-7. Original images from Yerkes, Honolulu, and Beirut spanning 16 hours. Photograph compliments of M. Niedner (c.v. Niedner and Brandt, 1978).

tudes, it would be impossible to relate features in successive images. I note in passing that the problem of determining the rotation period of cometary nuclei from photometric observations suffers from precisely the same limitations but there is not to my knowledge a single instance in which the required coordination has been achieved either deliberately or inadvertently.

My third, and perhaps most dramatic, example of the need for coordinated observations comes from wide-field photographs of the tails of comets. Figure 3 shows a disconnection event, in which the comet's ionic tail "breaks off" and is blown away by the solar wind while a new tail forms in the coma. This entire process takes place on a time scale of hours -- too slowly to be observed in one night from a single observatory but too fast to be observed by the the same observatory on successive nights. There are conflicting theories about how this process takes place (magnetic reconnection, flute instability, shock-shock interaction, and differential acceleration, c.f. Mendis et al. 1985) and it is impossible to resolve these discrepancies without good sequences of images showing the development of the process. Furthermore it is essential that data on the solar wind be available for the same period of time in order to ascertain whether or not there are substantial effects in the solar wind, such as crossings of sector boundaries, which are associated with the process.

The final type of coordinated observation is that between observatories doing different types of work. The particular example that I will cite uses data from large telescopes but the principle is equally applicable to small telescopes. Several groups have recently

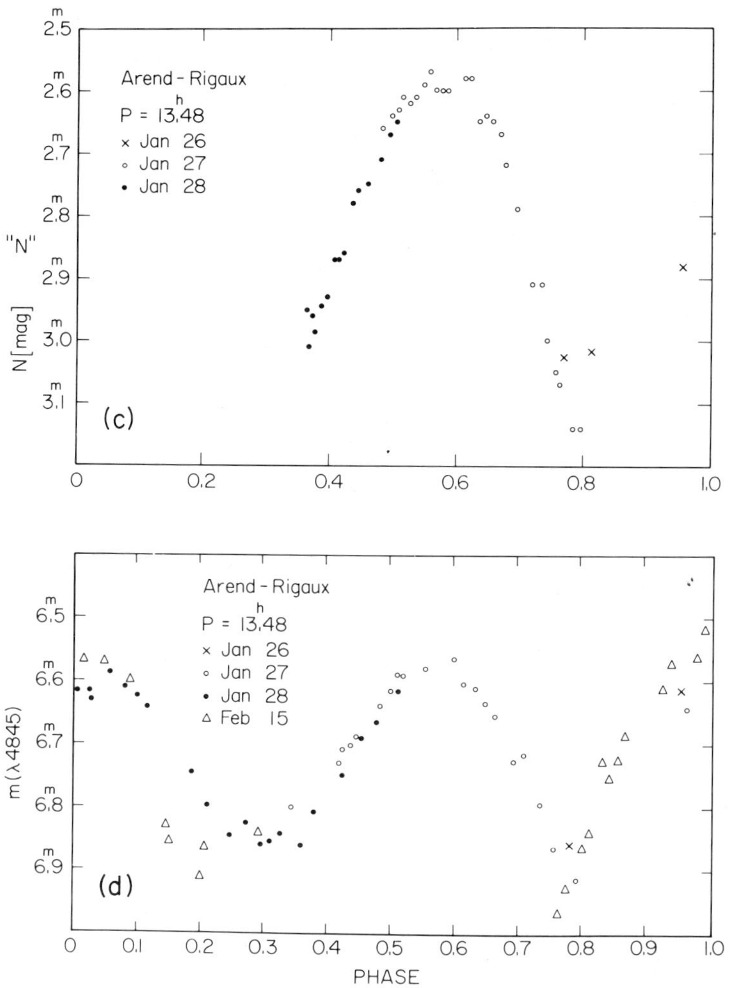

Figure 4. Light curves of the nuclear region of Comet Arend-Rigaux in the thermal infrared and the optical. Data taken on same nights using 88" telescope and Infrared Telescope Facility at Mauna Kea Observatory. From Millis et al. 1986.

studied the photometric variations of comets in which it was thought possible to see the nucleus. Photometric variations have been reported for Comet Arend-Rigaux by all groups and all agree that the period is a multiple of 6 3/4 hours (the choice of which multiple is in dispute due to lack of coordinated observations at different longitudes). A substantive question arises regarding the interpretation: is the variation due to variations in cross-section or in albedo? In the case of Comet Arend-Rigaux this was unambiguously resolved by measuring the variation simultaneously, with two different telescopes, in the reflected optical radiation and in the thermal infrared radiation (Millis et al., 1986). Both light curves were perfectly in phase, as shown in Figure 4, thus requiring that the variation be due to changes in cross-section and enabling us to determine the size and albedo of the nucleus. This type of coordination, among telescopes doing different types of observations, is in fact likely to lead to some of the most exciting scientific results on comets in the future.

3. PLANS FOR COMET HALLEY

Due at least in part to a concern for problems like those cited above, NASA is supporting the International Halley Watch (IHW) to coordinate observations of Halley's comet at this apparition. The three official purposes of this organization are advocacy, coordination, and archiving. The pertinent point for us is that the IHW is explicitly attempting to encourage observers at many different sites to coordinate their observations. Because Halley's is the only bright, periodic comet, it presents an ideal case for coordination. One aspect of the coordination has been to encourage many observers who normally observe galaxies or nebulae or stars to turn their attention to Halley's comet. Success in this will immediately help the problem of finding observers at a variety of different longitudes. This is not enough, however, since the coordination also requires that they be observing on the same nights. In order to encourage this, a series of Halley-Watch days has been identified as the prime times for observations. These days were selected to be at favorable lunar phases, to cover the "interesting" times in the comet's development, and also to maximize the coordination between ground-based observations and the spacecraft that will fly through the coma.

The primary contact with observers is through Discipline Specialists, experts in each of eight observational techniques who are charged with encouraging observers to use those techniques, providing all the relevant information on standardization, providing up to date information to the observers, and ultimately collecting all the observational data so that a complete archive of all data on Halley can be published after the apparition, presumably in 1989. This archive, incidentally, will contain only data, not interpretations; it is assumed that the results and their interpretation will be published in the usual manner by the individual investigators as articles in the astronomical journals. Naturally all of the discipline specialists are experienced cometary observers and will be observing Halley's comet as part of their own research programs and are encouraging collaborative analyses where these are appropriate.

Figure 5 shows as an example the locations where observers are expected to participate in the net carrying out narrow-band photometry. Obviously many observers are interested in making these observations but even these many turn out to be too few for some of the interesting problems. In the first place, bad weather will eliminate a significant fraction of the observers on any given night. More importantly,

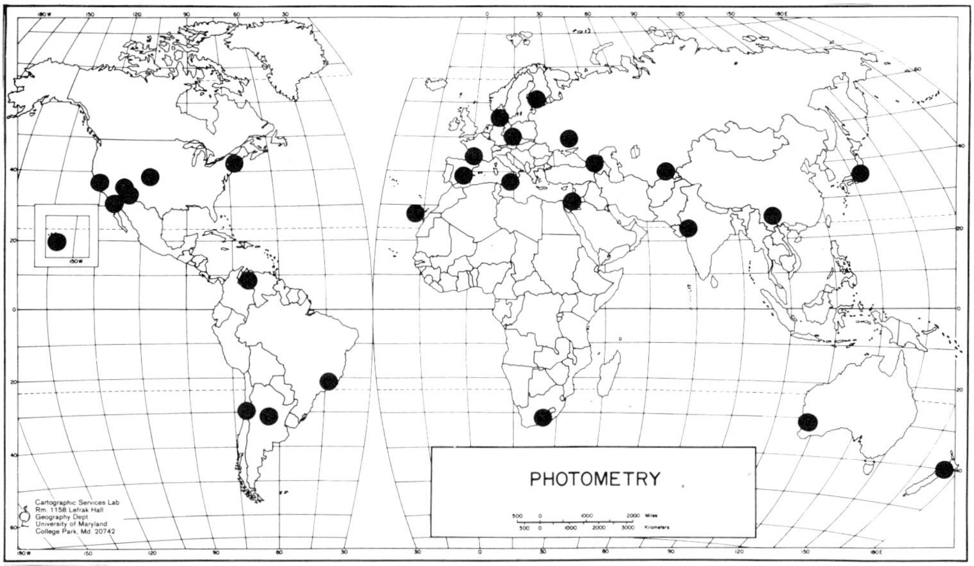

Figure 5. Map showing location of observers planning narrow-band photometry of Comet Halley in 1985-86.

however, Halley's comet at this apparition is observable almost exclusively from the southern hemisphere after perihelion (Halley's comet is always intrinsically brighter after perihelion than before) and, unfortunately for our purposes, much of the southern hemisphere is covered by oceans. Figure 6 shows the hours of the day during which the comet is observable by any member of the Large Scale Phenomena Net in December and during the week in March when several spacecraft will fly through the coma. It is clear that, even though many observers are participating, there are many hours of the day during March when none of the likely observers wil be able to see the comet. By contrast, the comet is observable from five or more participating observatories at all times in early December. These gaps, particularly in the south, severely limit our ability to study rapidly changing phenomena such as disconnection events in the plasma tail. Similar problems exist for all nets where rapidly changing phenomena are of interest.

As implied by the discussion of previous coordinated observations of comets, the small telescopes being discussed in this symposium will be used primarily for observations relevant to the Astrometry, Large Scale Phenomena, Near Nucleus Studies, and Photometry and Polarimetry Nets. The Infrared Net and the Spectroscopy and Spectrophotometry Net will use primarily larger telescopes while the Radio and Meteor Nets will generally use facilities totally outside the scope of this symposium. To illustrate the types of assistance which can be provided by the discipline specialists, I will just list some ways in which each of these four nets is attempting to provide assistance to its members. The Astrometry net has developed a catalog of reference stars with accurately known positions along the expected track of the comet, has distributed to a significant number of observers a precise, two-body

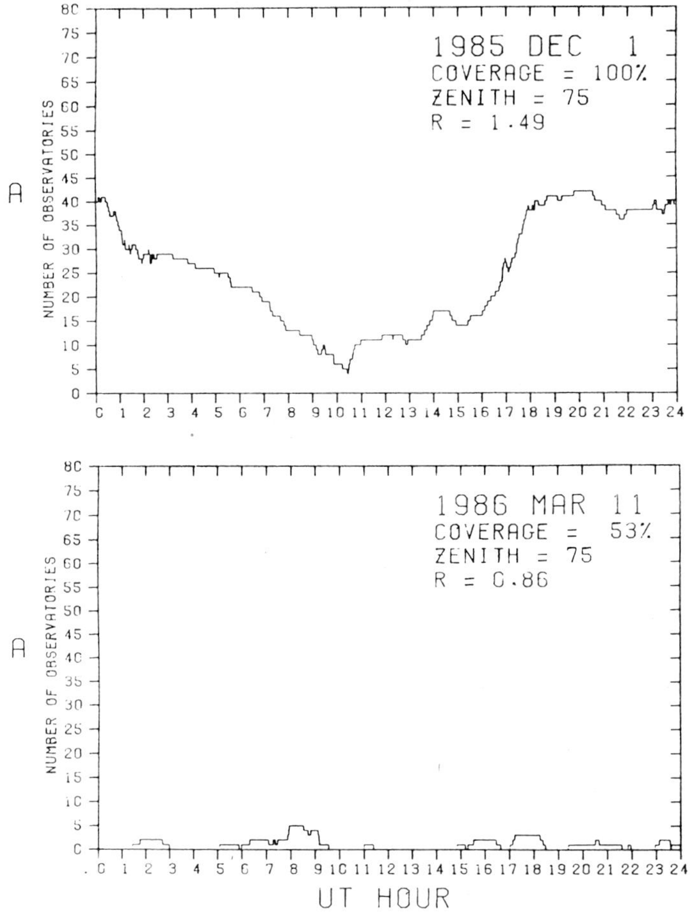

Figure 6. Number of observatories in Large Scale Phenomena Net which can observe Comet Halley (altitude > 15°, beyond twilight) as function of UT. Comet can be observed continuously in early December but not in March even if there are no clouds. Diagram compliments of M. Niedner.

ephemeris program, and regularly provides current orbital elements and ephemerides to all who want these data. The Large Scale Phenomena Net has concentrated on developing methods for photometrically calibrating plates taken with telescopes having large fields of view, typically but not exclusively Schmidt telescopes. The Near Nucleus Studies Net is first of all encouraging the transition from photographic techniques to CCD detectors. While the comet is bright, say from October 1985 to May 1986, virtually all the imaging can be done from telescopes less than one meter in size using a CCD with suitable transfer optics. Larson and

I, for example, each have a CCD system being used nearly full time on 20- and 24-inch telescopes respectively. This use of small telescopes considerably eases the problem of obtaining telescope time but more importantly gives us a larger field of view. This net has also coordinated the purchase and distribution of more than a dozen sets of narrow-band, image-quality interference filters. The Photometry and Polarimetry net has developed a complete photometric system using narow-band filters to isolate the emission bands and continuum of comets and distributed 75 sets of filters around the world. A set of standard stars has been chosen and an interim set of standard magnitudes is being presented as a separate, contributed paper at this meeting. Calibration of the standard magitudes into absolute fluxes is currently in progress and will also be distributed to the members of the net. Although most of the early photometric observations of Halley's comet were from large telescopes, some were obtained with a CCD on a 1.0 meter telescope and virtually all the observations when the comet is closer than 1.5 AU will be done with small telescopes.

As mentioned above, one of the key goals is to have simultaneous data from different nets. One example of such a planned program is the simultaneous use of CCD cameras and photometers on adjacent telescopes at the same observatory. We are, for example, testing whether it will be possible to completely replace photometers with CCD cameras at some time in the future. Because the calibration problems are so different, particularly when narrow-band filters are used with CCDs, it is not obvious _a priori_ how rapidly this can be achieved. Perhaps more importantly, simultaneous use of photometers and CCDs relieves one of the observers, presumably the one using the CCD, of the need to measure extinction coefficients each night. This is a particularly important point when the comet is very close to the sun (January and March) and only a short time is available for observations at large air mass. In another area, having simultaneous ultraviolet, radio, photometric, and spectroscopic data is critical for understanding the chemical processes in the cometary coma since the different techniques are typically used to derive abundances of different but possibly related species. A typical example would be the comparison of optically deduced CN abundances with the abundances of HCN deduced from radio observations as was done for Comet IRAS-Araki-Alcock (Irvine et al. 1984). In yet another area, astrometric and photometric data are needed simultaneously in order to understand occultations of stars by the cometary coma from which we can directly deduce the optical depth and albedo of the dust particles. These results will be even further enhanced if simultaneous infrared data are available.

At this time it is still premature to discuss the results of most of the programs on Halley. Nevertheless we can point out that all of the groups that observed Halley in 1983 and 1984 reported variability. Several groups tried inconclusively to establish periods based on their own data but a combination of the data from various groups showed that all the deduced periods were incorrect and that most of the variation was not periodic. It was only through this combination of data from several observers that it was possible to show the non-periodic nature of the variation.

4. COMETS AFTER HALLEY

Although Comet Halley is the only periodic comet routinely visible to the unaided eye, there are many other comets which can be studied with small telescopes. In fact, when a bright comet is unexpectedly discovered, it is often the case that only a small telescope can be made available rapidly enough to be useful. In the case of narrow-band photometry, there is an average of at least one comet observable every year, some periodic and others newly discovered. The coordination among different telescopes is much harder to arrange for a newly discovered comet but it is for precisely these comets that the most dramatic gains will be achieved by coordinated observations. Because comets like this are so bright, small telescopes can be used for many purposes but the coordination is most important.

5. REFERENCES

A'Hearn, M.F., P.V. Birch, P.D. Feldman, and R.L. Millis, 1985, 'Comet Encke: OH Production and Light Curve.' Icarus, **64**, 1-10.

Irvine, W.M. et al., 1984, 'Radio-Astronomical Observations of Comets IRAS-Araki-Alcock (1983d) and Sugano-Saigusa-Fujikawa (1983e).' Icarus, **60**, 215-220.

Larson, S.M. and Z. Sekanina, 1985, 'Coma Morphology and Dust-Emission Pattern of Periodic Comet Halley. III. Additional High Resolution Images Taken in 1910.' Astron. J., **90** 823-826.

Mendis, D.A., H.L. Houpis, and M.L. Marconi, 1985, 'The Physics of Comets', Fund. Cosmic Phys. **10**, 1-380.

Millis, R. L., M. F. A'Hearn and H. Campins, 1986, 'The Nucleus of Comet Arend-Rigaux' in preparation.

Niedner, M.B. and J.C. Brandt, 1978, 'Interplanetary Gas XXIII. Plasma Tail Disconnection Events in Comets: Evidence for Magnetic Field Line Reconnection at Interplanetary Sector Boundaries', Ap.J. **223**, 655-670.

Sekanina, Z. 1986, 'Effects of the Law for Nongravitational Forces on the Precession Model of Comet Encke', Astron. J., in press.

Whipple, F.L. and Z. Sekanina, 1979, 'Comet Encke: Precession of the Spin Axis, Nongravitational Motion, and Sublimation.' Astron. J. **84**, 1894-1907.

DISCUSSION

Moffet: What measurements will the various space probes be making on Halley's Comet?

A'Hearn: 1) Japanese probe (Planet A renamed Suisei) : Lyman-α camera and a solar wind monitor.
2) ESA (Giotto) and Russian (Vega 1 and 2) probes : neutral-and ion-mass spectrometers, dust impact counter, dust vaporisation mass spectrometer, plasma analyser, plasma wave measurer and a few others.

Hearnshaw: We have a 1 m telescope, which may be ready for the comet, and an échelle spectrograph. What wavelength intervals or features should we observe? At what resolution?

A'Hearn: One needs to resolve the rotational structure of the individual bands ($\Delta\lambda \sim 0.5$Å)

- CN λ 3883Å
- CO^+ λ 4260Å (looking at the acceleration of the ions)
- NH λ 3365Å
- C_2 bands are hard to observe.

Phillips: One shouldn't get fanatical about small telescopes. There are programs (on the comet) which can only be done with large telescopes.

Jacoby: Could you elaborate on the difficulty of doing photometry with a CCD as opposed to a photomultiplier?

A'Hearn: It's not more difficult but you have to be sure you do your extinction corrections with the CCD and not get carried away with the imaging.

RESEARCH PROGRAMMES, PRODUCTIVITY, AND USAGE STATISTICS FOR THE 0.9M TELESCOPES AT KITT PEAK

G. H. Jacoby
NOAO, Kitt Peak National Observatory
P.O. Box 26732
Tucson, AZ 85726

ABSTRACT: Current research programmes using the smaller telescopes at Kitt Peak include spectroscopy, photometry, and imaging of clusters, variable stars, nebulae, and extragalactic objects. Their scientific productivity is competitive with the 4m and 2.1m telescopes. Subscription rates indicate that when advanced instrumentation is made available on a small telescope, the demand will increase significantly.

1. THE KITT PEAK TELESCOPES

There are currently 7 primary telescopes for night-time use at KPNO, with apertures of 4, 2.1, 1.3, (two) 0.9, 0.6 meters and Coude feed. The two 0.9m telescopes have an instrument complement consisting of a CCD camera as the principle instrument at the #1-0.9m, and the Intensified Reticon Spectrograph (IRS) at the #2-0.9m.

The most significant attractions offered by the 0.9m telescopes are (1) modern instrumentation, (2) relatively generous time allocations, (3) efficient operation with reduced complexity, (4) a lower pressure environment, and (5) reduced scale to image larger fields or utilize large apertures and long slits in spectrophotometry.

Average dark-time oversubscription rates for the KPNO telescopes are given in Table I.

TABLE I

Telescope	4m	2.1m	1.3m	#1-0.9m	#2-0.9m	CF
1978 - 1985	3.0	2.1	1.4	1.9	1.4	1.4
1983 - 1985	2.8	1.9	1.5	2.1	1.5	1.5

Statistically there has been no real change in the subscription rate over the long term. There are smaller scale changes as shown in Figure 1 for the #1-0.9m telescope. This figure charts the demand for the telescope relative to the 4m thereby reducing seasonal effects. The demand has varied by a factor of nearly three, due in part to the availability of new instumentation.

J. B. Hearnshaw and P. L. Cottrell (eds.), Instrumentation and Research Programmes for Small Telescopes, 237–239.

The presence of the IRS had little effect, but the impact of the CCD direct camera was obvious. Two semesters after availability, the subscription rate exceeded that of the 4m; apparently projects previously requiring 2.1m and 4m telescopes were now feasible at the #1-0.9m. The high demand prompted transfer of the IRS to the #2-0.9m so that the CCD could be simultaneously scheduled. The subscription rate dropped thereafter, and was reduced further when the CCD camera became available at the 2.1m.

2. PRODUCTIVITY

Abt (1980, P.A.S.P., **92**, 249) showed that the scientific paper rate was roughly proportional to telescope aperture using data for the KPNO telescopes from 2.1m and smaller. Figure 2 shows the result for papers published during the past 2 years. The nearly linear trend is still evident. Abt explains the relationship as a result of large telescope usage directed toward fainter objects, and that the number of observer teams follows a roughly linear aperture relationship. One must also consider that because the larger telescopes are more heavily subscribed, allocation committees are more selective toward productive investigators.

Although, paper rates may not be a particularly good indicator of scientific productivity, current data support the trends found by Abt which suggest that small telescopes are scientifically cost effective.

3. RESEARCH PROGRAMMES

KPNO receives roughly 140 proposals for the two 0.9m telescopes each year. About half propose extragalactic studies. The major areas of recent research are optical properties of radio galaxies, surface photometry of galaxies, active galactic nuclei, populations in nearby galaxies, studies of variable stars, active chromospheres of stars, star formation and young stellar objects, lunar occultations, and recently, comet investigations.

One effect of heavy CCD and IRS scheduling has been a reduction in classical photometry. We are studying ways to use CCD cameras as photometers to relieve the demand for instrument changes, and to provide a much fainter limit resulting from the greater sensitivity and the areal properties of CCDs.

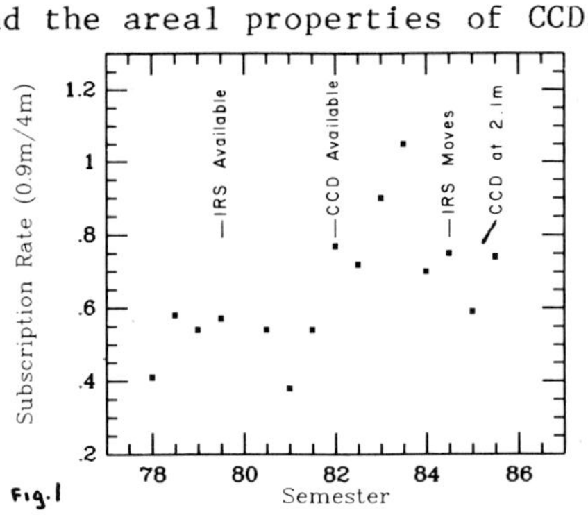

Fig.1

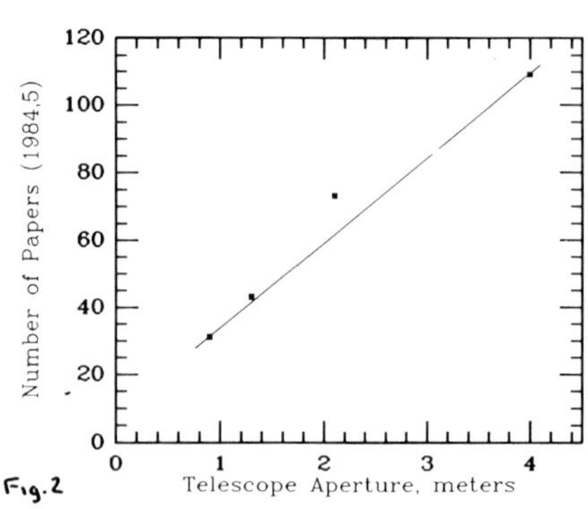

Fig.2

DISCUSSION

Kumar: Can you comment on the fringing effects with CCDs and narrow band filters?

Jacoby: We find the TI chips relatively immune, the RCA chips have some fringing effects which in many cases can be removed by flat fields. Problems arise when two strong emission lines fall within one bandpass, e.g. NII and Hα.

Warner: It is not yet possible to phase out photomultipliers completely (in favour of CCDs) because of the continuing need for high time resolution photometry. CCDs are limited by their read-out time (~1 second).

Jacoby: To a large entent this is true but we have been experimenting with scanning the CCD in rapid mode to get a resolution of ~ 1 second.

A PROGRAM OF NEW ACCURATE PHOTOMETRY FOR BRIGHT STARS

Dr. David L. Crawford
Kitt Peak National Observatory
National Optical Astronomy Observatories
P. O. Box 26732
Tucson, AZ 85726 USA

ABSTRACT. A new program of accurate photoelectric photometry for the brightest stars is getting underway, using the excellent small telescopes and talents of amateur astronomers.

The extensive use of new detectors and pulse counting electronics has produced an interesting fact that accurate photometric data on well-known standard systems is lacking for the brightest stars. The incoming flux from these stars overloads existing instrumentation for even relatively small telescopes. Yet these stars are often the ones for which the best astrophysical information is known, such as chemical abundances, rotation, surface magnetic fields, and so forth. It is important that very accurate photometric data should exist on all the standard photometric systems for these stars, so that the brightest stars with this detailed astrophysical information can be well linked to fainter stars via intermediate and wide-band photometry.

A number of amateur astronomers with first-class observing facilities and I have begun a program to correct this deficiency. We expect, over the next few years, to produce new, accurate, and homogeneous data for the brightest stars ($V < 5$ mag or so) on several of the most used photometric systems. Observing has begun at Braeside Observatory in Flagstaff, AZ, by Bob Fried, with its telescope diaphragmed down to approximately 3-inch aperture for some of the work. A second telescope (APTS at Mt. Hopkins) will begin operation, part time, on this program early in 1986. Others are welcome, but the quality of the telescope and photometer must be excellent, and the data reduction of the highest quality, with techniques and coordination adequate to guarantee homogeneous data.

The main goal is to have extensive data for these important and well known stars tied into the existing data for fainter stars on the same systems. Very useful by-products will be the long-term monitoring of these stars for variability, and analysis of the stability of the photometric systems and the sites in use.

J. B. Hearnshaw and P. L. Cottrell (eds.), Instrumentation and Research Programmes for Small Telescopes, 241–242.

DISCUSSION

Kurtz: I don't believe that using small apertures and neutral density filters are equivalent because of scintillation problems.

Crawford: I agree, and that's why we do both.

Gray: Will you rotate your aperture plate when it is used so as to sample a range of your optics, and thereby make the small aperture measurements more compatible with your full 16 inch aperture ones?

Crawford: I think not. On occasions we will run tests with the aperture plate in different positions.

SURFACE PHOTOMETRY SURVEYS OF GALAXIES WITH SMALL TELESCOPES

S. Djorgovski
Harvard-Smithsonian Center for Astrophysics
60 Garden St.
Cambridge, MA 02138
U.S.A.

ABSTRACT. Surface photometry of relatively bright galaxies is now becoming almost an exclusive province of small telescopes equipped with CCDs. I give a brief review of the scientific possibilities and problems of interest in this field. Small telescopes become tools of observational cosmology: they can provide large amounts of data necessary for the further understanding of fundamental properties of galaxies, clues to their formation, and their interactions with environment.

It is the high quantum efficiency of CCDs that granted to small telescopes this new power. Availability of large amounts of small telescope time makes it possible to do surveys of large numbers of objects, producing homogeneous, good-quality data bases, never available before, and of the quality equal or better than any of the previous, large-telescope photographic work. Small telescopes equipped with CCD cameras exist at KPNO, CTIO, and ESO, and many other places, and thus should be in principle available to anybody with a good scientific project. Appropriate reduction facilities exist at all observatories which operate this kind of equipment, and most major astronomical institutions. Examples of modern surface photometry surveys can be found in papers by Kent (1984, 1985) (105 galaxies), Lauer (1985b) (42 galaxies), and Djorgovski (1985) (262 galaxies). These papers, and references therein, also address the methods of modern surface photometry of galaxies. I will list briefly the possible useful scientific products of galaxy surface photometry. They roughly come in three groups:

First, there are "simple", or catalog-type measurements, such as magnitudes, diameters, colors, mean ellipticities and position angles, characteristic densities (a.k.a. mean surface brightness), and morphological classifications. Everybody who had to do anything in extragalactic astronomy knows how poorly determined such quantities are for all but the few brightest galaxies. The situation is particularly dismal with the magnitudes and diameters of galaxies, which are more often than not just simple eye estimates. There is much space for improvement here: diameters (and consequently standard apertures and

J. B. Hearnshaw and P. L. Cottrell (eds.), Instrumentation and Research Programmes for Small Telescopes, 243–245.

magnitudes) should be determined in a well-defined, quantitative manner from the surface brightness profiles.

In the second group are the measurements of quantities which can *only* be obtained from surface photometry. Radial density profiles of luminous material: it is now clear that the elliptical galaxies are *not* well described by any of the "standard" formulas. It is not clear how close to exponential are the disks galaxies, either. There is a great variety of surface brightness profile shapes and intensity scales, and quantifying and explaining this variety is the first step toward the better understanding of galaxies in general. Related to this are bulge-to-disk decompositions and measurements of bulge properties; see the paper by Kent (1985). This is necessary for dynamical modelling of disk galaxies, and relates directly to the problems of galaxy formation: are the bulges and the disks coeval? Are the bulges a direct continuation of the family of ellipticals? Ellipticity gradients and isophotal twists in elliptical galaxies carry direct information on galaxy shapes, not deducible in any other way; they may be used to constrain the velocity anisotropy in galaxies. Structure and kinematics of the inner parts of elliptical galaxies and bulges are "fossils" of galaxy formation. Core properties of galaxies are currently not well known. The quantities to be measured include the central surface brightness core radii, and central surface brightness slopes. Derived quantities, such as central luminosity densities, are also of considerable interest. The central surface brightness slopes are directly related to the velocity anisotropy; Lauer (1985a,b) has found that many (most?) ellipticals have non-isothermal cores. The dynamical structure of cores is also important for our understanding of the origin and fueling of nuclear acitivity in galaxies. Color gradients carry direct information on stellar populations and their distribution, and may provide direct constraints for the M/L ratios as functions of radius, and the distributions of dark matter in galaxies.

A quantitative parametrisation of morphological and kinematical properties of galaxies should open the possibility of reliable multivariate analysis of galaxies. Luminosity (or mass) is certainly one of the principal parameters, but what are the others? The number and identity of such quantities may give us some direct clues on the recipe for making galaxies, and some record of when and how that process happened. We may then be able to design better distance-independent luminosity indicators. For example, there is too much scatter in luminosity - velocity dispersion - metallicity relations for early-type galaxies to be accounted by the error bars alone. Reducing this scatter should lead us to an improvement in the mapping of the Local Supercluster velocity field, and perhaps a better measurement of the far-field Hubble constant.

Finally, there is a third, miscellaneous group of applications, drawing mainly on the *imaging*, and not necessarily on the photometry. One example is a search for subsystems in galaxies, e.g., dust lanes and stellar disks in ellipticals, bars and dynamical resonances in disks, etc.; see the paper by Djorgovski and Ebneter elsewhere in this volume. There is also a possibility of discovering new active and/or unusual nuclei. Repeated observations of a large number of galaxies should constrain the supernova rates. Tidal distortions in galaxy pairs provide a way to select a "clean" sample of physical binary galaxies.

REFERENCES

Djorgovski, S. 1985, *Ph.D. Thesis*, University of California at Berkeley.
Kent, S. 1984, *Astrophys. J. Suppl. Ser.* **56**, 105.
Kent, S. 1985, *Astrophys. J. Suppl. Ser.* **59**, 115.
Lauer, T.R. 1985a, *Astrophys. J. Suppl. Ser.* **57**, 473.
Lauer, T.R. 1985b, *Astrophys. J.* **292**, 104.

DISCUSSION

Garrison: I'm interested in the different surface photometry results for you E galaxies. Are these CD's and E's mixed together and do you see systematic results? Is there a luminosity dependence?

Djorgovski: There are only a couple of small galaxies in my sample. So far, there is a small indication that the shapes of surface brightness profiles correlate with luminosity.

DESIDERATA FOR NEARBY-STAR QUANTITIES OBSERVABLE WITH SMALL TELESCOPES

W. Gliese
Astronomisches Rechen-Institut
Mönchhofstrasse 12-14
D-6900 Heidelberg 1
F.R. Germany

At the General Assembly of the IAU at New Delhi van Altena reported on the new General Catalogue of Trigonometric Stellar Parallaxes (GCTSP) which was completed recently at the Yale Observatory. Time seems to be ripe now for the compilation of a Third Catalogue of Nearby Stars which will include all objects known to be nearer than 25 parsecs. This catalogue will contain positions, proper motions, radial velocities, spectral types, broad-band photometry, and parallaxes - quantities won by observations.

A preliminary version of the new GCTSP, available in 1984, yielded 1957 stars with parallaxes exceeding 0".0395 (singles and systems with altogether nearly 2 400 components). However on the basis of the luminosity function derived by Wielen et al. (1983) the total number of stars in a 25-pc sphere around the Sun is estimated at more than 7 000 objects.

Upgren and Armandroff (1981) have shown that our nearby star sample is statistically complete for the objects absolutely brighter than $M_V = +8$. Beyond this limit incompleteness of our knowledge is rapidly increasing and we cannot hope that the situation will be remarkably improved by trigonometric programs alone in the near future. Significant progress can be expected from observations of second-order distance indicators as spectral types, luminosity-sensitive quantities, colours. Today we know already more than 700 objects with spectroscopically and/or photometrically determined distance moduli $m-M < +2.02$.

Some programs observed with small telescopes which have contributed remarkably to our knowledge of objects in the solar neighbourhood will demonstrate potentialities, value and benefit of such series.

To overcome the "high-proper-motion selection effect" in trigonometric parallax programs Vyssotsky et al. (1943, 1946, 1952, 1956) searched for K and M dwarf stars on the collection of objective prism plates at McCormick Observatory. The supplement in southern declinations was observed with the Michigan Curtis Schmidt Telescope at Cerro Tololo (Upgren et al., 1972). Both series are restricted to stars brighter than 10th or 11th mag. Up to now the potential of the data in the Michigan Spectral Catalogues (Houk, 1975, 1978, 1982) is not yet fully analysed.

A certain weight and importance has to be assigned to objective prism

J. B. Hearnshaw and P. L. Cottrell (eds.), Instrumentation and Research Programmes for Small Telescopes, 247–249.

programs for red stars down to $m_V = 13$. Several series were observed with the Schmidt telescope (61/91 cm) at Warner and Swasey Observatory. The spectral classes of the individual objects seem to be uncertain but in combination with their proper motions the discrimination between dwarfs and giants is possible in many cases. Dwarf stars and suspected dwarf stars can be investigated later on by photometric programs.

In recent years the importance of photometric parallaxes for distance determinations in the solar neighbourhood has grown steadily. Broad-band photometry (UBV, RI) is observable with small telescopes also for stars fainter than 10th magnitude.

Eggen has found numerous nearby objects down to V = 15 from photometric observations with a 40-inch reflector.

Large and valuable sources for further detections are the proper motion catalogues of Luyten. The four NLTT volumes (Luyten, 1979 - 1980) contain probably about eighty percent of the stars existing within 25 pc of the Sun. Series with a 0.9-m telescope at Kitt Peak (Weis, 1984) and with the 70-cm Swiss telescope at La Silla (Grenon, 1984) are in full activity now, but further programs with small telescopes will find many more NLTT objects for observation.

Reliable radial velocity measurements with small telescopes seem to be possible only for objects of $m_V \leq 10$. However Griffin (1971 and following years) showed how spectroscopic binaries can be observed with a 36-inch reflector with photoelectric equipment.

The programs above were mentioned as examples how observations with small telescopes may support the compilation of data of objects in the solar neighbourhood. Further catalogues of nearby stars can be supplemented with new data and additional objects especially in the magnitude range around $m_V = 10$ observed with small telescopes.

REFERENCES

Grenon, M. 1984, private communication.
Griffin, R.F. 1971, *Mon. Not. R. Astr. Soc.* 155, 1.
Houk, N. and Cowley, A.P. 1975, *Univ. of Michigan Catalogue of Two-dimensional Spectral Types for the HD Stars, -90° to -53°0.* Ann Arbor, Michigan.
Houk, N. 1978, vol. 2, *-53°0 to -40°0;* 1982 vol. 3, *-40°0 to -26°0.*
Luyten, W.J. 1979 - 1980, *NLTT Catalogue,* vols. I-IV. Univ. of Minnesota, Minneapolis, Minnesota.
Upgren, A.R. and Armandroff, T.E. 1981, *Astron. J.* 86, 1898.
Upgren, A.R., Grossenbacher, R., Penhallow, W.S., MacConnell, D.J. and Frye, R.L. 1972, *Astron. J.* 77, 486.
Vyssotsky, A.N. 1943, *Astrophys. J.* 97, 381.
Vyssotsky, A.N., Janssen, E.M., Miller, W.J. and Walther, M.E. 1946, *Astrophys. J.* 104, 234.
Vyssotsky, A.N. and Mateer, B.A. 1952, *Astrophys. J.* 116, 117.
Vyssotsky, A.N. 1956, *Astron. J.* 61, 201.
Weis, E.W. 1984, *Astrophys. J. Suppl. Ser.* 55,289.
Wielen, R., Jahreiss, H. and Krüger, R. 1983, *IAU Colloquium No.76, The Nearby Stars and the Stellar Luminosity Function,* A.G.D. Philip and A.R. Upgren eds., L. Davis Press, Schenectady, N.Y., p.163.

DISCUSSION

McCarthy: The Schmidt telescopes are migrating around the world. The one to which you refer is the Burrell Schmidt (used by Stephenson and Sanduleak) in Cleveland, now at Kitt Peak, whereas the Curtis Schmidt was at Ann Arbor and is now at Cerro Tololo.

Penhallow: Most of the stars you mentioned had their parallaxes determined by small telescopes. What comment do you have about this?

Gliese: This is true, most telescopes for trigonometric parallaxes have apertures less than one metre.

Djorgovski: What do you think about the hypothesis that the Sun has a companion star which killed the dinosaurs?

Gliese: It may in some way have been confirmed by observation and it is now a matter of believing it or not.

ASTEROSEISMOLOGY WITH SMALL TELESCOPES

D.W. Kurtz
Department of Astronomy
University of Cape Town
Rondebosch 7700, South Africa

ABSTRACT. Study of the solar interior through analysis of solar oscillations, a field now referred to as *helioseismology*, has generated interest in the development of *asteroseismology*, the study of the interior of pulsating stars through the analysis of their oscillations. Some progress has been made in the study of the Rapidly Oscillating Ap stars, cool magnetic Ap stars which pulsate in high overtone, low degree non-radial p-modes with periods between 4 and 15 minutes. Since all of these stars found so far are bright ($V < 10$), their light variations can be studied with small telescopes; most of the work done has been with a 0.5-m telescope. Their light variations are extremely complex, however, which means that they must be intensively observed, something which can only be done with small telescopes. Due to aliasing problems, multisite observing programs of the Rapidly Oscillating Ap stars are often necessary in order to decipher their frequency patterns. A collaborative observing program is suggested.

Daily aliasing is a problem in observing all of the Rapidly Oscillating Ap stars so that collaborative observing projects from more than one observatory are always desirable. For some of them, it is mandatory. HR 1217, for example, oscillates in at least five independent modes which just happen to be separated by about 3 day^{-1} (Kurtz 1982, Kurtz & Seeman 1984, Kurtz, Schneider & Weiss 1985). Each of the frequencies associated with these modes is amplitude modulated with the rotation period of HR 1217, 12.4564 day. According to the oblique pulsator model, each of these pulsation modes will be split into a triplet (for an $\ell = 1$ mode), a quintuplet (for an $\ell = 2$ mode), a septuplet (for an $\ell = 3$ mode), etc. In order to disentangle all of these frequencies in the amplitude spectrum of HR 1217 and determine the spherical harmonic degree of the oscillations, it will be necessary to obtain observations over two consecutive rotation cycles from at least two, and preferably several, observatories well spaced in longitude. HR 1217 is at $\alpha = 4$ hr, $\delta = -12°$ which means that it transits at midnight in late November and can be observed from both northern and southern hemispheres, although southern hemisphere observations will clearly be more easily made.

J. B. Hearnshaw and P. L. Cottrell (eds.), Instrumentation and Research Programmes for Small Telescopes, 251–253.

I plan to observe this star from Sutherland during the period from 11 November to 8 December 1986 and would like collaboration from as many other observers as possible. In particular, observations from Chile and Australia and/or New Zealand are particularly desirable. Because of the very short southern hemisphere nights (7 hr) at that time of year, and because of the possibility of cloudy weather, observations from all other observatories at all longitudes are desirable. The 60-cm telescope on Mauna Kea is particularly well situated for this project as it is at a longitude of only 20°N and is nearly antipodal to South Africa. Observations from Kitt Peak, McDonald, the Canary Islands, and Israel would also be most useful. All of these observatories have small telescopes for which observing runs of two to four weeks are possible.

All observations of HR 1217 will be made using continuous 20-s integrations through a Johnson B filter with pulse counting photometers. Interruptions for sky observations should be made as often as necessary; I observe the sky background often enough that the change from one sky observation to the next is never more than 0.1 to 0.2% of the count rate on HR 1217. Since HR 1217 has V = 6, this is not much of a problem even during full moon. Observations should always be made with diaphrams of 30 to 40 arcsec. Because of telescope tracking errors, vibrations in the telescope due to wind, seeing, and scintillation, the diffraction spikes and rings (which the observer cannot see in the telescope eyepiece) may spill off the edge of the diaphragm with smaller diaphragms. Photometry of the accuracy shown in the light curves in this paper is difficult or impossible to obtain with the use of small diaphragms. Furthermore, for bright stars like HR 1217 there is no need whatsoever for using a small diaphragm; sky background is not a major problem.

REFERENCES

Kurtz, D.W., 1982. *Mon. Not. R. astr. Soc.*, **200**, 807.

Kurtz, D.W., Schneider, H., & Weiss, W.W., 1985. *Mon. Not. R. astr. Soc.*, **205**, 11.

Kurtz, D.W., & Seeman, J., 1983. *Mon. Not. R. astr. Soc.*, **205**, 11.

DISCUSSION

Kumar: Myron Smith at KPNO has looked at line profile variations for B stars. Has anyone done that for this type of project?

Kurtz: Various people have tried but without success. Typically the amplitudes are a few thousands of a magnitude (maybe a factor of ten less). The profile variations are going to be very subtle. That job needs to be done with a big telescope.

Djorgovski: CCDs are, of course, redundant for high-time-resolution photometry. A good way to do it would be with a Runicon or some other microchannel plate imaging detector. Then you can save only the parts of the frame which are of interest: the main target star, and the comparisons. Then you can do relative photometry, PSF fitting, and improve on accuracy.

Kurtz: I agree, that's the way in the future.

Finkenzeller: Could you comment on which filters you used, and how the filter band-width would affect your signal strength and signal contrast?

Kurtz: The observations were taken in the Johnson B filter. I have looked at these objects in U, B and V. The amplitude in U and B is approximately the same but it is substantially down in V. The important thing is that there is a phase shift U to B to V (in the most extreme case it is 60°). You would get better observations with narrow band filters but the universality of Johnson B enables intercomparison of results from different observations. One at least wants a filter bandpass centered near that of B.

Sterken: If you look for monitoring from different sites, would it not be wiser to use a filter with a narrower bandpass (for example Strömgren b or v) than Johnson B?

Kurtz: Some problems with these : 1) b is ~ 300Å redward of B and amplitude is dropping very fast, 2) for fainter stars and small telescopes one will run into photon problems and 3) v is contaminated by the strong (and variable) Sr II 4077Å line which is stronger than Hδ (4101Å).

Sterken: Maybe some specialized narrow-band filter should be produced and distributed to observers (or observatories) in South Africa, Chile, Australia and New Zealand.

MONITORING OF VARIABLE STARS ON A LONG TIME-BASELINE

C. Sterken
Astrophysical Institute
University of Brussels
Pleinlaan 2
B 1050 Brussels, Belgium

ABSTRACT. The functioning of a project for systematic observations of variable stars on a long time scale is described.

Today photometric monitoring of variable stars during a time span of several years is almost impossible for observers who collect their data at the large visitor-operated observatories. The study of variable stars with periods of the order of many years for example can only be accomplished by the steady observation of these stars during several decades, and there is no way for a single observer or investigator to successfully undertake such an enterprise. Therefore joint research in a dedicated team seems to be fundamental.

In 1982 a dozen of active variable star observers belonging to the majority of the European member-countries of the European Southern Observatory started an international long-term collaboration with the intention to study a number of selected variables of different type over a timespan of at least one decade. Individual observing programmes were merged, and the most interesting stars were grouped into several separate research topics. For each field of research, one or two principal investigators were appointed. These persons are responsible for the coordination of the work within their own section.

Table I lists all sections, and the corresponding principal investigators.

TABLE I. Principal Investigators for each section

1.	Pre-main sequence stars	P.S. The, H. Tjin a Dijie
2.	Ap stars	H. Hensberge, J. Manfroid
3.	Eclipsing Binaries	H.W. Duerbeck, A. Bruch
4.	Be stars	D. Baade, P. Harmanec
5.	Supergiants	B. Wolf, M. de Groot
6.	X-ray sources	M. Burger
7.	Events of Opportunity	C. Sterken
8.	Peculiar Late-type stars	F. Querci, C. Zwaan

J. B. Hearnshaw and P. L. Cottrell (eds.), Instrumentation and Research Programmes for Small Telescopes, 255–256.

Section 7 consists of all objects which need immediate monitoring due to the occurrence of an unexpected event (e.g. flares) or due to exceptional observational circumstances (e.g. simultaneous ground-based and space observations).

The actual list of objects contains 150 stars; about 50% of them are observed with a frequency of one observation per night, and the remaining stars are observed less frequently. All measurements are performed in a differential way relative to one or two comparison stars. A 50 cm or a 60 cm telescope is used, and all observation is done in the Strömgren uvby system.

Since October 1982, 20 observing runs have already been granted by ESO. The observers are particpants who volunteer to carry out the measurements according to the adopted scheme, and each observing run has a typical length of about three to four weeks. After termination of each observing run, all data are sent on magnetic tape to Dr. J. Manfroid at the University of Liège, where the central reduction is done. The final results are mailed to the principal investigators, who in turn redistribute the results to the participants in their section. The publication of the scientific results is done by those participants who carried out the analysis, but the data are separately published by the observers.

The project offers several interesting possibilities. First of all an important contribution to collecting valuable data is made. Second, the program leads to a highly efficient and economic way of using a small telescope, and it also prevents inactivity of useful instruments.

In principle anyone can apply for data. Just send to the corresponding principal investigator your application with a clear description of why you need data on a particular object, which precision you expect, and how long and how often you wish it observed. Do not forget to send the finding charts, and coordinates (epoch 2000.0) for the program stars and the comparison stars.

Acknowledgement: The author wishes to acknowledge financial support from the Ministry of Education of Belgium.

DISCUSSION

Millis: You listed one of your categories as "Events of opportunity". Could you give examples of these?

Sterken: 1) If you were expecting a flare in a certain object, 2) simultaneous observations with IUE and/or another satellite, or 3) a possible expected eclipse.

STUDY OF BOK GLOBULES WITH SMALL TELESCOPES

P.V. Kulkarni, M.R. Deshpande, U.C. Joshi and
A.K. Kulshrestha
Physics Research Laboratory
AHMEDABAD-380 009, INDIA

ABSTRACT. Polarization of radiation in the UBVRI spectral bands of the stars in the vicinity of four Bok Globules/dark clouds is observed with 1.5 and 1 m telescopes. Moderate to high degree of polarization and alignment of polarization vectors indicate the possibility of frozen magnetic fields. Other possibilities for retarding the cloud collapse have also been pointed out.

With IRAS satellite discovering a few hot regions/young stars in the BOK Globule B-5 and in some more dark clouds it is now necessary to observe more such objects for their better understanding.

We have studied four such globules by measuring polarization in UBVRI spectral bands of stars in their vicinity. Table I summarizes the data.

Table I. Polarimetric observations on stars in the vicinity of Bok Globules

Globule	Observed in	Place, Telescope	Filters	No. of stars	Ref
B-361	Sep '82	Tenerife, 60"	UBVRI	10	R1
B-5	Jan '84	Mt.Catalina, 61"	UBVRI	24	R2
L-1534	Feb '85	Kavalur, 40"	(Ga-As)	28	-
L-134	Apr '85	Nainital, 40"	(Ga-As)	25	-

R1: Williams, I.P. et. al., *Mon. Not. R. astr. Soc.* (1985) 212, 181
R2: Joshi, U.C. et. al., *Mon. Not. R. astr. Soc.* (1985) 212, 275

Data acquisition systems attached to all polarimeters were microprocessor based. I, Q, U, V, the Stoke's parameters could directly be calculated. From these, percentage of plane polarization P and angle of polarization θ for each star along with their probable errors were printed on-line. Figure 1 shows the plots of Ps, the polarization

J. B. Hearnshaw and P. L. Cottrell (eds.), Instrumentation and Research Programmes for Small Telescopes, 257–259.

vectors, on stars. Length of P is proportional to % polarization and its inclination is East of North.

Results obtained on the four globules are shown in Table II.

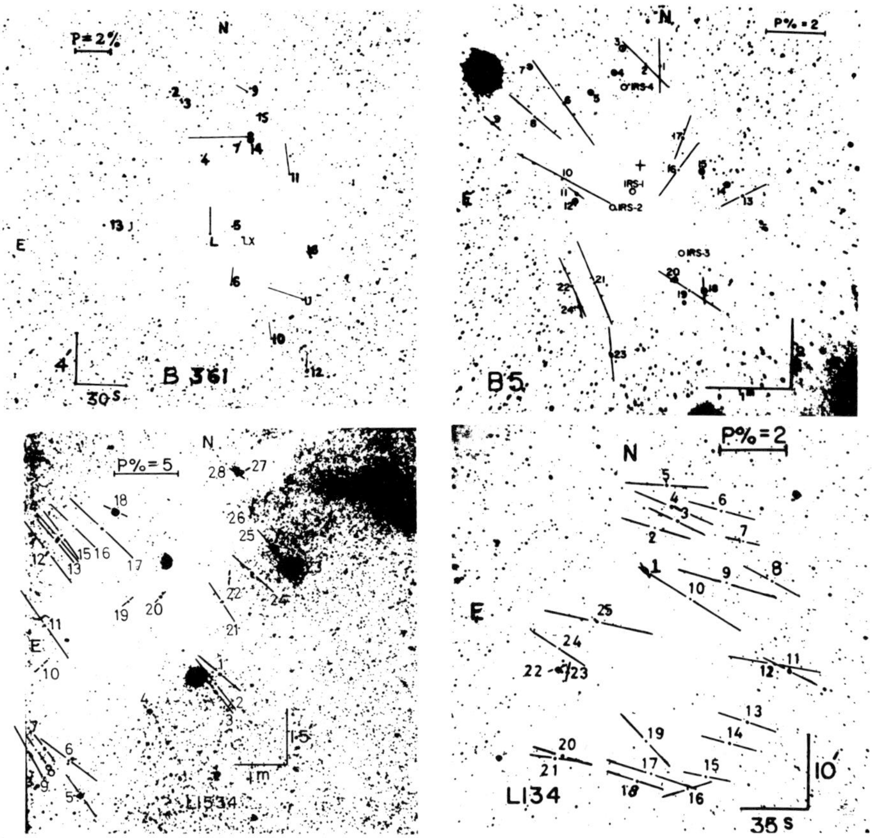

Figure 1.

Table II. Percentage of polarization and the alignment of Ps.

Globule	Polarization	Alignment	Direction*
B-361	maximum 2%	Poor	-
B-5	" 2%	Fair	-
L-1534	" 6%	Very good	0 ± 15°
L-134	" 2%	Very good	30 ± 10°

* Average direction of Ps with respect to the galactic plane.

Alignment of Ps in L-1534 and L-134 is quite high, however magnitude of P in L-1534 is unusually high. Four stars which may belong to another globule(s) in L-1534 show Ps at right angles to those of other stars. No hot spots are detected in L-1534 and L-134. The direction of Ps in L-1534 is along the galactic plane and in L-134 titled by 30° to it. The high degree of alignment suggests the frozen magnetic field aligning the grains.

Other factors which may retard the gravitational collapse are (a) rotation of the cloud, and if there is an embedded star near the cloud centre, (b) stellar wind with outward plasma flow or (c) a torus with a magnetic field due to the central star.

DISCUSSION

Graham: Have you observed any foreground stars outside the field of the globules to see if the alignment still persists? I would worry that this effect is a characteristic of the field and not the globules.

Kulkarni: There aren't many foreground stars (1 or 2) but they generally have less polarisation than the globules.

II. PHOTOMETRIC RESEARCH PROGRAMMES

b) Poster papers

uvby PHOTOMETRY WITH A CCD

Barbara J. Anthony-Twarog
Physics and Astronomy Department
University of Kansas
Lawrence, Kansas 66045
U.S.A.

Initial observations using uvby filters with a CCD detector were completed in June of 1985, using the Cerro Tololo 4m PFCCD system. It has been possible to derive all Strömgren indices (b-y, m_1 and c_1) for V<18 from frames with acceptable integration times. For fields in the open cluster IC 4651, averages of 2 to 3 frames with exposure times of 15, 20, 40, and 500 seconds for the y, b, v, and u filters, yielded high precision magnitudes and indices for V<15. Exposure times up to 5 times longer were required for the v and u filters in order to derive c_1 indices of sufficient precision for stars with V between 15 and 16. One result of our survey is the apparent absence of main sequence stars below V = 15.

The nearby globular cluster NGC 6397 was also observed. Frames with exposure times of 200, 400, 750 and 2500 seconds were used to construct the color-magnitude diagram shown. Analysis of the m_1 and c_1 indices in comparison to VandenBerg and Bell's (1985) isochrones indicate Z = 0.0003 and an age of 14 x 10^9 years; higher Z would imply an even younger age.

The support of the National Science Foundation through grant AST 84 20209 is gratefully acknowledged.

VandenBerg, D.A., and Bell, R.A. 1985. Astrophysical Journal Supplement, 58, 561.

J. B. Hearnshaw and P. L. Cottrell (eds.), Instrumentation and Research Programmes for Small Telescopes, 267–268.

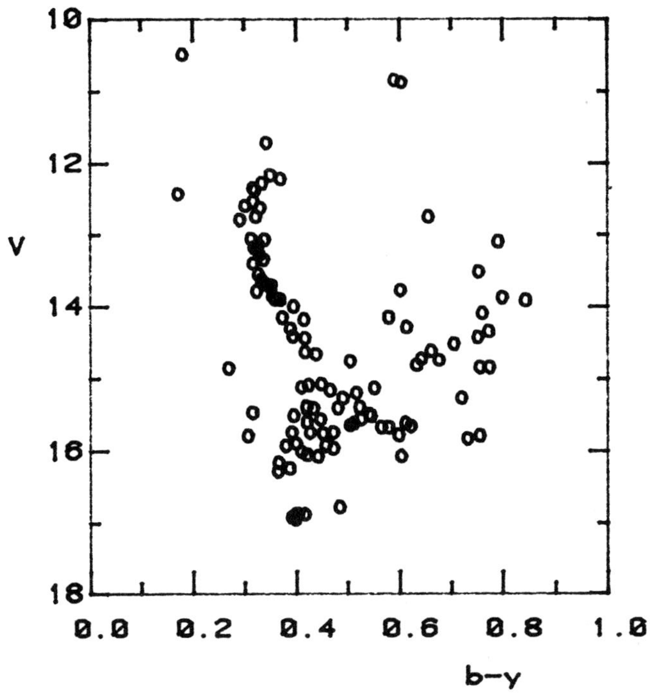

Figure 1. Color-magnitude diagram for the open cluster IC 4651.

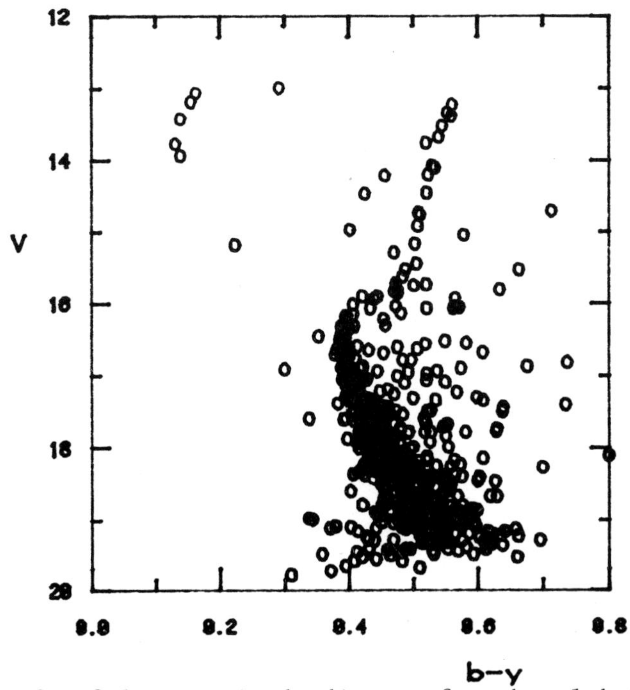

Figure 2. Color-magnitude diagram for the globular cluster NGC 6397.

PHOTOGRAPHIC MONITORING

Emilia Pisani Belserene
Maria Mitchell Observatory
Vestal Street
Nantucket, Massachusetts 02554
USA

ABSTRACT. The usefulness of photography in long term studies of variable stars is demonstrated by recent results on pulsation periods. Photography has been in use long enough for evolutionary changes to be detected, at least during stages that are relatively rapid, such as after core He exhaustion.

For many purposes photography is being supplanted by newer techniques, but there is an important role that it still plays. Photography has been in use for many decades. Continued monitoring with small telescopes can provide the long time base needed for detecting small changes in periods. Our Observatory uses a telescope of 18-cm aperture, scale 4 arcmin/mm, to follow variable stars. The limiting magnitude is 15 to 17 with exposures of 30 to 60 minutes.

There are useful plates taken with this telescope 60 years ago and more. The data set, sometimes supplemented by published magnitudes and by plates at the Harvard College Observatory, has become valuable for studies of small changes in the periods of pulsating variable stars. Table I summarizes results for cases where a reasonable fit to an O-C diagram is obtained by a curve no more complicated than a parabola (constant rate of change of period).

For 13 of the 25 stars, indicated by an asterisk, the change is no larger than about twice its mean error, indicating that a straight line fit to the O-C diagram (constant period) is still a reasonable hypothesis in these cases. Upper limits to the rates are given by the listed values and their mean errors.

For V759 Cyg and UZ Com the parabolas are purely formal fits, since there is evidence of somewhat more complex period behavior. The average rates of change, however, are significantly different from zero. Least-squares parabolas give a reasonable approximation.

The time scale of change for the BL Her stars is consistent with post-horizontal-branch evolution, after exhaustion of He in the core. UZ Com, an RR Lyr star whose amplitude is high for its long period, may be related.

The assistants who did the bulk of the work, most of them college students on summer vacation, are named in the last column. More

J. B. Hearnshaw and P. L. Cottrell (eds.), Instrumentation and Research Programmes for Small Telescopes, 269–270.

details on individual stars are or will be available in brief papers in the Journal of the AAVSO. Table I supplants a table which appeared on p. 41 of vol. 8 of that Journal. Work on the less well-behaved periods will be published separately. This project has received support from Earthwatch and from National Science Foundation grants AST 78-07405 A01, AST 80-05162 and AST 83-20491.

Table I. The rates of change of periods of pulsating variable stars monitored at the Maria Mitchell Observatory. Rates are given in fractions of the period per million years and are accompanied by mean errors obtained from least squares fits to the data. BL and CW identify population II Cepheid of types BL Her and W Vir, respectively. Cep I identifies the classical Cepheids.

STAR	TYPE	PERIOD	(1/P)(dP/dt)	OBSERVER
V759 Cyg	RRab	$0^d.360$	+1.67 ±.12	J.F.Waugh
TT Sct	RRab	0.453	-0.03 ±.05*	S.Cederbloom
IM Aql	RRab	0.457	-0.66 ±.05	E.Lada
FV Com	RRab	0.472	+1.29 ±.44	M.A.Reynolds
V801 Cyg	RRab	0.516	+0.53 ±.13	M.Yarlott
V943 Aql	RRab	0.519	-0.17 ±.13*	B.Meyer
EE Com	RRab	0.540	-0.80 ±.90*	A.Reeves
RR CVn	RRab	0.559	-0.02 ±.08*	W.Larson
V894 Cyg	RRab	0.571	+0.13 ±.04	J.Fahn
MY Sct	RRab	0.574	+0.07 ±.06*	S.Davis
V1510 Cyg	RRab	0.581	-0.03 ±.04*	L.Hoecker
CD Com	RRab	0.590	-1.32 ±.67*	D.Gilmore
V957 Aql	RRab	0.605	+0.24 ±.06	E.Street
V783 Cyg	RRab	0.621	-0.14 ±.04	A.Loser
EO Com	RRab	0.632	+0.67 ±.60*	M.A.Reynolds
DV Com	RRab	0.654	-0.33 ±.16*	D.Gilmore
UZ Com	RRab	0.737	+0.69 ±.06	D.Svetlov
V2022 Sgr	BL	1.533	+1.22 ±.17	J.Provencal
VZ Aql	BL	1.668	+0.70 ±.08	J.Christianson
V741 Sgr	CW	15.168	+7.0 ±4.6 *	M.Mitchell
XY Cas	Cep I	4.502	-0.35 ±.14	J.Fahn
V1954 Sgr	Cep I	6.179	-0.79 ±.34	M.Wooldridge
BX Sct	Cep I	6.411	-0.17 ±.33*	J.Sigismondi
V336 Aql	Cep I	7.304	+0.83 ±.65*	L.Bissonnette
TY Sct	Cep I	11.053	+3.0 ±1.5 *	C.Bailey

COMETARY ASTROMETRY WITH THE SYDNEY OBSERVATORY ASTROGRAPH

Col Bembrick, Jeff Byron & David Phipps
British Astronomical Association (New South Wales Branch)
c/- Sydney Observatory, Observatory Hill
Sydney, N.S.W., 2000
Australia.

ABSTRACT. The New South Wales Branch of the British Astronomical Association is currently making use of the Sydney Observatory astrograph for cometary astrometry. Astrometric plates (16 x 16 cm) of P/Halley are being taken at a plate scale of 116 arcsec/mm with a field of 5 x 5 degrees. Plate measuring is accomplished by means of a mechanical single screw Hilger measuring machine. Plate reduction utilises the program supplied by the International Halley Watch organisation. Measurement and reduction is facilitated by the machine readable catalogue and plotting routines supplied by the United States Naval Observatory. Currently the BAA observing team is working towards achieving a 24 hr turnaround for plate reductions. This will enable useful contributions to be made during the critical pre-encounter periods.

1. INTRODUCTION

The technique of astrometry involves the taking of astrophotos of a precisely known scale for the purposes of accurately measuring the positions of stars, minor planets and comets. The southern hemisphere is particularly poorly covered by astrometric observers, and here the amateur has a chance to make significant contributions.

2. CONCEPTS

To measure accurately the positions of objects in the sky, three essential steps are involved:-

i) taking a photograph (glass plate or sheet film),
ii) measuring the distances on the plate between various reference stars and the object whose position is required,
iii) scaling of these measurements to obtain the object's position in terms of Right Ascension and Declination.

Thus, the following are fundamental to the practice of astrometry:-

i) a large image scale & distortion free field in the focal plane,

J. B. Hearnshaw and P. L. Cottrell (eds.), Instrumentation and Research Programmes for Small Telescopes, 271–272.

ii) a dimensionally stable photographic film or plate,
iii) sufficiently accurate measurement and reduction methods.

The astrograph also needs a relatively wide field so that sufficient catalogued reference stars may be found.

3. TECHNIQUES

When taking an astrometric plate exposures should be only long enough to record the "nucleus", plus stars down to mag 9-10 for reference purposes. The telescope is normally guided in such a way as to maximise the chance of recording a measurable comet image on the plate. Thus, the stars are recorded as short trails.

Measurement of these astrometric plates depends on the use of a simple mechanical (single screw) measuring machine. This consists of a travelling microscope on a stage which is moved by an accurately machined screw. Distances along this screw may be read to one micron by means of vernier scales.

Using a microcomputer, the reduction of the plate measurements (in microns) to accurate positions (in RA and Dec) is a relatively straightforward, though tedious procedure.

4. THE SYDNEY OBSERVATORY ASTROGRAPH

The Sydney Observatory astrograph is actually two astrographs and a 26cm guidescope mounted together. The small astrograph which we use has a 23cm, f/7.7 lens with a field of 5 x 5 degrees. Guiding is accomplished by means of a special guiding eyepiece which is mounted on an X-Y stage and can be offset from the optic axis of the guidescope.

5. INITIAL OBSERVATIONS BY BAA OBSERVERS

Initial trials (comets P/d'Arrest and P/Kopff) eventually yielded positions to an accuracy of about 1 arcsec.

Prior to the IHW trial run in March, 1984, an astrograph operating manual was compiled which includes programs in BASIC and HP41 programmable calculator notation.

6. CONCLUSIONS

During the 1985/86 return of P/Halley, the BAA (NSW Branch) observers will be involved with the IHW astrometry network and will operate the Sydney Observatory astrograph at the observatory site during this period. The relatively few southern hemisphere astrometric observers will play an important part in spacecraft encounters with Halley in 1986.

THE EMPIRICAL PERIOD-RADIUS RELATION FOR PULSATING STARS : A SYNTHESIS BASED ON PHOTOMETRIC AND RADIAL VELOCITY MEASUREMENTS

G. Burki
Observatoire de Genève, Switzerland

The relation existing between the radius and the period for the pulsating stars of a given class constitutes a powerful test for the theory of stellar evolution and for the identification of the pulsation modes. In recent years, several authors have determined the mean radius of a lot of pulsating stars of various classes by applying the Baade-Wesselink method. Fig. 1 presents the resulting general logP - logR diagram grouping these determinations. The sources for the radii are given by Burki and Meylan (1986). The variable stars in known binaries have been excluded since the presence of a companion biases the radius calculation (Burki, 1984). The determinations marked by arrows are based on the radial velocities by CORAVEL (1m telescope at the Haute-Provence Observatory, France) or/and on the photometry in the Geneva system (40cm and 70cm telescopes at La Silla Observatory, Chile).

For the classical cepheids, two linear regressions have been calculated:

$$\log R = 1.16 + 0.72 \log P \qquad P<10d \qquad (1)$$
$$\log R = 1.03 + 0.77 \log P \qquad P>10d \qquad (2)$$

The difference between these two relations, if really significant, could be due to the mass loss process affecting the evolution of massive stars (Burki, 1985). For the cepheids with P<10d, the various theoretical models yield a mean relation $\log R = 1.18 + 0.69 \log P$ (Fernie,1984), which is quasi-identical with (1). This allows one to derive, from the definition $Q = P(\overline{\rho}/\overline{\rho}_\odot)^{1/2}$, an equation relating Q, R and M (mass) :

$$Q = 0.0238\ M^{0.5}\ R^{-0.1} \qquad P<10d \qquad (3)$$

For the population II cepheids (BL Herculis and W Virginis stars), the relation is:

$$\log R = 0.87 + 0.54 \log P \qquad (4)$$

and it is remarkable to see that this relation satisfies globally well the RR Lyrae, δ Scuti and SX Phoenicis stars. By adopting 0.6 $M_\odot$ for the typical mass of the population II cepheids and RR Lyrae stars (Harris, 1985), the following relation can be deduced :

J. B. Hearnshaw and P. L. Cottrell (eds.), Instrumentation and Research Programmes for Small Telescopes, 273–274.

$$\log Q = -1.4 + 0.2 \log P \qquad (5)$$

In Fig. 1 is also shown the location of FG Sge, the nucleus of a planetary nebula for which Mayor and Acker (1980) derived a value of the mean radius for the years 1978-1979.

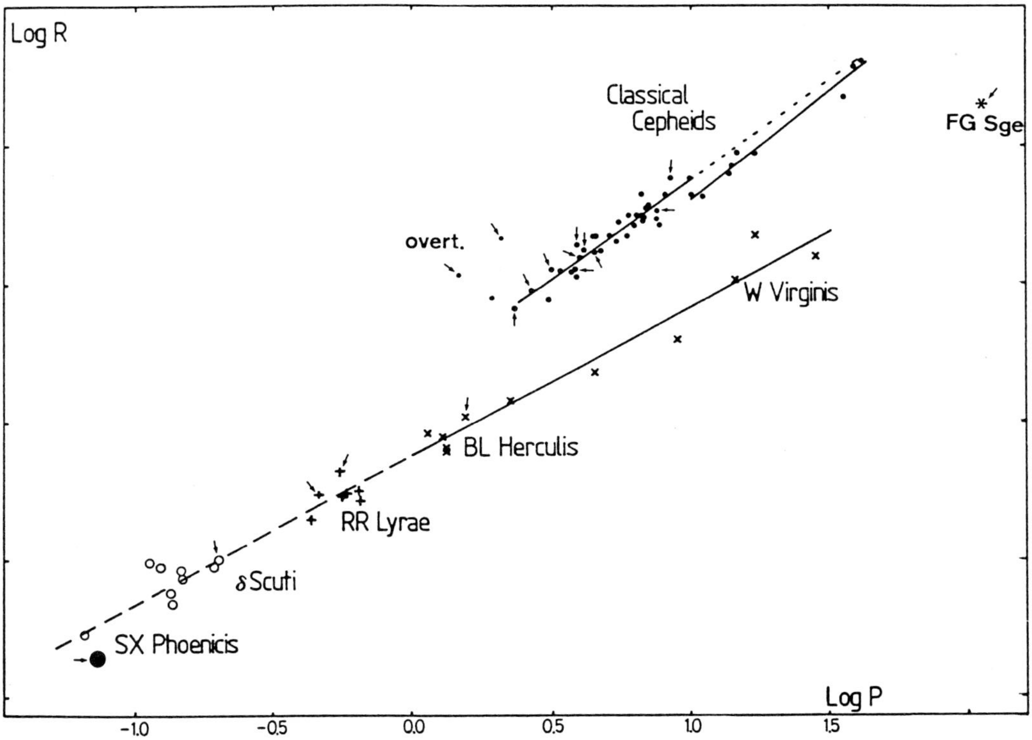

Figure 1. The logR - logP relations for pulsating stars.

REFERENCES

Burki, G.: 1984, Astron. Astrophys. 133, 185

Burki, G.: 1985, Duplicity, Mass Loss and the Cepheid Mass Anomaly, in Cepheids: Observations and Theory, IAU Coll. no 82, Ed. B.F. Madore, Cambridge Univ. Press, Cambridge, p. 71

Burki, G., Meylan, G. : 1986, Astron. Astrophys., in press.

Fernie, J.D.: 1984, Astrophys. J. 282, 641

Harris, H.: 1985, Population II Cepheids, in Cepheids: Observations and Theory, IAU Coll. no 82, Ed. B.F. Madore, Cambridge Univ. Press, Cambridge, p. 232

Mayor, M., Acker, A.: 1980, Astron. Astrophys. 92, 1

STARSPOT IMAGING USING VRI PHOTOMETRY

A. Collier Cameron
Institute of Astronomy
Madingley Road
Cambridge CB3 OHA
England

ABSTRACT. Recent progress on maximum-entropy imaging of stellar surface brightness distributions using VRI photometry and Doppler-broadened absorption line profiles is reported. VRI photometry generally provides tighter constraints on starspot configurations than does Doppler imaging (which requires the use of large telescopes). Observers carrying out long-term photometric monitoring programmes on spotted stars are urged to work in the VRI system, as well as the UBV.

1. INTRODUCTION

The RS CVn and related late-type active chromosphere stars have attracted much well-deserved attention from photometrists in recent years and - judging from the number of groups at this conference who are actively engaged in long-term monitoring programmes - will continue to do so. Such programmes will ultimately yield important information concerning the evolution of individual stellar active regions, surface differential rotation and activity cycles in stars with a wide range of rotation rates and envelope structures. However, it is important to consider at the outset just what type of observation should be made in order to maximise the scientific return, in the light of recent developments in the interpretation of such data, which are described here.

2. IMAGE RECONSTRUCTION METHODS

The most difficult problem to be surmounted in the interpretation of RS CVn light-curve observations is that of uniqueness. Generally, any number of stellar surface brightness distributions can be devised which for a given axial inclination (often unknown a priori) can reproduce the observed V light-curve.

Vogt (1981) introduced an important refinement which used the colour-surface brightness relations of Barnes, Evans and Moffett (1978) to resolve the ambiguity between the temperature of the starspots and their total covering fraction on the visible stellar disk at each

J. B. Hearnshaw and P. L. Cottrell (eds.), Instrumentation and Research Programmes for Small Telescopes, 275–276.

rotation phase. The success of this method relies on the availability of VRI photometry, as line-blanketing in the U and B bands causes the colour-surface brightness relations at the blue end of the spectrum to break down at the temperatures typical of starspots.

The application of the maximum entropy method to the reconstruction of stellar surface brightness distributions from VRI photometry and/or Doppler imaging (Vogt and Penrod 1983) has been described by Cameron and Horne (1986). This technique involves the setting-up of a transformation from a two-dimensional image of the stellar surface to a data set consisting of flux observations in a number of different wavelength bands at different phases throughout the rotation cycle. The image is then adjusted iteratively using the maximum-entropy algorithm of Skilling and Bryan (1984) to yield that image which minimises the amount of configurational information in the starspot distribution on the image, subject to the constraint that it must reproduce the observed data at a satisfactory level of the χ^2 statistic.

Our early studies with both synthetic and real data (Cameron et al 1986) indicate that it is feasible to reconstruct stellar images from VRI photometry, provided that (1) complete phase coverage is available; (2) internal errors on individual data points are less than 10% of the total amplitude of variation; and (3) the inclination of the rotation axis to the line of sight is less than 70° or so. At higher inclinations, the lightcurve contains little information to constrain the spot latitude. At lower inclinations, artificial elongations of the spots in either the latitudinal or longitudinal direction occur if a reconstruction is attempted at the wrong inclination; given data of sufficiently high quality, the axial inclination may be determined to within about 10° from the lightcurve alone.

3. FUTURE WORK

Perceptible changes in lightcurve morphology in the RS CVn and related systems have been observed to occur on timescales of weeks to months. Some of these changes appear to be caused by shifts in the relative positions of spots at different latitudes through differential rotation, others by the growth and decay of entire active regions. Image reconstruction methods of the type described here provide a powerful tool for the analysis of these processes on spotted stars, provided that high-quality lightcurves in the VRI system can be obtained at regular intervals for the systems under study.

4. REFERENCES

Barnes TG, Evans DS and Moffett TJ 1978 *Mon Not R astr Soc* 183 285

Cameron AC and Horne KD 1986 } in *Fourth Cambridge Workshop on Cool*
Cameron AC et al 1986 } *Stars, Stellar Systems and the Sun* Springer-Verlag in press

Skilling J and Bryan RK 1984 *Mon Not R astr Soc* 211 111

Vogt SS 1981 Astrophys J 250 327

Vogt SS and Penrod GD 1983 *Pub astr Soc Pacific* 95 565

A SEARCH FOR FEATURES IN EARLY-TYPE GALAXIES

S. Djorgovski and K. Ebneter
Astronomy Department
University of California
Berkeley, CA 94720
U.S.A.

ABSTRACT. We are conducting a search for dust lanes, incipient stellar disks and bars, and other deviations from elliptical symmetry in a sample ~ 250 early-type galaxies. Division by a model galaxy image is a very powerful enhancement technique for this type of work.

A classical image of elliptical galaxies contains no dust or stellar disks: these should be the attributes of spirals. Modern data do not support this simple picture. Many *bona fide* ellipticals are now known to contain dust (cf. Ebneter & Balick 1985, or Salder & Gerhard 1985). The presence of dust in elliptical galaxies is most likely a signature of of past mergers, and it provides some clues about the origin of nuclear activity in galaxies. The morphological and kinematical separation of ellipticals and SO's is getting more and more blurred. A continuity of galaxian properties between different morphological types is of great importance for theories of galaxy formation and evolution.

We are currently completing a large search for morphological features in early-type galaxies, *viz.*, all deviations from the elliptical symmetry of the underlying light. The data come from the CCD surface photometry by Djorgovski (1985), and cover ~ 200 elliptical and ~ 50 SO galaxies. This work is fairly similar to the surveys of Lauer (1985), and Sparks *et al.* (1985), and employs the same digital processing technique, which we describe below. The difference is primarily in the sample size, as we are covering several times more objects than the previous investigators; the sample selection effects are well under control.

The main problem here is essentially the suppression of the strong radial intensity gradient. This can be accomplished by a polar-coordinates gradient technique (radial stretch or squeeze, possibly a slight rotation, and subtraction from the original image). We tried that method, and found it satisfactory, but another technique worked better, *viz.*, division by the purely elliptical galaxy model image. The method works as follows:

From the surface photometry of a galaxy (best fit surface brightness, ellipticity, and position angle as functions of semimajor axis), we reconstruct a model galaxy image which obeys those profiles, but has

J. B. Hearnshaw and P. L. Cottrell (eds.), Instrumentation and Research Programmes for Small Telescopes, 277–279.

purely elliptical isophotes; it helps to smooth the ε and PA profiles somewhat. The model is smoother than the original data frame, because of the isophotal averaging, and smoothing of the ε and PA profiles. We then divide the original image by the smoothed image. The reason for superiority of this technique over the simple polar gradient is that we are adding some *a priori* information in the image restoration process, *viz.*, the exact form of the galaxy light distribution, plus the demand that the "unperturbed" galaxy has purely elliptical isophotes.

REFERENCES:

Djorgovski, S. 1985, *Ph. D. Thesis*, University of California at Berkeley.
Ebneter, K., and Balick, B. 1985, *Astron. J.* 90, 183.
Lauer, T.R. 1985, *M.N.R.A.S.* 216, 429.
Sadler, E.M., and Gerhard, O.E. 1985, *M.N.R.A.S.* 214, 177.
Sparks, W.B. *et al.* 1985, *M.N.R.A.S.* 217, 87.

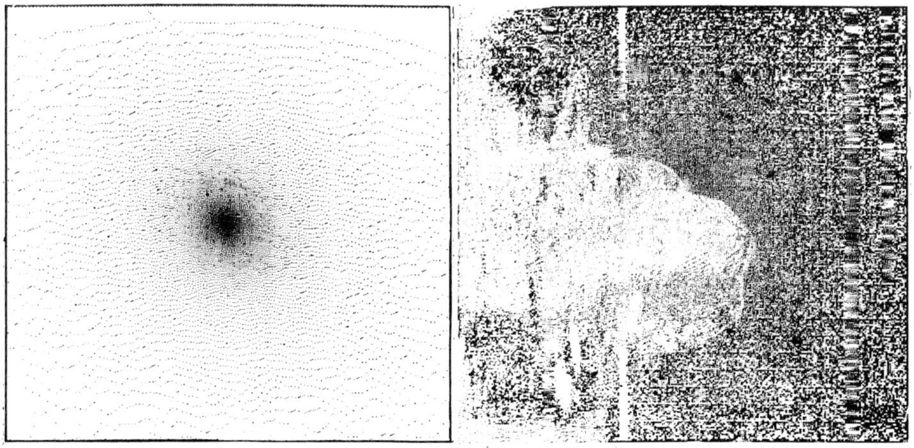

Figure 1: An example of feature enhancement. The original image is on the left and the model-divided image is on the right. The object is NGC4374 which reveals a dust lane (a ring?); note the limb-darkening of the ring: the dust is optically thin!

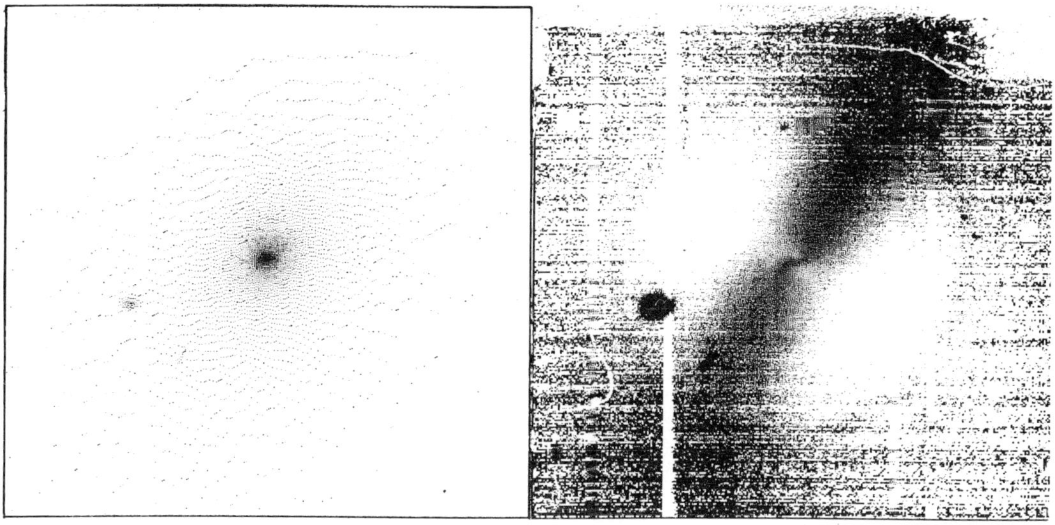

Figure 2: Another example of feature enhancement, this time for NGC 584 which shows a thick stellar disk, and a smaller, tilted one in the nucleus.

THE STRUCTURE OF GLOBULAR CLUSTERS

S. Djorgovski, I.R. King, C. Vuosalo, A. Oren and H. Penner
Astronomy Department
University of California
Berkeley, CA 94720, U.S.A.

ABSTRACT. We report the preliminary results of a CCD surface photometry survey of a large fraction of all known Galactic globular clusters. About 1/5 of all surveyed clusters show a characteristic post-core-collapse (PCC) morphology. The PCC clusters are on average closer to the Galactic center than the King-model-like clusters.

We have completed a multi-color CCD surface photometry survey of some 113 Galactic globular clusters. Many of the observed clusters had no previous photometry, or density profiles of any kind. The first scientific goal of this survey was to do a complete census of cluster cores, and determine the fraction of clusters which show a post-collapse morphology. Several good reviews of this topic can be found in the volume edited by Goodman & Hut (1985).

This work is a follow-up to the earlier study (Djorgovski & King, 1984). Most data for the present survey were obtained with the CTIO 1.5-m telescope and RCA and GEC CCDs, and with the *UBVRI* filters. Some of the data were obtained with the Lick 1-m telescope and TI and GEC CCDs, and with a special red filter. Several hundreds of cluster exposures were obtained. Most of the data are already reduced to one-dimensional surface brightness profiles, but with little absolute calibration. The full account of the survey will appear in future papers. We supplement our data with reliable surface density profiles from the literature.

There are 21 definite PCC clusters, 6 possible ones, 3 uncertain ones, 97 definite or probably King-model-like clusters. Thus, PCC clusters are about 1/5 of the total.

The PCC clusters are generally not very rich; this may be an additional evidence for their advanced dynamical state: the central, stabilizing binary pumps the energy into the cluster, and thus speeds up its dissolution.

One interesting result emerged from this study; the PCC clusters are on the average much more concentrated toward the Galactic center. For a set of 20 definite or possible PCC clusters, the mean and the median galactocentric radii are 4.2 and 2.7 kpc respectively; for a

J. B. Hearnshaw and P. L. Cottrell (eds.), Instrumentation and Research Programmes for Small Telescopes, 281–282.

set of 91 King-model-like clusters, the corresonding numbers are 14.8 and 7.6 kpc. The difference is highly significant, and it suggests that the tidal shocks (from disk passages) play an important rôle in the *internal* dynamics of globular clusters, *viz.*, by accelerating their dynamical evolution. This effect was predicted by Chernoff, Kochanek & Shapiro (1985, preprint).

There is one remaining puzzle, however: there are a number of very concentrated clusters with very dense cores, and thus very short relaxation times (a few x 10^7 yr). They had plenty of time to collapse, but they look like perfect King models. Either they managed to recover completely from the collapse (faster than the current theories would predict), or they *do* have collapsed cores, but they are hidden in the invisible heavy stellar remnants (neutron stars, stellar black holes).

REFERENCES

Djorgovski, S. and King, I.R. 1984, *Ap. J. Lett.* 277, L49.
Goodman, J. and Hut, P., (eds.) 1985, *Proceedings of the I.A.U. Symposium #113, "Dynamics of Star Clusters"*. Dordrecht: D. Reidel.

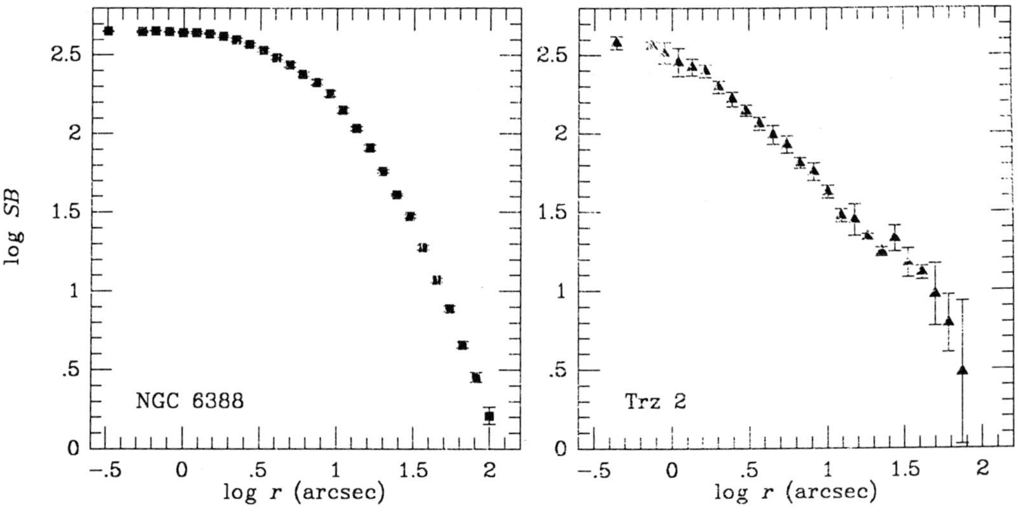

Figure 1. Examples of cluster surface brightness profiles: NGC 6388 (left) is an example of a highly concentrated, King-model-like cluster: note the flat core, and a steep envelope. Trz 2 (right) is an example of a PCC cluster: the light distribution is almost a pure -1 power law. Both profiles were derived from CCD images taken in a B band, at CTIO 1.5-m telescope.

STELLAR ACTIVITY

D.K. Duncan
Mt Wilson and Las Campanas Observatory
813 Santa Barbara Street
Pasadena CA 91101, United States of America

ABSTRACT. We have been monitoring the CaII H and K emission in late-type giant stars using the Mt Wilson 60-inch telescope. It is a continuation of work that Olin Wilson began almost 20 years ago. As a result not only have starspot cycles been detected (equivalent to the solar sunspot cycle), but rotation periods have now been measured.

The most recent interesting result is the detection of the rotation of red giant stars. Also, differential rotation is starting to show up, that is there is more than one period in the data at the same time.

At this symposium we are trying to emphasize three major advantages of this type of work for smaller telescopes and institutions : (1) regular access to the telescope (we have made 100,000 observations over the last 19 years), (2) there are not large amounts of data to reduce (two points in the spectrum of a star) and (3) the method uses a simple instrument and detector.

J. B. Hearnshaw and P. L. Cottrell (eds.), Instrumentation and Research Programmes for Small Telescopes, 283.

CCD SCANNING WITH A SMALL TELESCOPE

Tom Gehrels
Lunar & Planetary Laboratory and Steward Observatory
University of Arizona
Tucson, AZ 85721

We use a scanning CCD for acquisition and astrometry of new and recovered comets, asteroids and other objects. The CCD is an RCA SID 53612 thinned, buried channel array of 512 x 320 30-micron square pixels that are back-illuminated and refrigerated to -60 C in a vacuum housing. The readout noise is ± 200 electron-hole pairs (ehp) per pixel per readout, the thermal dark current is 50 ehp/pixel/sec, and the scale of our 12-bit analog-to-digital converter is 25 ehp/AD unit. The Newtonian focus of a 91-cm telescope has been modified from f/5 to f/3.85 with a relay lens to give a platescale of 1.73 arcsecs per pixel, a scale chosen for efficient coverage of sky. The CCD is operated in the scanning mode with the telescope drive off, and the rate of transfer of signal charges is tuned to correspond to the rate at which the star images drift across the focal plane. The exposure time (the time the images take to transit the "512" dimension of the CCD) is 60 seconds at the celestial equator, giving a "six sigma" limiting (visual) magnitude of 19.5. A typical scan covers 30 minutes of time in right ascension by 0.156 deg in declination and is stored as a digital array of 14848 x 320 CCD pixels. A set of three 30-minute scans near the opposition point along the ecliptic nets about 5 new main-belt asteroids. Potential moving objects are revealed by determining the positions of all the stellar images in each scan and comparing the results of the three scans. Several potential reference (SAO) stars are in one scan. The reduction from pixel coordinates to apparent topocentric equatorial coordinates of date involves only three free parameters, making our astrometric reductions simpler than the classical affine transformation of plate coordinates. There is no rotational transformation and the scale in right ascension is defined by the clock. The declination scale is determined from the reference stars. The zero points of R.A. and Dec. are given by the average differences between the pixel coordinates and catalog positions of the reference stars.

J. B. Hearnshaw and P. L. Cottrell (eds.), Instrumentation and Research Programmes for Small Telescopes, 285–286.

REFERENCES

Frecker, J. E., Gehrels, T., McMillan, R. S., Merline, W. J., Perry, M. L., Scotti, J. V. and Smith, P. H. 1984. A CCD System for Photometry of Direct and Spectroscopic Images, Proc. of the Workshop on Improvements in Photometry, ed. W. J. Borucki and A. Young, NASA CP-2350, 137-151.

Gehrels, T. 1981. Faint Comet Searching, Icarus **47**, 518-522.

Gehrels, T. 1984. Comparison of a 1.8-m CCD Scannerscope with a 1.2-m Schmidt Camera, Proc. of the 2nd Asian-Pacific Reg. Mtg. on Astron., B. Hidajat and M. W. Feast, eds. (Jakarta: Tira Pustaka), 335-340.

Gehrels, T. 1984. Fundamental Studies of Asteroids, Bull. A.A.S. of India **12**, 16-39; also in l'Astronomie **98**, 115-132 and 159-170 (in French); also in Astrum **54**, 5-18 (in Spanish).

Gehrels, T. 1985. Asteroids and Comets, Physics Today **38**, No. 2, 32-41.

Gehrels, T., Frecker, J. E., McMillan, R. S., Perry, M. L., and Scotti, J. V. 1984. Four Modes of CCD Operation, Bull. A.A.S. **16**, 986.

McMillan, R. S. and Stoll, C. P. 1982. Software Simulations of the Detection of Rapidly Moving Asteroids by a Charge-Coupled Device, Proc. Soc. Photo-Opt. Instrum. Eng. **331**, 104-112.

Scotti, J. V., Frecker, J. E., Gehrels, T., and McMillan, R. S. 1984. Astrometry using the CCD System of the Spacewatch Camera, Bull. A.A.S. **16**, 902.

SEARCH FOR VARIABLE STARS IN GLOBULAR CLUSTERS

M. L. Hazen-Liller
Harvard-Smithsonian Center for Astrophysics
60 Garden Street
Cambridge, MA 02138
U. S. A.

ABSTRACT. A program to search poorly studied southern globular clusters for variable stars is being carried out on the 1-m Yale telescope at CTIO. Recently two galactic bulge clusters of moderately high metallicity have been found to contain RR Lyrae variables; a cluster of similar metallicity a little farther from the galactic center apparently has none.

For the past decade, the 1-m Yale telescope at CTIO has been used in a program to photograph poorly studied southern globular star clusters in order to search for and to study their variable stars. Many of these clusters lie near the direction to the galactic center and thus are in crowded fields, making other types of studies of horizontal branch (HB) morphology very difficult.

The 1-m telescope at the Ritchey-Chretien focus (19 arc sec mm^{-1}) can reach a limiting magnitude of B $\sim$ 18.5-19.0 in a reasonably short exposure time ($\sim$30-45 min) on sensitized 103a-0 plates. Such a limit reaches possible HB variables in 90 per cent of the globulars listed in the compilation of Zinn (1985).

Data from the 1-m on the variables (or absence of variables) in some 15 globular clusters have been published over the last several years. The most recent work has involved studies of three clusters of moderately high metallicity: NGC 6569, [Fe/H] = -0.86; NGC 6388, [Fe/H] = -0.74; and NGC 6652, [Fe/H] = -0.89. The clusters are listed in order of distance from the galactic center: R $\sim$ 0.8:, 3.8, and 5.2 kpc respectively. The [Fe/H] and R are taken from Zinn (1985).

The study of NGC 6569 (Hazen-Liller 1985) reveals eight RR Lyrae stars that are probable members and four that are possible members of the cluster. (Membership criteria include proximity to the cluster center and similarity of mean brightness.) No other cluster of such a high metallicity has been previously found to contain HB variables. The result suggests that NGC 6569 has an unusually blue HB for its

J. B. Hearnshaw and P. L. Cottrell (eds.), Instrumentation and Research Programmes for Small Telescopes, 287–288.

metallicity. Recent, as yet unpublished, data for NGC 6388 indicate that this cluster, too, has at least five RR Lyrae stars that may be members. NGC 6388 has even higher metallicity than NGC 6569. On the other hand, preliminary investigation of NGC 6652 has failed to find any RR Lyrae stars in that cluster.

The results here agree with the study by Zinn (1980) that concluded that clusters lying near the galactic center have bluer HBs than clusters of similar metallicity lying farther from the galactic center. One possible explanation, following Zinn, is that the clusters with smaller R are older than those farther out.

REFERENCES

Hazen-Liller, M. L. 1985, Astron. J. **90**, 1807.
Zinn, R. 1980, Astrophys. J. **241**, 602.
Zinn, R. 1985, Astrophys. J. **293**, 424.

EX HYDRAE TIMING DATA

K.M.Hill, D.P.Sharma and R.D.Watson
Physics Department
University of Tasmania
P.O.Box 252C, Hobart, 7001
Australia

The 98 min eclipsing cataclysmic variable EX Hya possesses a strong 67 min modulation in its light. This has led to a discussion of EX Hya in the context of intermediate polar variables. The observed $P/\dot{P}$ of the 67 min modulation provides a useful constraint on EX Hya models. Here we report additional timing data obtained over the interval 1982 to 1985 which bears on this matter.

Table I lists 35 moments of 67 min maximum. Data were derived from observations made with the 1 m reflector of the Kavalur observatory, South India, (denoted by an asterisk) and with the 1 m reflector of the University of Tasmania. As well, the Tasmanian observations yielded 21 mid-eclipse times, which are given in Table II, where Heliocentric Julian date is corrected to Ephemeris Time. Here the data in Table I is combined with that in Cordova *et al.* (1985), Jablonski and Busko (1985) and all previous data, as summarized in Hill and Watson (1984). The resulting quadratic ephemeris in Figure 1 is

$$HJD(max) = 2437699.8901(\pm 6) + 0.046546516(\pm 13)E - 0.84(\pm 7) \times 10^{-12}E^2.$$

From the E^2 term a $P/\dot{P} = -3.5(\pm 0.3)\times 10^6$ yrs is deduced. Following Ritter (1985), this permits limits on the primary mass accretion rate to be inferred. The value lies in the range 10^{-10} to $10^{-11}\ M_\odot\ yr^{-1}$. This is consistent with the $1\times 10^{-10}\ M_\odot\ yr^{-1}$ obtained by Patterson (1984) using an independent technique which relies on the observed disk flux. Eclipse times in Table II, combined with previously published data, show O-C residuals from a constant period which are similar to those of the 1960's, after a systematic shift in the following decade. Similar drifts are seen in other cataclysmic variables.

REFERENCES

Cordova, F., Mason, K.O. and Kahn, S.M.: 1985, *M.N.R.A.S.*, 212, 447.
Hill, K.M. and Watson, R.D.: 1984, *Proc.Astron.Soc.Aust.*, 5, 532.
Jablonski, F. and Busko, I.C.: 1985, *M.N.R.A.S.*, 214, 219.
Patterson, J.: 1984, *Ap.J.Suppl.*, 54, 443.
Ritter, H.: 1985, *Astron.Ap.*, 148, 207.
Vogt, N., Krzeminski, W. and Sterken, C.: 1980, *Astron.Ap.*, 85, 106.

J. B. Hearnshaw and P. L. Cottrell (eds.), Instrumentation and Research Programmes for Small Telescopes, 289–290.

TABLE I

Heliocentric Times of 67 min Maximum

HJD (2440000+)	E (100000+)	HJD (2440000+)	E (100000+)
5031.316	57508*	5825.213	74564
5031.361	57509*	5825.253	74565
5042.309	57744*	5855.969	75225
5042.354	57745*	5875.008	75634
5043.326	57766*	6149.168	81524
5043.371	57767*	6149.219	81525
5344.438	64235*	6149.266	81526
5344.484	64236*	6150.009	81542
5389.349	65200*	6150.055	81543
5389.395	65201*	6150.098	81544
5792.212	73855*	6150.149	81545
5792.258	73856*	6150.201	81546
5792.303	73857*	6150.247	81547
5792.357	73858*	6153.032	81607
5793.239	73877*	6153.126	81609
5823.154	74520	6153.180	81610
5825.075	74561	6153.224	81611
5825.118	74562		

TABLE II

Heliocentric Times of Mid-Eclipse

HJD (2440000+)	E (100000+)	HJD (2440000+)	E (100000+)
5823.1132	19049	6149.2023	23828
5823.1820	19050	6149.2704	23829
5825.0924	19078	6150.0214	23840
5825.1600	19079	6150.0890	23841
5825.2283	19080	6150.1574	23842
5856.0015	19531	6150.2259	23843
5856.0694	19532	6153.0229	23884
5871.0133	19751	6153.0921	23885
5871.0821	19752	6153.1600	23886
5875.0385	19810	6153.2282	23887
5885.0687	19957		

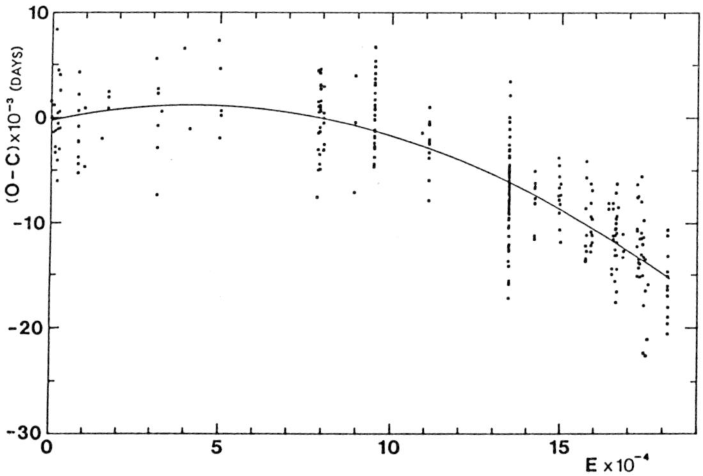

Figure 1. The residuals, O–C, in units of 10^{-3} day, versus cycle number E in the 67 min modulation of EX Hya. These residuals are derived relative to the ephemeris of Vogt *et al.* (1980). A least squares quadratic fit is shown, which is the basis of the ephemeris presented here.

RECENT TRIGONOMETRIC PARALLAXES FROM TWO HEMISPHERES

Philip A. Ianna
Department of Astronomy
The University of Virginia
Charlottesville, VA 22903-0818
U.S.A.

ABSTRACT. Both southern and northern hemisphere parallax results are reported for a number of stars including several white dwarfs, the central star of NGC7293, Barnard's star, and several other red dwarfs. The 1 meter astrometric reflector of the McCormick Observatory is described and some preliminary results are briefly discussed.

1. INTRODUCTION

Stellar parallax programs have almost exclusively been carried out with long-focus refractors, once among the largest telescopes in the world, but small aperture telescopes by current standards. Modern ground-based astrometric programs have now been extended to reflectors, several of specialized design, and some effort is also being directed toward implementing non-photographic detector systems of high quantum efficiency for astrometric applications. The new detectors are resulting in high accuracy results to magnitudes much fainter than heretofore practical.

2. SOUTHERN HEMISPHERE RESULTS

We first present a number of new parallaxes from our program with the 66 cm Yale-Columbia photographic refractor at the Mt. Stromlo Observatory. Recent measurements with the PDS 1010A at MSSSO yield a positional accuracy of about $\pm$ 0.0015 mm (s.e.) and parallaxes with standard errors $\pm$ 0″003 for series with about 100 exposures. Relative parallaxes reported include NGC7293 (0″004 $\pm$ 0″003), EG141 (0″039 $\pm$ 0″003), CD-42° 14462 (0″006 $\pm$ 0″003), CD-38°10980 (0″065 $\pm$ 0″004), BPM 24601 (0″028 $\pm$ 0″003), Gr406 (0″072 $\pm$ 0″004), and LHS 288 (0″220 $\pm$ 0″008). The binary LHS 1047, resolved by speckle interferometry, is found to have one of the lowest mass companions known. These results will be reported in more detail elsewhere.

J. B. Hearnshaw and P. L. Cottrell (eds.), Instrumentation and Research Programmes for Small Telescopes, 291–292.

3. NORTHERN HEMISPHERE RESULTS

3.1 Fan Mountain Reflector

The 1 meter astrometric reflector was built under the direction of L. W. Fredrick. It is an f/13.5 flat-field Schmidt cassegrain Baker-design with a 1.1 m f/7.5 ULE primary, a 0.45 m ULE secondary, and a 1 m UBK-7 corrector. Both primary and secondary are aspherical. The unvignetted field is 40 arcmin in diameter; with useable field of 50 arcmin. There is no detectable aberration flare over at least a one degree field.

Primary-secondary separation is maintained by three invar spacing rods. Tests show the collimation to be maintained to less than a few arcseconds of secondary tilt in all practical orientations of the instrument, and to be stable over several years.

The 200 program stars fall into several categories: Hyades members, white dwarfs, subdwarfs, and very low luminosity red dwarfs. The stars have a wide range in apparent magnitude, the bulk falling between 11th and 16th magnitude. Unhypersensitized Kodak 103a-G + GG495 plates reach a threshold apparent magnitude of 16.4 in 40 minutes; we use a variety of plate-filter combinations.

3.2 Preliminary 1 Meter Results

The plate scale was determined from 7 exposures of 36 stars in the Russell Praesepe test region on IIIa-J plates behind a GG455 filter. The plates were measured on the MSSSO PDS and fit to the standard coordinates of the cluster stars corrected for proper motion. No significant magnitude or coma terms were present and the plate adjustment used a three constant model. The reductions performed by B. McNamara yield a plate scale of $15''.003 \pm 0''.003$ per mm.

A preliminary study of Barnard's star by McNamara used 39 plates measured on the McCormick Grant engine. The mean error of unit weight was $\pm$ 0.0022 mm in x and y, typical of the Grant machine. The solution employed 3 linear plate constants, as no significant additional terms were found. We find a relative parallax of $0''.556 \pm 0''.012$ and an annual proper motion of $10''.327$ in good agreement with other results. Most images were obtained with parallax factors 0.5 or less, leading to the low weight for the parallax.

4. BARNARD'S STAR RESULTS FROM THE McCORMICK 67 cm REFRACTOR

The reality of the small amplitude perturbation for Barnard's star remains questionable. Measurement of a selected sample of the best McCormick plates on the MSSO PDS does not support the Sproul perturbation.

This work is supported by the NSF, the Estate of Leander J. McCormick, and the Mount Stromlo and Siding Spring Observatories.

THE MONASH UNIVERSITY OBSERVATORY: EQUIPMENT AND RESEARCH

J.L. Innis, K. Thompson and D.W. Coates
Department of Physics, Monash University, Wellington Road,
Clayton, Australia

1. INTRODUCTION

The Monash Observatory, located at Mount Burnett, Victoria, seventy kilometres east of the city of Melbourne, at a latitude of 38° S has a 0.45-m Cassegrainian and a 0.25-m Newtonian telescope. Both are equipped for photoelectric photometry, with microcomputer-based data logging systems. The 0.45-m has recently replaced the observatory's original 0.4-m Newtonian. Our site is near the coast, and has less than 20% photometric weather. Access to the telescopes and facilities of Mount Stromlo and Siding Spring Observatories (MSSSO) increases our observing time and the scope of our research. For a description of the Monash Observatory at mid-1984 see Coates et al, (1984).

2. RESEARCH

While our original work was on short period pulsating stars, our current research interest is the study of chromospherically active stars. We commenced work on these stars in 1979, initially concentrating on southern RS CVn candidates, searching for broadband variations due to starspots, determining rotation rates, and searching for light curve variability due to changes in spot sizes and/or locations. In 1983 we began spectroscopic observations with the telescopes of MSSSO to study the binary nature of these stars and the changes in the Ca II and Hα emission lines. In 1984 we began collaborative work with radioastronomers from the Australian Commonwealth Scientific and Industrial Research Organisation (CSIRO), Division of Radiophysics (Drs. O.B. Slee, G.J. Nelson and A.E. Wright).

Light curves and other data which we have obtained illustrate the rapidity of the changes in the properties of spots on these stars (eg: Innis et al, 1984, and work to appear in Proc. Astron. Soc. Aust.). For example in figure 1 we show V light curves for the rapidly rotating single star AB Dor, for 1984 February-March, 1984

J. B. Hearnshaw and P. L. Cottrell (eds.), Instrumentation and Research Programmes for Small Telescopes, 293–294.

November, and 1985 February. The first two curves were obtained with the 0.25-m telescope, while the third, which shows a small flare near phase 0.8, was obtained at Siding Spring Observatory. The changes in the light curves are quite marked. Rucinski (1983) reported even faster changes for this star, on the order of days.

Our recent work on AB Dor suggests that it is a young star, just arriving on the main sequence. It appears to be very similar to the rapidly rotating Pleiades K dwarfs (Van Leeuwen and Alphenaar, 1983). We are continuing the analysis.

3. ACKNOWLEDGEMENTS

We acknowledge the contributions of former graduate students Len Halprin and Terry Moon to the development of the Monash Observatory. We thank the staff of the Department of Physics Electronic and Mechanical Workshops for excellent technical support. We also thank MSSSO for access to their telescopes and facilities. The Monash Observatory is supported by the Australian Research Grants Scheme, a Monash University Special Research Grant, the Monash/CSIRO Collaborative fund, and the Department of Physics. John Innis is supported by an Australian Commonwealth Postgraduate Research Award.

4. REFERENCES

Coates, D.W., Thompson, K., Innis, J.L. and Moon, T.T., 1984. In Advances in Photoelectric Photometry Vol. 2, p. 74, eds. Wolpert, R.C., and Genet, R.M., Fairborn, Ohio.

Innis, J.L., Coates, D.W. and Thompson, K., 1984. Proc. Astron. Soc. Aust. 5, 540.

Rucinski, S.M., 1983. Astron. Astrophys. Suppl. Ser., **52**, 281.

Van Leeuwen, F. and Alphenaar, P., 1983. IAU Coll. No. 71, Activity in Red Dwarf Stars, p. 189, eds. Byrne, P.B. and Rodono, M., Dordrect, Reidel.

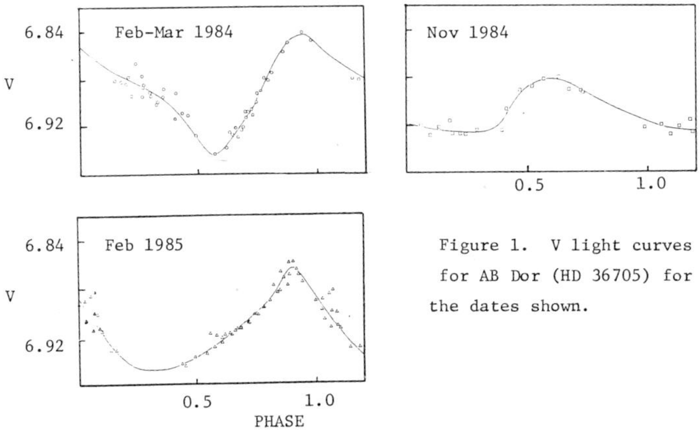

Figure 1. V light curves for AB Dor (HD 36705) for the dates shown.

LONG TERM PERIOD BEHAVIOUR OF CONTACT BINARIES

H.D. Kennedy
Astronomy Centre
Darling Downs Institute of Advanced Education
P.O. Darling Heights
Toowoomba, Queensland, Australia 4350

ABSTRACT. Determination of the rate of mass flow in contact binary systems is one of the most important motives for their observation. Many of these binaries are in over-contact configuration with a common equipotential surface located somewhwere between the inner and outer Lagrangian zero-velocity surfaces. Most are active systems displaying frequent lightcurve and period changes resulting in residuals behaviour depictable by intersecting linear segments.
Period changes are predominantly caused by mass loss or mass transfer. Period behaviour is frequently complicated by rotation of the line of apsides (resulting in out-of-phase periodic variations of minima) or orbital motion about a third component (resulting in periodic phase variations). Residuals diagrams have to be rectified for apsidal motion and lighttime effect.

Twenty four contact binaries have been under observation during a nine-year period. For sixteen of these, sufficient data relating to period changes, have been accumulated to enable plotting of residuals versus Heliocentric Julian dates as (O-C) diagrams. Such diagrams reveal that the long term period behaviour of most of these systems is of similar if not identical nature.

Decreases in period range from $.05^{d}-.55^{d} \times 10^{-5}$. This range is likely to reflect different stages of binary evolution in terms of mass ratio and degree of contact. A plot of period changes versus mass ratios confirms that period changes range between the above values which correspond to mass ratios between .4 and .9.

This clustering must be partly contributable to observational selection. Considering however, the range in colour and period of the systems selected, there appears to exist a relationship between the period changes, δP (in $d \times 10^{-5}$) and the mass ratios, q, of the components in the form: $\delta P = 0.48 - 0.47q \pm 1.3 \cdot 10^{-5}$). A number of period change values appears to fall outside the region covered by the above equation. Most likely this is attributable to surface activity (starspots).

A full paper on the above topic is in preparation and will be published in 1986.

J. B. Hearnshaw and P. L. Cottrell (eds.), Instrumentation and Research Programmes for Small Telescopes, 295.

PHASE-NIGHT DIAGRAM FOR PERIODIC VARIABLE STARS

Z. Kvíz
School of Physics
University of New South Wales
P.O. Box 1, Kensington, 2033
Australia.

A program has been developed for the Hewlett-Packard Computer 9845 to represent in a graphical form the period of observability, for individual nights, of periodic phenomena of stars with respect to the phase. One version of the program (Fig. 1) gives the observability with respect to the phase for the whole year. The other version (Fig. 2) gives the beginning and end of the suitable observing interval for several observatories around the world thus facilitating preparation of a program for 24-hour continuous observation.

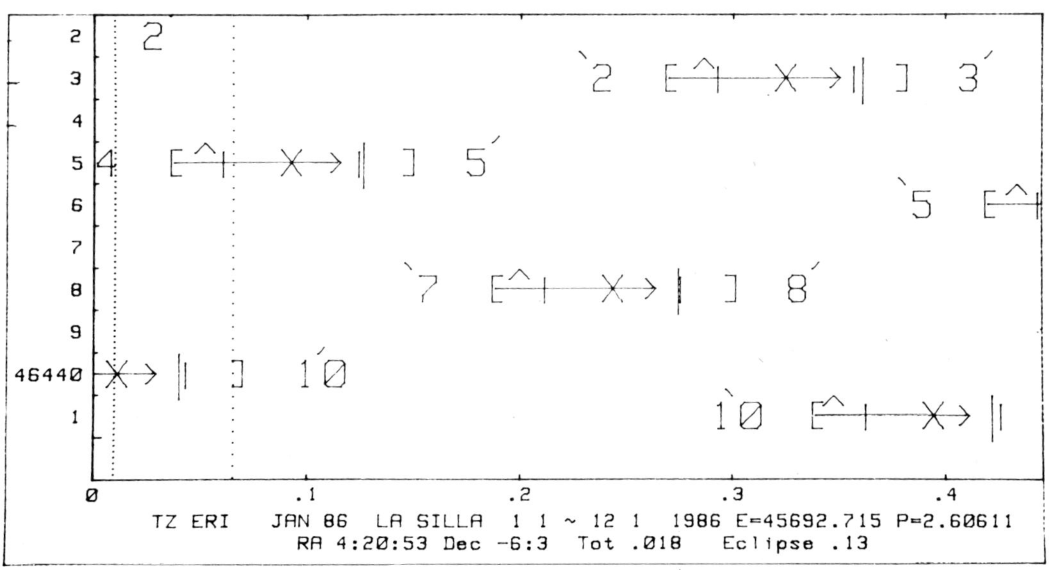

Fig. 1

J. B. Hearnshaw and P. L. Cottrell (eds.), Instrumentation and Research Programmes for Small Telescopes, 297–298.

EXPLANATION OF SYMBOLS USED IN THE DIAGRAM

____ - Full horizontal line - Phase of the period when the star is at air mass less than 2 during the night.

X - Phase at Midnight.

^ - Phase at Culmination.

< > - Phase at the times when the star is at air mass 2, rising or setting respectively.

[ ] - Phase at the end or beginning of astronomical twilight.

` ´ - Phase at 6 p.m. and 6 a.m.

| - Phase at 10 p.m. and 2 a.m. local solar time.

| - Phase at the time when the star is at air-mass 3 (not marked if in daytime or twilight).

Numbers are evening and morning dates, the inner edges of the numbers represent the phase at sunset and sunrise.

Vertical Axis - Julian Day.
Horizontal Axis - Phase (0 - primary minimum for eclipsing stars).

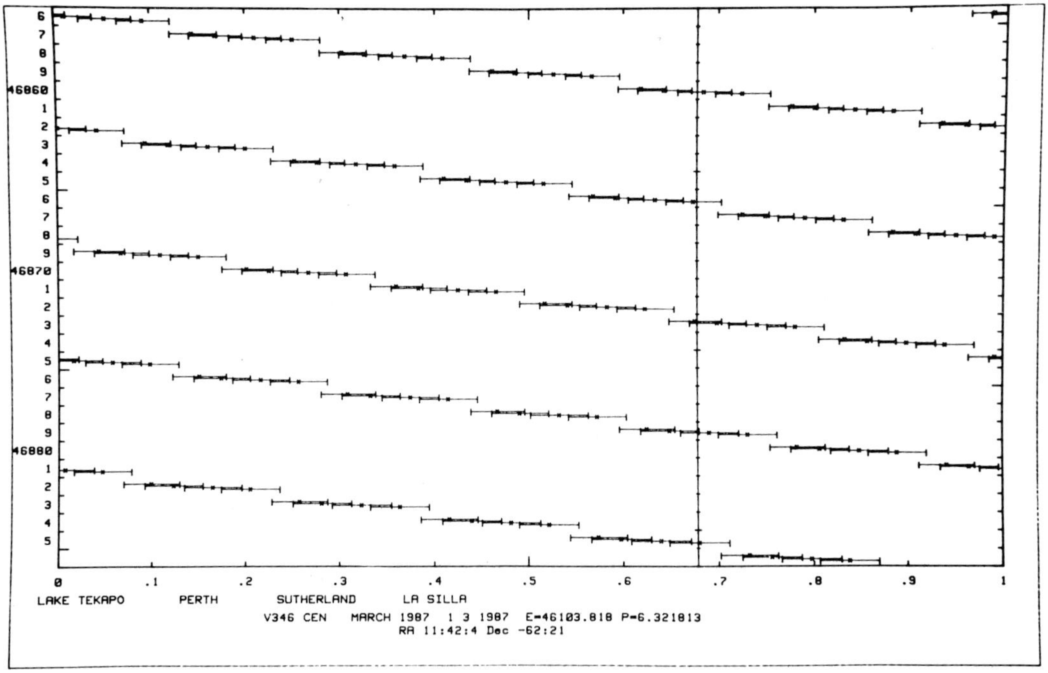

Fig. 2

PHOTOMETRY OF RZ OPHIUCHI - AN INTERNATIONAL COLLABORATION

L.B.G. Knee and C.D. Scarfe, University of Victoria, Canada
B.W. Baldwin, Fanshawe College, London, Canada
S.J. Meatheringham, MSSSO, Canberra, Australia

ABSTRACT. UBV observations, plus a few in R and I, were obtained during the 1984 eclipse of RZ Ophiuchi. Bolometric corrections and temperature calibration, applied to the magnitudes and colours of the stars, were used to derive the ratio of the stellar radii, and a light curve solution was obtained with this parameter fixed. Neither component fills its Roche lobe, but the system may be at a late stage of case C mass transfer.

The eclipsing binary RZ Ophiuchi consists of a small F type star and a K star that is so much larger and cooler that a) secondary eclipse is undetectable and b) primary eclipse cannot be made to yield all three of r_g, k and i independently (Forbes and Scarfe 1984). In his pioneering study of the system, Baldwin (1978) chose to fix the inclination i at 90°. He concluded that neither star fills its Roche lobe, and thus left unexplained the system's circumstellar matter and evolutionary state. Smak (1981) reinterpreted Baldwin's data on the assumption that the radius of the larger star r_g equals that of its Roche lobe. Forbes and Scarfe showed that either model was consistent with new data from the 1981 eclipse, but suggested that a value of k similar to Baldwin's might be obtained from the stars' bolometric magnitudes and effective temperatures.

However, better photometry was needed to decide the matter finally. To obtain good coverage of the partial phases UBV observations of the 1984 June eclipse were obtained from western North America, Sicily and Australia. In addition a few R and I observations were obtained during totality and outside eclipse. The comparison stars were the same as used by Baldwin, Olson and Hickey (1983) and Forbes and Scarfe and the observations were transformed to the standard system of Landolt (1983).

Colour indices were derived for each star from the eclipse depths, on the assumptions that variations due to the disk eclipse detected by Olson and Hickey are small, and that the disk does not affect the colours significantly. These colours were corrected for reddening using the visual absorption $A_V = 0.4$ found by Olson and Hickey and were

J. B. Hearnshaw and P. L. Cottrell (eds.), Instrumentation and Research Programmes for Small Telescopes, 299–300.

then used to estimate effective temperatures and bolometric corrections for the stars. For the hotter star B-V and V-R were matched to Kurucz (1979) model atmospheres, assuming the star to have a supergiant's surface gravity consistent with its F5Ib spectral type and its Stromgren c_1 index. The cooler star's V-R and R-I were converted to the Johnson system and its effective temperature and bolometric correction derived from the tables of Johnson (1966).

The bolometric magnitudes and effective temperatures were used to derive the ratio of the stellar radii, k, and this was imposed on Russell-Merrill solutions of the U, B and V light curves. The results obtained by this procedure were as follows:

Star	Hotter	Cooler
Effective Temperature	6600° ± 250°	3600° ± 200°
Fractional Radius	0.017 ± 0.002	0.13 ± 0.02
Orbit Inclination	90° ± 4°	

Further insight into the physical nature of RZ Oph requires a radial-velocity solution of greater accuracy than the preliminary one of Baldwin. Fortunately M. Mayor has derived a new orbit from CORAVEL observations which he has kindly made available to us. It leads to masses of 5.7 ± 0.2 $M_\odot$ and 0.7 ± 0.1 $M_\odot$ for the hotter and cooler stars respectively. Despite the small size of its Roche lobe, the cool star's radius is only 0.6 that of the lobe.

Thus RZ Oph seems now to be detached, although it has almost certainly undergone considerable mass transfer in the past. It may be that it is an example of Case C mass transfer, for which Lauterborn (1970) predicted a quiescent phase between the major transfer stage and a later lesser one. We believe RZ Ophiuchi may be in this quiescent stage.

We are greatly indebted to the directors of Pine Mountain Observatory, Lick Observatory, Osservatorio di Catania and Mt. Stromlo and Siding Spring Observatories for the use of their telescopes, and to Dr. M. Mayor for providing us with his unpublished data. A more complete account is in preparation jointly with him.

References

Baldwin, B.W. 1978. Ap. J. 226, 937.
Forbes, D.W. and Scarfe, C.D. 1984. Pub. Ast. Soc. Pac. 96, 737.
Johnson, H.L. 1966. Ann. Rev. Ast. Ap. 4, 193.
Kurucz, R.L. 1979. Ap. J. Supp. 40, 1.
Landolt, A.U. 1983. Ast. J. 88, 439.
Lauterborn, D. 1970. Ast. Ap. 7, 150.
Olson, E.C. and Hickey, J.P. 1983. Ap. J. 264, 251.
Smak, J. 1981. Acta Ast. 31, 25.

TIMES OF MINIMA AND PERIOD CHANGES OF ECLIPSING BINARIES AS A PROGRAMME FOR SMALL TELESCOPES

J. M. Kreiner
Astronomical Department, Institute of Physics,
Higher Pedagogical School
ul. Podchorazych 2, Krakow, Poland

Investigations of times of minima in eclipsing binaries give us important clues concerning a number of quite different and frequently unrelated phenomena such as the apsidal motion, mass-loss, mass-transfer or gravitational radiation. It is hard to over-estimate the importance of these relatively easy observations. However, simple statistics show that much remains to be done to assure proper data for even a moderate-size sample of eclipsing binaries. Here we present essential results of such statistics made for stars brighter than 12 magnitude in minima (this limit assumed arbitrarily as accessible to small-telescope observers).

For selection of stars brighter than 12 magnitude in minima we used the 3-rd edition of the General Catalogue of Variable Stars with its three Supplements. These stars were checked for entries of times-of-minima in the Card-Index Catalogue of the Cracow Observatory (unpublished). These data were supplemented by visual minima published by B.B.S.A.G., A.A.V.S.O., B.A.V., and others. In addition a collection of about 20 thousand minima and corresponding O-C curves for W UMa systems (Z.Glownia,unpublished) was used.

The number of eclipsing binaries in both hemispheres was:

N: $\delta > 0°$	549	systems
S: $\delta < 0°$	539	systems
Total	1088	systems

of which

Algol-type (EA)	670	systems
Beta Lyrae-type (EB)	269	systems
W UMa-type (EW)	125	systems
unknown type (E)	24	systems

The ephemerides are better established for northern systems: those published in the Rocznik Astronomiczny Obserwatorium Krakowskiego, International Supplement (SAC) are available for:

N: 398 systems (72 percent)
S: 280 systems (52 percent).

The amateurs provide a large fraction of minima for all 1088 systems and again the difference between hemispheres is rather obvious:

J. B. Hearnshaw and P. L. Cottrell (eds.), Instrumentation and Research Programmes for Small Telescopes, 301–302.

N: 374 systems (68 percent)
S: 112 systems (21 percent)

An additional difference is in the uniformity of observations. If we divide the stars according to the year of the last observed minimum, we see a striking lack of recent observations of southern stars. In addition, for 40 percent of southern stars (and for 24 percent over the whole sky) we have no data at all!

THE LAST OBSERVED MINIMUM
(photoelectric or visual or photographic)

Years	Northern systems		Southern systems	
	number of stars	percent	number of stars	percent
1984-1985	172	31	28	5
1981-1983	170	31	45	8
1971-1980	91	17	123	23
earlier than 1971	69	13	127	24
lack of minima	47	8	216	40

Two additional comments:
1. Only for about 85 binaries have minima been observed systematically (photoelectrically and visually) during the recent 30 years, this mostly being work of amateurs; the contributions of professional astronomers have been rather small.
2. The secondary minima, which are very important for apsidal-motion investigations, and stars with amplitudes less than 0.4 mag. are not observed by amateurs. These very important gaps should be filled by photoelectric observations.

PHOTOMETRIC RESEARCH WITH A SMALL TELESCOPE (BAMBERG 37-cm. REFRACTOR) AT BOSSCHA OBSERVATORY, INDONESIA

Hakim L. Malasan * ; Akira Okazaki **
Atsuma Yamasaki *** and Gregorius Hadiyanto *
* Bosscha Observatory ITB, Lemband, Java, Indonesia
** Tsuda College, Kodaira, Tokyo 187, Japan
*** Department of Earth Science and Astronomy, University of Tokyo, Meguro-ku, Tokyo 153, Japan

In 1979 Partoharsodjo (see Kitamura et. al., 1985) constructed a new photoelectric photometer with an RCA IP21 photomultiplier, so that it : 1) can be coupled easily with the Bamberg 37-cm. refractor at the Bosscha Observatory, and 2) can be used both for direct current and photon counting modes. Since its installation at the telescope in the early 80's, photometric work at Bosscha Observatory was resumed.

In view of meteorological and telescope conditions, and optical limitations, we put emphasis on the observations of bright variable stars ($\acute{m} \leqslant 9$) south of declination +17°. In this programme differential, rather than absolute, photometry can be applied. For variable star research we have selected eclipsing binaries with periods of around 24 hours whose light curve are not yet well determined or not well understood.

Photometric activities since 1981 include:

1. Observations of eclipsing binaries
   Eight pairs (DV Aqr, δ Cap, DX Aqr, Y Aql, EE Peg, ζ Phe, 35 UU Psc, and V 822 Aql) are on our current observing programmes. DV Aqr has been studied in 1983. Analysis of its light curve using the synthesis method developed by Yamasaki (1981) lead us to conclude that this object is a short period non-contact close binary system (Okazaki et. al., 1985). In 1985, the light curve of δ Cap was completed at Bosscha Observatory and part of it had been obtained at Okayama Observatory (Yamasaki, 1985, private communication). A preliminary study shows that this system thus belongs to a few of such systems of binary stars. Observations of the other six pairs are being continued.
2. Atmospheric extinction study
   In order to obtain the extinction coefficients, UBV standard stars (kindly prepared by Nishimura, 1981) were used, both in absolute photometry and differential photometry. The coefficients of extinction (k_v and k_b) were derived from observations of these stars at various zenith distances. As a side study, we found high values of the annual average of k_v and k_b for the year 1982. This is attributed to fine dust originating from the eruption of Mt.

J. B. Hearnshaw and P. L. Cottrell (eds.), Instrumentation and Research Programmes for Small Telescopes, 303–304.

Galunggung and floating in the stratosphere (Malasan et. a., 1984; Hidayat and Malasan, 1985).

3. Be stars photometry
This work has just started both at Bosscha and Okayama Observatories (Dawanas et. al., 1985).

For the future, observations of variables with periods of around 24 hours, occultation of stars by planets, the moon or satellites, bright flare stars and comet heads may be best carried out in coordination with several observatories located along a considerable length of geographic longitude. This coordinating work would ensure better time and spatial resolutions, and thus higher precision. As for Indonesia, a plan for a 60-cm. reflector for modern astronomical photometry is underway.

Our thanks to Profs. B. Hidayat and M. Kitamura for their constant interest in our work. To Messrs. C. Nurwendaya and A. Rosana we owe much valuable technical advice.
The Institute of Technology, Bandung, has put the observational facilities at our disposal, for which we would like to record our gratitude. Last but not least, one of us (HLM) would like to thank the Leids-Kerkhoven-Bosscha Foundation, whose travel support has made it possible for him to attend this meeting.

REFERENCES

Dawanas, D.N., Hirata, R.; Kogure, T. : 1985, in 'Galactic Structure and Variable Stars', Proc. 6-Year Cooperation between Indonesia-Japan in Astronomy, T. Kogure and B. Hidayat (eds.), Tokyo, 1985 p. 95.

Hidayat, B.; Malasan, H.L. : 1985, 'Coefficients of Extinction at Lembang', IAU Comm. no. 50 meeting in India.

Kitamura, M.; Yamasaki, A.; Okazaki, A.; Malasan, H.L. : 1985, in 'Galactic Structure and Variable Stars', Pro. 6-Year Cooperation between Indonesia-Japan in Astronomy, T. Kogure and B. Hidayat (eds.), Tokyo, 1985, p.78.

Malasan, H.L.: Nurwendaya, C.; Hidayat, B. : 1984, Proc. of the Institute of Technology, Bandung, vol. 17, p. 11 (in Indonesian).

Okazaki, A.; Yamasaki, A.; Nurwendaya, C.; Malasan, H.L. : 1985, Publ. Astron. Soc. Pacific, 97, 62.

Yamasaki, A. : 1981, Astrophys. Space Sci., 77, 75.

A NEW B-SUBDWARF ECLIPSING BINARY WITH AN EXTREMELY SHORT PERIOD

J W Menzies and F Marang
South African Astronomical Observatory
P O Box 9
Observatory 7935
South Africa

ABSTRACT. The B-type Subdwarf BD-07° 3477 has been found to be an eclipsing binary with a period of about 2^h48^m. A preliminary analysis of the light curve reveals the system to be fully detached. It appears to be similar in many respects to the O-Subdwarf binary AA Dor which is thought to be the remnant of a system which has undergone a common envelope phase of evolution.

1. INTRODUCTION

During a UBV Survey of UV-bright objects from the S2/68 catalogue (Carnochan and Wilson, 1983) we found BD-07° 3477 (α 1241.8, δ - 0824 (1950)) to be an eclipsing variable. This star has been classified as a possible B subdwarf by Berger and Fringant (1980).

2. ANALYSIS

(a) Period

From 27 primary minima covering an interval of 493 days, we find the following ephemeris:-

$$\text{HJD (primary minimum)} = \underset{\pm .000017}{2445730.556057} + \underset{\pm .0000000098}{0.116719651}.n$$

The remarkably short period (2^h48^m) is only slightly longer than that found for the central star of Abell 41 (2^h43^m) by Grauer and Bond (1983).

b) Velocity Curve

The star is a single-lined spectroscopic binary and from the spectra we have derived the following parameters:-

$k_1 = 87.9 \pm 4.8$ km/sec $\qquad \gamma = -12.5 \pm 3.6$ km/sec

$a_1 \sin i = 1.41 \times 10^5$ ($\pm .08 \times 10^5$) km $\qquad f(M) = 0.0082 \pm .0013$ Mo

$= 0.20\ R_\odot$

J. B. Hearnshaw and P. L. Cottrell (eds.), Instrumentation and Research Programmes for Small Telescopes, 305–306.

c) Light Curve

The best $UBV(RI)_c$ light curves available to us were used for analysis. All observations were made with the 0.5m or 1.0m telescopes in Sutherland. Neither eclipse is total, and there is clearly a very pronounced reflection effect.

An exploratory investigation revealed that the inclination of the system is high, that both stars must have high gravities for any reasonable value of the mass ratio and that the secondary contributes very little light except through the reflection effect. We used the parameters determined from this simple analysis as starting values for the Wilson-Devinney differential correction program. The temperature of the primary was set at 26000K on the basis of the UBV colours and the Q-θ relation of Greenstein and Sargent (1974) for high gravity objects. Theoretical limb darkening coefficients were used for the hot star, while for the secondary it was found necessary to use a value of 1.0 for all colours. A geometric albedo of 1.0 was used for the secondary. We found T_2 = 4500K to give a reasonably good fit though this may be uncertain by $\pm$ 500K or more. Photometry in the infrared is needed before a satisfactory temperature can be found for the secondary.

A solution was performed for all 5 light curves simultaneously, so that the parameters that were derived are the best compromise. This resulted in polar radii (in units of the separation) of r_1 = 0.203 and r_2 = 0.207.

A plausible combination of masses is found to be m_1 = 0.25 $M_\odot$ and m_2 = 0.12 $M_\odot$ in which case the primary could be a hydrogen shell burning star with a degenerate core while the secondary could lie on the lower zero age main sequence. The system could have passed through a common envelope stage of evolution when the primary became a red giant, and then have lost a substantial fraction of its mass. This is the model proposed for the O-subdwarf binary AA Dor by Paczynski (1980), though in that case the primary is hotter and the secondary less massive than in BD-07° 3477. Other similar objects in a different evolutionary stage may be the central star of Abell 41 and UU Sge.

Since BD-07° 3477 is so relatively bright, it might be possible to see evidence of the secondary in high dispersion spectra. Infrared photometry would be useful in determining the secondary temperature.

REFERENCES

Carnochan, D.J. and Wilson, R., 1983. Mon. Not. R. astr. Soc., **203** , 317.
Berger, J. and Fringant, A.-M., 1980. Astron. Astrophys., **85** , 367.
Grauer, A.D. and Bond, H.E., 1983. Astrophys. J., **271** , 259.
Greenstein, J.L. and Sargent, A.I., 1974. Astrophys. J. Suppl., **28** , 157.
Paczynski, B., 1980. Acta Astron., **30** , 113.

PHOTOMETRY OF SELECTED CEPHEIDS

Thomas J. Moffett
Department of Physics
Purdue University
West Lafayette, Indiana 47907
USA

ABSTRACT. Cepheids with periods around 15 days sometimes show a reversal, or dip, in their light curves near phase 0.85. The theoretical interpretation of this feature is still in doubt. We present a list of suspected dip Cepheids, accessible to small telescopes, which need additional photometry to verify the dip.

Ever since 1912 when Henrietta Levitt published the first period-luminosity relation, Cepheids have been the cornerstone of the cosmological distance scale. Over the past 75 years, Cepheids have received so much attention from observers that one might ask if there is anything new to be gained from additional Cepheid photometry. The answer is a resounding yes! There are features in the light curves of Cepheids which need better observational definition.

A new feature in the light curve of the Cepheid X Cygni was reported by Davis, Moffett and Barnes (1981). On the rising branch of the light curve a dip in luminosity occurs near phase 0.85. This dip was first observed by the amateur astronomer H.J. Landis (1973) but went unnoticed by the professional community. The physical cause for this reversal on the rising branch is still uncertain but two possibilities have been suggested.

Theoretical Cepheid light curves sometimes show dips in the computed curves associated with the passage of a compression wave through the photosphere. Since real Cepheids did not show this dip, most theorists argued that the dip was an artifact of the approximate treatment of shocks in the models. The photometry of X Cygni showed a dip at the phase predicted by theory.

Under the assumption that the theoretical dips were real, Davis, Moffett and Barnes (1981) were able to fit theoretical light curves to the observed curve of X Cygni and thereby determine a mass for the Cepheid. The mass of X Cygni, determined in this manner, was found to be nearly the evolutionary mass. This may be a fortuitous result so it must be tested on other dip Cepheids. If this interpretation turns out to be correct, then we have a new means of attacking the Cepheid mass problem.

Klapp, Goupil and Buchler (1985) put forth a different interpret-

J. B. Hearnshaw and P. L. Cottrell (eds.), Instrumentation and Research Programmes for Small Telescopes, 307–308.

ation which does not depend on the Cepheid's mass. They suggest that the dips may be related to an amplitude saturation mechanism resulting from an interaction between the linear driving and the nonlinear quadratic coupling terms.

The theoretical interpretation will probably remain uncertain until more examples of dip Cepheids are found by observers. Using the light curves published by Moffett and Barnes (1980, 1984) and Pel (1976) we have selected Cepheids which show some indication of a dip (see Table I).

This work was supported in part by NSF grant AST-8417844. This financial assistance is gratefully acknowledged.

TABLE I
DIP CEPHEID CANDIDATES

CEPHEID	α (1900)	δ (1900)	PERIOD (DAYS)	<V>	COMMENTS
SZ Aql	18 59 35	+01 09.4	17.141	8.675	SUSPECT
TT Aql	19 03 09	+01 08.5	13.754	7.182	SUSPECT
RW Cam	03 46 10	+58 21.3	16.415	8.720	DEFINITE
RW Cas	01 30 43	+57 14.9	14.798	9.281	SUSPECT
TX Cen	14 27 36	-60 33.0	17.090	10.530	SUSPECT
SZ Cyg	20 29 38	+46 15.6	15.110	9.469	SUSPECT
TX Cyg	20 56 26	+42 12.4	14.710	9.572	SUSPECT
CD Cyg	20 00 37	+33 49.7	17.074	9.015	DEFINITE
β Dor	05 32 45	-62 38.3	9.842	3.754	SUSPECT
AA Gem	06 00 22	+26 20.3	11.302	9.741	SUSPECT
Z Lac	22 36 55	+56 18.4	10.886	8.451	SUSPECT
SV Mon	06 16 04	+06 30.9	15.233	8.302	DEFINITE
SZ Mon	06 46 24	-01 15.2	16.375	10.286	SUSPECT
AD Pup	07 43 53	-25 19.7	13.594	10.965	SUSPECT
WZ Sgr	18 11 06	-19 06.6	21.850	8.090	DEFINITE

REFERENCES

Davis, C.G., Moffett, T.J., and Barnes, T.G. 1981, Ap. J., 246, 914.
Klapp, J., Goupil, M.J., and Buchler, J.R. 1985, Ap. J., 296, 514.
Landis, H.J. 1973, A.A.V.S.O., 2, 38.
Moffett, T.J., and Barnes, T.G. 1980, Ap. J. Suppl., 44, 427.
1984, Ap. J. Suppl., 55, 389.
Pel, J.W. 1976, Astr. Ap. Suppl., 24, 413.

PHOTOMETRY OF THE RS CVn BINARY II PEG

S. Mohin, A.V. Raveendran and M.V. Mekkaden
Indian Institute of Astrophysics
Bangalore 560034
India

B and V photometry of II Peg obtained at Kavalur on 39 nights during 1980-81, 1981-82 and 1984-85 observing seasons is presented in Fig. 1. From an analysis of the available data on this object we find that no two light curves agree in any of the following : shape, amplitude, phases of the light maxima and minima, mean light level, or brightness at the light maxima and minima.

In Fig. 2 we have plotted the phase of light minimum against the mean Julian Day of observation. The values are visual estimates from the respective light curves. Photometric data on II Peg are taken from Chugainov (1976), Rucinski (1977), Vogt (1981), Nations and Ramsey (1981), Raveendran, Mohin, and Mekkaden (1981), Bohusz and Udalski (1981), and Rodono, Pazzani, and Cutispoto (1982).

It is interesting to see that during the interval from JD 2444175 to JD 2444595, the light curves had two minima indicating the simultaneous presence of two major active regions, say A and B, separated in longitude. The solid lines represent the 'best fit lines' passing through well defined minima. The slightly different slopes indicate different photometric periods of 6.672 ± 0.010 and 6.694 ± 0.002 days for these two regions A and B respectively. This implies that the respective regions were located at different latitudes also. The minima observed at JD 2444878 and JD 2444988 were also included in group B for the solution of the photometric period, since both lie close to the line passing through the minima of group B.

Further, we find that both the minima observed by Rucinski in 1976 and Vogt in 1977 lie on the line corresponding to group B. Probably the active region had its origin sometime in 1976 and finally disintegrated sometime in 1981. This puts a limit of about 5-6 years for the life time of the active region. The minima belonging to group A which appeared in 1979 could not be traced after 1981 and hence probably had a shorter life time. Similar values for the life time for the centres of activity have been found in the case of DM UMa, another active member of the RS CVn class (Mohin et al. 1985).

The minima observed by Chugainov in 1974 follow a completely different path indicating that the corresponding activity had a different latitude and longitude. Again the recent photometry shows that the two

J. B. Hearnshaw and P. L. Cottrell (eds.), Instrumentation and Research Programmes for Small Telescopes, 309–311.

minima observed do not fall on any of the two lines and hence probably are due to different active regions formed later.

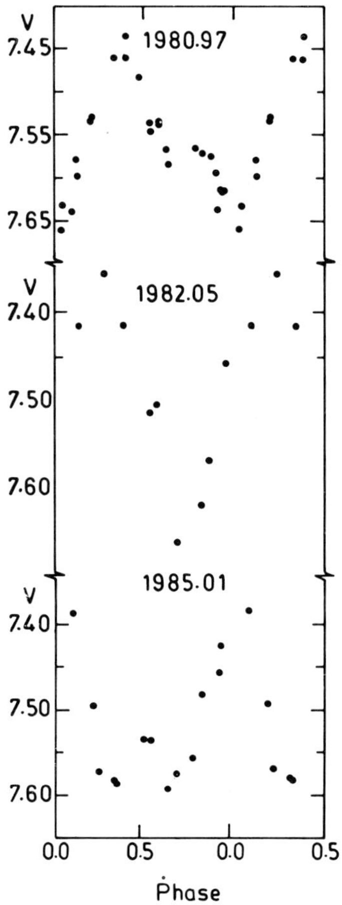

Figure 1

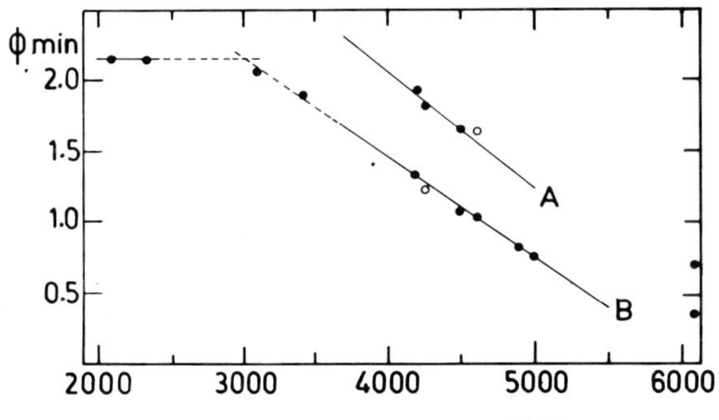

Figure 2

It is also found that there is apparently no correlation between the colour (B-V) and the visual magnitude, but there is a large scatter with a total range of ~ 0.1 mag, much larger than the expected observational error.

An extended version of this paper will be submitted to Bulletin of the Astronomical Society of India.

We are grateful to Dr J.C. Bhattacharyya and the late Dr M.K.V. Bappu for their kind encouragement during the course of this work.

REFERENCES

Bohusz, E. & Udalski, A., 1981. Acta Astra., 31, 185.

Chugainov, P.F., 1976. Bull. Crim. Astrophys. Obs., 54, 89 (Russian).

Mohin, S., Raveendran, A.V., Mekkaden, M.V., Hall, D.S., Henry, G.W., Lines, R.D., Fried, R.E., Louth, H. & Stelzer, H.J., 1985. Astrophys. Space Sci.

Nations, H.L. & Ramsey, L.W., 1981. Astr. J., 86, 433.

Raveendran, A.V., Mohin, S. & Mekkaden, M.V., 1981. Mon. Not. R. Astr. Soc., 196, 289.

Rodono, M., Pazzani, V. & Cutispoto, G., 1982. Activity in Red-Dwarf Stars IAU Colloq. No.71, p.179, Reidel, Dordrecht, Holland.

Rucinski, S.M., 1977. Publs. Astr. Soc. Pacif., 89, 280.

Vogt, S.S., 1981. Astrophys. J., 247, 975.

STANDARD STARS FOR COMET PHOTOMETRY

Wayne Osborn
Physics Department, Central Michigan University
Mt. Pleasant, Michigan 48859, U.S.A.

Peter Birch
Perth Observatory
Bickley, Western Australia 6076, Australia

Michael Feierberg
Astronomy Program, University of Maryland
College Park, Maryland 20742, U.S.A.

ABSTRACT. Photoelectric photometry of comets is a feasible and worthwhile research program for small telescopes. Nine filters designed to isolate selected emission features and continuum regions in cometary spectra have been adopted by the IAU as those recommended for comet photometry. Magnitudes of standard stars for the IAU filter system have been derived. The results have been compared with previous observations and transformations to place all observations on a common system obtained.

1. INTRODUCTION

A profitable research program for small telescopes is the photoelectric monitoring of comets. Because comets appear suddenly and most studies require measurements made over an extended period of time, observations with larger, heavily scheduled instruments are often impossible. Furthermore, comets are extended objects and for projects involving the comet's integrated brightness a small scale is preferable. We have established a system of photoelectric standards intended for comet photometry.

2. COMET PHOTOMETRY

Comet photometry can be divided into two types: studies of the continuum and studies of emission lines. Continuum observations are directed toward understanding the dust component of the comet, and ideally only involve measures of reflectivity of the solar spectrum. The emission line studies are concerned with the production of the components of the

J. B. Hearnshaw and P. L. Cottrell (eds.), Instrumentation and Research Programmes for Small Telescopes, 313–314.

gas, and ideally involve only measures of the emitted flux of a particular species as a function of solar distance. In practice, the two types of studies are intertwined.

In order that observations be standardized, IAU Commission 15 has adopted a set of filters recommended for photoelectric observations of comets. Briefly, the set consists of five primary and four secondary filters selected to measure various emission features and continuum points. A large number of IAU filter sets have been distributed to observers worldwide through the International Halley Watch in preparation for the current appartition of Comet Halley.

3. STANDARD STARS

The adoption of a filter system for comet photometry necessitates the establishment of a system of photometric standards. Two classes of standard stars are required. First, one must have G-type stars that mimic as closely as possible the solar spectrum for use in the continuum studies. Second, one needs flux standards, usually chosen to be B stars because of the few spectral lines, to calibrate the emission line fluxes. Obviously, the standard stars must be distributed in right ascension, and it is preferable if they are equatorial stars accessible from observatories in both hemispheres. At the same time, comets often are observable only for a few hours near dawn or dusk and at high air masses. This mandates standards close to the comet in the sky, and it is therefore usually also necessary to have a set of secondary standards located along the path of the comet.

A set of primary standard stars for comet photometry has been suggested by A'Hearn (Solar System Photometry Handbook, 1983). Seven solar analogs were selected as the G standards and twelve early B stars chosen as flux standards. Secondary standards, of course, need to be chosen for each comet. Examples of secondary standards are the list of Comet Halley stars selected by A'Hearn, Vanysek, and Campins (International Halley Watch Newsletter No. 4, 1984).

4. OBSERVATIONAL MATERIAL

We have obtained observations of A'Hearn's suggested primary standards and Comet Halley secondary standards using the IAU filters. The northern Hemisphere observations were obtained at Lowell Observatory employing the 0.6-meter and 1.1-m reflectors. The southern hemisphere observations were made at the Perth Observatory with a 0.6-meter reflector. These data are available from the authors.

A comparison of our results with the IAU comet filter observations of Wisniewski and Zellner (Icarus 63,333, 1985) shows good agreement, averaging .025 mag. after correction for different zero points. The agreement with the preliminary results of Vanysek and Wolf is less satisfactory, averaging 0.08.

This work was made possible by an International Halley Watch contract and by a Central Michigan University research leave.

ASTROMETRY WITH SMALL TELESCOPES

W. S. Penhallow
Physics Department
University of Rhode Island
Kingston, Rhode Island, 02881
U. S. A.

ABSTRACT. At the Quonochontaug Observatory of the University of Rhode Island, we specialize in the development of simple seeing limited optical telescopes. Our main instrument is a heavy Serrier trussed f/10.7, .4-m astrometric reflector which provided the last-minute astrometry in the successful prediction of the occultation of SAO 120774 by Herculina in 1978. It is similar in design to the U. S. Naval Observatory 1.5-m reflector. Its scale of 47.1"/mm makes it potentially more powerful than the Lick .5-m, f/7 astrograph. However, its field of view of 1.3° x 1.7° is insufficient to capture enough AGK3R reference stars which are needed to consistently provide tenth-of-a-second-of-arc accuracy. To solve this problem, a .24-.32-m, f/6.3 lensless Schmidt was developed to provide a 3.8° field at 135"/mm and is mounted on the .4-m telescope. It can be used by itself and/or to provide a secondary reference frame for the main instrument. The measures from the plates and film are reduced by a suite of interactive computer programs which allows one to choose the proper reduction model based on the Eichhorn criterion. Over 100 positions of stars, asteroids, comets and supernovae have been provided with these instruments, and a regular observing program including participation in the IHW is underway.

1. RESULTS

1.1. Table I - Residuals for Comet Observations (Astrometric Reflector)

	1978 U.T.	$\Delta\alpha\cos\delta$	$\Delta\delta$	
		"	"	
Wild 2 (1978b)	Feb.9.08	+0.8	+0.6	
	10.11	-0.3	+0.8	
	12.08	-0.5	0.0	
Meier (1978f)	May 3.12	+0.7	[+2.1]	[disc.]
	RMS	0.6	0.6	

J. B. Hearnshaw and P. L. Cottrell (eds.), Instrumentation and Research Programmes for Small Telescopes, 315–316.

1.2. Table II - Residuals for P/Giacobini-Zinner (Schmidt)

	1985 U.T.	$\Delta\alpha\cos\delta$	$\Delta\delta$
		"	"
P/Giacobini-Zinner	July 24.19955	+ .77	-1.24
	25.25833	- .45	- .33
	Aug. 3.25833	-1.33	+1.15
	5.26771	[-5.21][disc.]	-1.30
	7.21965	+ .32	- .60
	18.34722	+1.55	- .73
	RMS	.99	.96

1.3. Table III -Internal Residuals for Amphitrite (Schmidt)

	1985 U.T.	$\Delta\alpha\cos\delta$	$\Delta\delta$
		"	"
Amphitrite	June 21.08057	+0.07	-0.17
	21.08542	-0.12	+0.30
	21.09167	+0.05	-0.13
	RMSF	0.08	0.21
	RMSO	0.05	0.12

1.4. Table IV - A Star That Might Be Occulted by P/Halley (Schmidt)

No.	hr	m	s	RMSO	°	'	"	RMSO
1511	2	4	21.494	(±.23)	+17	42	5.94	(±.19)
			.497				4.39	

2. CONCLUSIONS

It has been demonstrated that simple reflecting telescopes of small size and low cost can be used to make useful astrometric observations when suitable measuring engines are available. The rapid turn-around time, particularly with asteroid, comet and planetary occultations, provides for a powerful means of evaluating computer errors, reduction models, telescope optics, measuring techniques, star catalog positions, and general procedures. Accuracies better than 0".1 can be obtained with the .4-m (16-inch), f/10.7 astrometric reflector with four exposures under favorable conditions. Corresponding accuracies for the .24-.32-m, f/6.3 lensless Schmidt are about 0".2. A linear six constant reduction model is adequate to attain the latter accuracy with the Schmidt when the field is kept to 3°.8.

3. RECOMMENDATIONS

Occultations of stars by solar system bodies have become a very active and fruitful field of endeavor. The need for more astrometry is pressing. It is recommended that (a) more observatories, both professional and amateur, be encouraged to attempt this type of work; (b) a survey be made of the old and little-used measuring engines that might be made available for possible long-term loan to observatories interested in doing this type of work; (c) that some type of amateur-professional organization be formed to guide and nurture the reawakening of the ancient art of doing good astrometry.

A DESCRIPTION OF A CATALOGUE OF BRIGHT uvbyβ STANDARD STARS

Charles L. Perry
Department of Physics & Astronomy
Louisiana State University Observatory
Baton Rouge, Louisiana 70803, USA

David L. Crawford
Kitt Peak National Observatory
P. O. Box 26732
Tucson, Arizona 85726 USA

ABSTRACT. A description is given of an all-sky catalogue of bright standard stars for the uvbyβ photometric systems.

During the planning stage of a long-term photometric investigation of the spatial distribution of interstellar reddening in the southern hemisphere, one of us (CLP) has composed an all-sky master list of bright uvbyβ standard stars for use with small telescopes. The list was compiled from uvby and/or beta photometric survey papers of bright stars published by various Copenhagen and Tucson observers during the past 25 years. Galactic cluster papers were excluded. The only criterion for inclusion in the master list was that each star have a minimum of ten uvby and/or beta observations in a given reference. The photometric references were then searched for all uvby and/or beta observations for the stars in the list. Weighted uvby and beta averages were determined for each star; these data became the basis of the present catalogue. The final catalogue tabulates, in succession, the HR numbers, the 2000-year coordinates, and the spectral types (from the Bright Star Catalogue (Hoffleit and Jaschek, 1982)); the visual magnitudes (to be discussed below); and the uvby and beta indices. The total number of observations and the sources of the data are listed with each type of photometry.

The situation is not so simple for the visual magnitudes. When available, the visual magnitudes were taken from the intermediate-band references, but no minimum was imposed on the number of observations. These visual magnitudes were derived from observations made with the intermediate-band y filter transformed to the UBV system. For the remaining stars, visual magnitudes were assigned from the wide-band photometric references. Because the half-width of the y filter of the uvby system differs from that of the V filter of the UBV system, the visual magnitudes derived from the uvby photometry may deviate slightly from those derived from UBV photometry. The visual

J. B. Hearnshaw and P. L. Cottrell (eds.), Instrumentation and Research Programmes for Small Telescopes, 317–318.

magnitudes for many catalogue stars are not firmly established due to: (a) the half-width differences in the intermediate- and wide-band photometric systems, and (b) the lack of a sufficiently large number of observations per star. Clearly, more data are required even for naked-eye stars. Crawford will present a paper at this symposium which proposes a solution to this problem.

The stars in the master list were checked against the Index Catalogue of Visual Double Stars (Jeffers and van den Bos). Those standard stars which are members of visual binary systems are listed in a separate table which tabulates all components within 10 arc-sec of a given primary. A check was also made on the external consistency of the photometric references. Mean values of the photometric indices were determined for the stars in the "standard" references. Next, the residuals in the indices were found for all stars common to a given reference and the "standard" references. For all indices, the mean difference and its mean error (one star) between a given reference and the "standard" references indicate that no significant deviations exist between the different references. Finally, stars noted in the references as being photometrically variable are indicated as such in the catalogue. Additional remarks concerning the duplicity, variability and other pecularities of individual stars may be found in the Bright Star Catalogue. There exists quite a good distribution of stars of different spectral types over the sky, both north and south.

In conclusion, a comprehensive catalogue of bright standard stars for uvbyβ photometries is now available for observers using small telescopes. The catalogue data and references will be published in the near future. Additional standard stars of later spectral types would be desirable, as Olsen (1984) has extended the uvby calibrations into the G and K spectral range. Lastly, the authors wish to thank the many observers whose efforts contributed to this catalogue.

PHOTOMETRY OF APSIDAL MOTION STARS - A PROGRESS REPORT

C.D. Scarfe
Climenhaga Observatory, Department of Physics
University of Victoria
Victoria, B.C.
Canada

ABSTRACT. Times of minima of apsidal motion binaries have been obtained at this observatory and at Siding Spring and Cerro Tololo Observatories. This paper discusses the results for CW Cephei, DR Vulpeculae, V477 Cygni and the seldom-observed southern system YY Sagittarii.

A long-term project ideally suited to small telescopes, timing the minima of eclipsing binaries, has been in operation at the Climenhaga Observatory since 1969. All observations are photoelectric, and made differentially with respect to nearby comparison stars. They are reduced by the method of Kwee and van Woerden (1956).

Among the systems which have been studied are several with eccentric orbits which show advance of the line of apsides. In addition several similar objects have been observed from Siding Spring Observatory in 1978 and from Cerro Tololo Interamerican Observatory in 1980. In this paper I shall discuss the three most-frequently observed northern systems, CW Cephei, DR Vulpeculae, and V477 Cygni, and the southern system YY Sagittarii.

In order to treat the apsidal motion separately from other causes of period variation, use was made of its different effects on primary and secondary minimum. For each system, each season in which both eclipses were observed yields a value of the left side of the equation

$$\frac{\pi}{P_s}\,(t_2 - t_1 - \tfrac{1}{2}\,P_s) = (2 + \cot^2 i)\; e \cos\omega - \tfrac{1}{3}\, e^3 \cos 3\omega \qquad (1)$$

In this equation, which is valid for small e and $i \simeq 90°$, P_s is the mean sidereal period, t_1 and t_2 are times of adjacent primary and secondary minima, e and i are the orbital eccentricity and inclination, and ω is the instantaneous longitude of periastron. This equation was used to determine e, the apsidal period U, and the time of conjunction of the apsidal line (ω = 90° or 270°). From these an ephemeris was calculated to fit all the available times of minima, adjusting P_s and initial epoch by least-squares.

For CW Cephei we have obtained observations of six primary and

J. B. Hearnshaw and P. L. Cottrell (eds.), Instrumentation and Research Programmes for Small Telescopes, 319–320.

seven secondary eclipses. Together with data from the literature these provide nine epochs spread over thirty years for use with equation (1). The values of e and U derived are given in Table I, the latter being slightly longer than that found by Soderhjelm (1976). The r.m.s. residual for all known photoelectric minima from the resulting ephemeris is $0\overset{d}{.}0019$. Thus there is no evidence for any period change other than that due to apsidal motion.

For DR Vulpeculae we have obtained observations of ten primary and seven secondary minima; there are eight epochs, spread over 21 years, at which $t_2 - t_1$ has been determined. Equation (1) yields the values of e and U given in Table I, which are in reasonable agreement with those of O'Connell (1972) . However when the apsidal terms are subtracted from the individual times of minima, it is clear than the mean period has increased over the last two decades. The data are insufficient to determine whether this increase has been continuous or abrupt.

For V477 Cygni, we have obtained observations of nine primary and four secondary minima since those listed by Scarfe et al (1976). There are now a dozen seasons in which $t_2 - t_1$ has been determined; for these equation (1) gives the results in Table I, which agree well with those of O'Connell. The r.m.s. residual in this case is $0\overset{d}{.}0012$.

For YY Sagittarii one primary minimum was obtained at SSO and primary and secondary minima on consecutive nights at CTIO. The latter may be combined with the data of Keller and Limber (1951) to obtain U = 238 yr, which is too uncertain to yield a reliable apsidal constant.

Apsidal constants, corrected for the general relativistic effect, are included in Table I. For DR Vul it was necessary to estimate masses from the light ratio and spectral type, since no spectroscopic solution has been published. Possibly DR Vul's extra period change is a light-time effect, and if so the third body may contribute to the apsidal motion, and k may thus be incorrect.

Table I

Star	CW Cep	DR Vul	V477 Cyg	YY Sgr
P_s (days	2.729	2.251	2.347	2.628
e	0.0285	0.0947	0.3045	0.163
U (years)	45.0	36.1	350	238:
k	0.0086	0.0092	0.0056	

References

Keller, G. and Limber, D.N. 1951. Ap. J. 113, 637.
Kwee, K.K. and van Woerden, H. 1956. B.A.N. 12, 327.
O'Connell, D.J.K. 1970. Vistas in Ast. 12, 271.
O'Connell, D.J.K. 1972. Specola Vaticana 8, 319.
Scarfe, C.D., Barlow, D.J., and Niehaus, R.J. 1976. Ap. and Sp. Sci 39, 129.
Soderhjelm, S. 1976. Ast. and Ap. Supp. 25, 151.

THE USE OF MACHINE-READABLE ASTRONOMICAL CATALOGS AT SMALL OBSERVATORIES

Wayne H. Warren Jr.
Astronomical Data Center (ADC)
National Space Science Data Center, Code 633
NASA Goddard Space Flight Center
Greenbelt, Maryland 20771 U.S.A.

ABSTRACT. The development of computer controlled telescopes at small observatories has dramatically increased the demand for and potential usefulness of astronomical catalogs in machine-readable form. The compilation and storage of catalogs containing program and standard stars are obvious necessities for the operation of an automatic telescope, but to date most observers have been collecting their own data and manually entering them into microcomputer disk storage. (This is clear from the small number of machine catalogs distributed by the ADC to smaller observatories.) Astronomical data centers located in several countries around the world currently archive, maintain and disseminate a wide variety of machine catalogs in virtually every discipline of astronomy, and these facilities can provide observers with nearly any kind of data needed for controlling telescopes (positional catalogs), reducing data (catalogs of all types of photometry, spectroscopy, etc.) and providing access to fundamental quantities needed for the interpretation of observations (catalogs of binaries, variables, radial and rotational velocities, etc.). The ADC presently has approximately 450 machine catalogs in its archives and these are available to observatories upon request. Procedures for obtaining data from the ADC and policies for distribution are described in this paper, while a list of all catalogs available can be obtained by contacting the ADC.

1. USING EXISTING CATALOGS

The large collection of existing machine-readable catalogs offers small observatories the possiblility of compiling specialized lists of program objects in almost any discipline. Astrometric catalogs such as the FK4 (Fricke and Kopff 1963) can be used to extract accurate positions for the automatic acquisition of standard and program stars, while existing catalogs of variable stars (Kukarkin *et al.* 1982) can assist greatly in setting up observing programs for photometry of suspected variables.

Compilation catalogs for most photometries already exist in machine form. Most of these have been compiled by specialists at the photometric data center in Geneva and include, in addition to all individual

J. B. Hearnshaw and P. L. Cottrell (eds.), Instrumentation and Research Programmes for Small Telescopes, 321–322.

observations, files of homogeneous mean values based on weights determined by reference to system standards. Examples are the general *UBV* catalogs (Mermilliod and Nicolet 1977, Nicolet 1978, Mermilliod 1983) and *uvbyβ* compilation (Hauck and Mermilliod 1985). A complete list of available catalogs can be obtained by writing to the author. These machine-readable catalogs can also be ordered from the Centre de Données Stellaires in Strasbourg, France, the Soviet Center for Astronomical Data in Moscow, the Japanese Astronomical Data Center at the Kanazawa Institute of Technology, and the Zentralinstitut für Astrophysik in Potsdam, DDR.

2. OBTAINING CATALOGS FROM THE ADC

The ADC at NASA is a joint effort of Goddard's Laboratory for Astronomy and Solar Physics (J. M. Mead) and the NSSDC (W. H. Warren Jr.). While both groups prepare and document catalogs for distribution to the astronomical community, the latter group actually disseminates the data through NSSDC (domestic) and the World Data Center A for Rockets and Satellites (international). Although charges are sometimes assessed to cover the cost of processing a request, the Head of NSSDC and/or the Director of WDC-A-R&S may waive charges, as resources permit, for modest numbers of data when they are to be used for scientific and educational purposes. Further information and ordering materials may be obtained by contacting the ADC at the address above.

REFERENCES

Fricke, W. and Kopff, A. 1963, *Fourth Fundamental Catalogue (FK4) and Supplement, Veröff. Astron. Rechen-Inst., Heidelberg,* Nos. 10, 11.

Hauck, B. and Mermilliod, M. 1985, *Astron. Astrophys. Suppl.* 60, 61.

Høg, E. and von der Heide, J. 1976, *Perth 70, Positions of 24900 Stars, Abh. der Hamburger Sternwarte,* Band IX.

Kukarkin, B. V., Kholopov, P. N., Artiukhina, N. M., Fedorovich, V. P., Frolov, M. S., Goranskij, V. P., Gorynya, N. A., Karitskaya, E. A., Kireeva, N. N., Kukarkina, N. P., Kurochkin, N. E., Medvedeva, G. I., Perova, N. B., Ponomareva, G. A., Samus', N. N. and Shugarov, S. Yu. 1982, *New Catalogue of Suspected Variable Stars* (Moscow: Publishing House of the Academy of Sciences of the U.S.S.R.).

Mermilliod, J.-C. 1983, *Bull. Inform. Centre Données Stellaires,* No. 25, p. 79.

Mermilliod, J.-C. and Nicolet, B. 1977, *Astron. Astrophys. Suppl.* 29, 259.

Nicolet, B. 1978, *Astron. Astrophys. Suppl.* 34, 1.

SMALL TELESCOPES AND RESEARCH FROM SPACE

W. Z. Wisniewski
Lunar and Planetary Laboratory
University of Arizona
Tucson, Arizona 85721
U.S.A.

The number of artificial satellites dedicated to astrophysical research is increasing rapidly. Nearly 30 satellites currently under development or in the early planning stages will be in orbit within 10-12 years and will have more sensitive detectors and better data-processing technology as a result of current research. Many of the galactic and extragalactic objects discovered by the new technology are variable on surprisingly short time scales ranging from sub-seconds to many months. The new variable objects include but are not confined to: neutron and binary stars; quasars (and associated active centers of galaxies); newborn infrared stars and associated clouds; the coronal activity of main sequence stars; cataclysmic variables (white dwarfs in binaries); and novae, supernovae and remnants. We now realize that the short time-scale variations of many unusual stars and active galactic nuclei demand that ground and space data be taken as close in time as possible and that they be carefully planned and coordinated.

Astronomers face a situation where precise and simultaneous observations are necessary in order to make further progress and to decide between competing models. Present worldwide astronomical research plans call for billions to be spent on space-based technology, but neglect the coordinated ground-based optical support program needed. This haphazard situation will continue indefinitely unless telescopes can be built that are dedicated to space astronomy. An important step forward in ground-based support of astronomy satellites would be to build six new telescopes of the 2.5 meter class (e.g., Michigan, Dartmouth, MIT or angle design) which would be located to provide time-continuous support of any satellite and cosmic source in either hemisphere.

The total 20-year budget is shown in Table I and may be compared with the worldwide astronomical satellite effort during the same period. This table shows that the total cost of building and maintaining the ground-based support program is less than 1% of the cost of astrophysical space research, less than a recent expenditure for repairing the solar maximum mission satellite or less than the cost of putting one of the large space observatories into orbit. If the ground-based support program is not provided for, then much significant science on the spectral energy variations of many cosmic sources will be lost. In some

J. B. Hearnshaw and P. L. Cottrell (eds.), Instrumentation and Research Programmes for Small Telescopes, 323–324.

cases astronomers would not be able to distinguish between competing models of the most exciting sources.

The question is what we can do now to alleviate the situation. I think that we can proceed in two steps:

1. By creating a clearing house for all observatories' schedules worldwide. Satellite observers in need of ground-based support could contact a clearing house in the hope that somewhere there is someone observing with the right instrument who is willing to cooperate.

2. With the advent of new detectors like the CCD and expected array detectors for the IR region 1.5 μm to 5 μm and 10 μm to 20 μm, the small telescope of near 1-meter class can play an important role in providing ground-based support and follow-up observations for astronomical satellites. It has been shown that a 20 mag star can be easily observed with a 0.9-meter telescope.

If some observatories would agree to dedicate their near 1-meter class telescopes to the project and maintain them, then through international effort those telescopes could be equipped with sets of identical instruments. These instruments could also be used for follow-up observations in support of surveys done by satellites. One is immediately reminded of an exciting program of the astrometric satellite HIPPARCOS. This satellite will gather very high precision astrometric data on 100,000 stars brighter than 12 mag. Additional, somewhat coarser accuracy observations will be generated by satellite auxiliary sensors on more than 400,000 stars. This data, if followed up by observations from the ground, will increase our knowledge of stellar distances, absolute magnitudes of stars, and the structure of HR diagram by the order of several magnitudes. What a perfect project for small telescopes!

This international organization, if formed, could also provide data for other observing programs requiring coordinated multi-site observations such as asteroids and comet nuclei rotation, or observations of oscillating astronomical phenomena as suggested by Sterken and Christensen-Dalsgaard (February 1985, "New IAU Commission?" IAU Bulletin 53, pp. 22-23). For logistical reasons, all of the involved observatories should be tied together and with a disposition center by telex.

Table I

Budget for Groundbased Support of Satellites

Items required over 20 years	Cost (millions 1985)
Six 2.5 meter angle telescopes	12
Six telescope domes	6
Instrumentation (exchanged every 5 years 6 X 500K X 4 computers)	12
Personnel (30K/yr X 20 yr X 6)	
Three observers (professional astronomers)	11
Three night assistants	11
Two engineers (BS degree)	8
Two technicians	8
Two data processors	8
Miscellaneous supplies (100K X 6 X 20)	12
Disposition center (100K X 20)	2
	90/20 years

ARC-SECOND INFRARED IMAGING OF THE BN-KL STAR FORMATION REGION

C.E. Woodward, A. Moneti, J.L. Pipher and W.J. Forrest
University of Rochester and The C.E.K. Mees Observatory
Rochester, New York 14627 U.S.A.

ABSTRACT: Infrared images of the BN-KL region in Orion obtained with the Univ. of Rochester IR Array Camera on the Mees 0.6-m (2.5"/pixel) and the Kitt Peak 1.3-m telescopes (1"/pixel) are presented. These images demonstrate the utility of small telescopes for infrared studies of star formation regions.

We have developed a near-IR (1-5 μm) array camera, based on a 32 x 32 InSb detector array read out by a Si CCD, and have used extensively both 0.6-m and 1.3-m telescopes for testing the camera system and for obtaining images of a variety of astronomical sources (Forrest *et al*. 1985 *Pub. Astr. Soc. Pac.* 97, 183). These observations are intrinsically very interesting and have also proven invaluable in guiding our research efforts on larger aperture telescopes. IR array detectors with good quantum efficiency ($\gtrsim 0.60$), excellent pixel to pixel uniformity, individual pixel, and non-destructive read-out, are now becoming available to the general astronomical community. The impact of this new technology on the study of active star formation regions, external galaxies, bipolar nebulae, etc., will be immense. Since large field-of-view, dark current limited ($\lambda < 3$ μm) and background-limited ($\lambda \gtrsim 3$ μm) observations at high spatial resolution ($\lesssim 3$"/pixel) are possible with small aperture telescopes, many interesting projects can be undertaken without the need for large (and generally over-subscribed) telescopes.

The high spatial resolution images of the BN-KL region presented here were obtained to identify new sources, and to deduce their physical nature. Astrometric positions were obtained from H (1.65 μm) and K (2.2 μm) images where mainly foreground stars and scattered light are detected. At longer bandpasses (L' [3.75 μm]; M' [4.67 μm]) the effects of extinction diminish. The spatial distribution of heated dust was revealed and highly obscured point sources were detected, including IRc2 (possibly the source of energetic flows in the core of the star formation region). At M', IRc2 is the second brightest source after BN. With our higher resolution we were able to determine which sources were point-like and which were extended. From the intensity

J. B. Hearnshaw and P. L. Cottrell (eds.), Instrumentation and Research Programmes for Small Telescopes, 325–326.

maps we derived color temperature maps which were used to deduce the probable source emission mechanism.

We have detected 28 sources: 20 of these can be identified with foreground stars, 8 have no visual counterparts. The IR sources are surrounded by extended nebulosity. Figure 1a shows the large scale structure of the BN-KL complex, especially the two extended lobes of nebulosity NW (P.A.∿350°) and E (P.A.∿90°) of BN. At H and K a trough in the diffuse nebulosity is seen (Figure 1b) which is identified with a circumstellar doughnut around IRc2, proposed as the disk collimating the outflows from IRc2. The total luminosity of IRc2 is ∿ $10^5 L_\odot$ (Downes *et al.* 1981 *Ap. J.* 244, 869) and it is probably responsible for nearly all of the observed luminosity of the BN-KL. The only other self-luminous sources are thought to be BN and IRc9 (Wynn-Williams *et al.* 1984 *Ap. J.* 281, 173). In our color maps (K-L', L'-M') these three sources are the prominent temperature peaks; however interpretation of these maps in conjunction with high resolution polarimetric observations (Werner *et al.* 1983 *Ap. J.* 265, L13) suggest that the sources IRc3 and 4 are not illuminated by IRc2 and could be self-luminous.

Our images also show interesting structure associated with BN. A wisp extending NE from BN (most prominent in the K and L' bandpasses) is identified as scattering off the inner surface of the doughnut surrounding IRc2. Further, in the K band images a jet-like structure consisting of several colinear sources whose apex is BN is found at a P.A. ∿ 132°. This may indicate that energetic flows are also associated with BN.

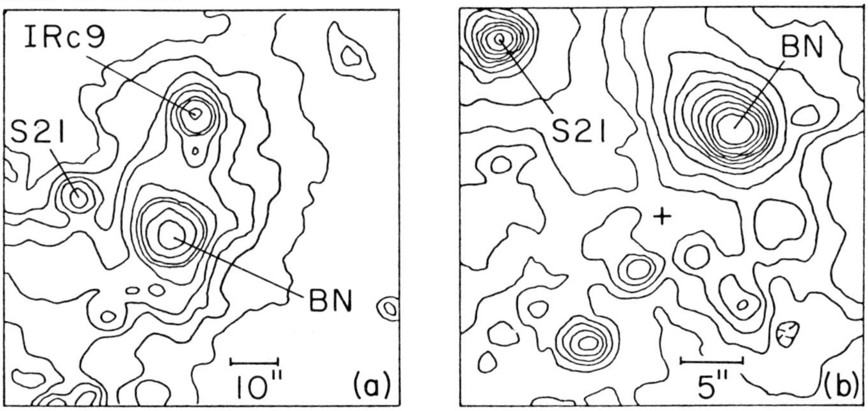

Figure 1: The Orion BN-KL region at 2.23 μm (a) from the Mees 0.6-m (2.5"/pixel). The contours are in logarithmic intervals. Here the large scale structure of the infrared complex is clearly seen. The total observing time for this image was 198 sec. (b) from the Kitt Peak 1.3-m (1"/pixel). The contours are 8.3 - 17.3 mags/pixel every 0.4 mags/pixel. At this resolution the myriad of individual IR point sources is evident. Note the wisp of emission extending NE from BN at a P.A. ∿ 30° and a jet of emission at P.A. ∿ 132°. The cross indicates the position of IRc2 determined from the M' image where it is clearly detected. The total observing time for this image was 425 sec. Prominent sources are labeled, and the scale is inset in each map. The orientation of each map is North up and East to the left.

THE RS CVN PROJECT AT CAPILLA PEAK OBSERVATORY*

M. Zeilik
Institute for Astrophysics
The University of New Mexico
800 Yale Blvd., N. E.
Albuquerque, N. M. 87131 U.S.A.

1. INTRODUCTION

The 61-cm telescope at the Capilla Peak Observatory of the University of New Mexico was used extensively from 1979 to 1982 for UBVR observations of selected RS CVn stars. (See Elston and Zeilik, 1982, for details of the system.) Complete coverage was obtained from most of the short-period systems (Hall, 1976)--RT and CG Cyg, XY Uma (the most active of these systems), UV Psc and ER Vul--with a precision of 0.01 mag or better for each datum. Our early experiences showed that these stars had rapid intrinsic variations (within a month).

2. DATA ANALYSIS AND RESULTS

The now-standard model for RS CVn activity (Hall, 1972) posits large spotted regions covering a considerable fraction of the active star's surface. Models to date have usually found that two spotted regions with effective temperatures about 1000 K below that of the photosphere mimic the observed photometric behavior reasonably well (Eaton and Hall, 1979; Poe and Eaton, 1985). Although such models are not unique, they can follow the long-term trends in the evolution of the spotted regions: their positions, areas, and temperatures.

We report here on the first effort to model starspots on the short-period systems using the Capilla Peak database. These time-serial observations are of high-precision, made with a dedicated instrument, and include the R-band essential to finding spot temperatures (which is otherwise a free parameter). We use UV Psc here as a typical example.

We have adopted the methods of Budding (1973, 1977) to fit the binary light curves and the distortion waves in them. The data was fit by an eclipsing binary model for at least 5 geometrical parameters optimized to a minimum chi-squared. Then the theoretical light curve was subtracted from the observed one to leave behind the distortion wave. This wave was modeled by two spots, assumed to be circular and fixed in latitude. We subtracted the computed distortion wave from the original data to leave a "clean" light curve. We finally fit a new binary model

J. B. Hearnshaw and P. L. Cottrell (eds.), Instrumentation and Research Programmes for Small Telescopes, 327–328.

to find the actual values of the system parameters.

Figure 1 shows a theoretical fit to the V-band data on UV Psc. It gives the distortion-wave model with two spots on the equator. The spotparameters are: spot 1, longitude 190°, size 9.8°, spot 2, longitude 286°, size 11.2°.

Given the results from these preliminary models, we conclude that high-quality photometric data can provide sufficient information for starspot models on RS CVn systems for which the distortion wave has a small amplitude. The photometry must have a precision of at least 0.01 mag (0.007 to 0.005 mag is better) in order to pull the distortion wave out of the eclipsing light curve for meaningful starspot modeling.

REFERENCES

Budding, E., 1973, Astrophys. Space Sci., 22, 87.
Budding, E., 1977, Astrophys. Space Sci., 48, 207.
Eaton, J. A. and Hall, D. S., 1979, Ap. J., 227, 907.
Elston, R. and Zeilik, M., 1982, PASP, 94, 729.
Hall, D. S., 1972, PASP, 87, 323.
Hall, D. S., 1976, in Multiple Periodic Variable Stars, edited by W. S. Fitch (Reidel, Dortrecht), 278.
Poe, C. and Eaton, J. A., 1985, Ap. J., 289, 644.

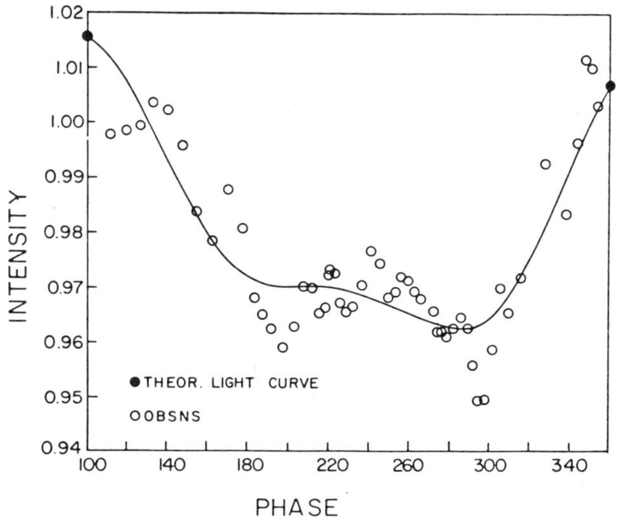

Fig. 1-Two spot fit for UV Psc.

DISCUSSION OF POSTER PAPERS : SECTION II

Anthony-Twarog - uvby photometry with a CCD

Graham: Did you have any trouble with interference fringes when using Strömgren filters with a CCD?

Anthony-Twarog: We had no trouble with interference fringes with such narrow filters, even for the y-filter where the night-sky line normally produces problems.

Belserene - Photographic monitoring

Graham: You work on period changes. Do you choose well-behaved RR Lyrae stars?

Belserene: We do not know in advance, but we have at most about 25 plates per year. If there is significant period behaviour within ~ 3 months, then we can not do much about the star.

Latham: How faint can you work with your system?

Belserene: 15 mag. in Milky Way. At 4 arcmin/mm the Milky Way then becomes an extended object. 17 mag at high galactic latitudes. These are one hour exposures on 103 aO plates.

Osborn: I would like to comment that for those with small telescopes and samll budgets that preclude CCD systems and the like, this type of work is very worthwhile. Although photographic photometry may have large zero point errors, this is not important for the timings of maxima or minima necessary for period change studies.

Warner: If one starts observing now one has to wait 15-20 years to detect period changes.

Garrison: Christine Clement has been doing work of this kind for 15 years with the Toronto 60cm telescope in Chile. The globular clusters pass directly overhead there and she has found quite a number of period changes and double periods. It is quite useful work that can be done with a small telescope and minimal reduction equipment.

Hazen-Liller: I would like to mention the Harvard patrol program being carried out in the southern hemisphere here at Mount John, and in the northern in Massachusetts. These plates have recently been used, for example, by Steve Shawl to determine recent

J. B. Hearnshaw and P. L. Cottrell (eds.), Instrumentation and Research Programmes for Small Telescopes, 329–341.

ephemerides for long period variables. The collection at the Harvard College Observatory is open to any astronomer wishing to use it.

Bembrick et al - Cometary astrometry with the Sydney Observatory.....

Millis: What astrometric accuracy are you able to achieve and do you have any interest in asteroid astrometry?

Bembrick: We hope to achieve $\lesssim {}^{1}/10$ arcsec accuracy but we do not have enough plates to say whether we are at this limit yet. One of our plans is to do minor planet work.

Sullivan: Being an Australian institution why do you call yourselves the British Astronomical Association?

Bembrick: We are a branch of the BAA, the last surviving one, I believe, in a once worldwide empire.

Burki - The empirical period-radius relation for pulsating stars

Latham: This is an outstanding effort which has been organised out of Geneva to do 50,000 precise stellar radial velocities (< 1 km s^{-1} accuracy of stars down to 14th mag) using small telescopes in both the northern and southern hemispheres. A great team effort, due largely to Mayor but with many collaborators around Europe. When are you going to go to the globular clusters and give us a good distance scale?

Burki: We need larger telescopes. We tried to observe RR Lyrae stars in the globular clusters with the 1.5 m Danish telescope at ESO and obtained marginal results.

Graham: Could you comment on how the Baade-Wesselink method works for giving absolute magnitudes? Do you think it is at all dependent on structural features in the atmosphere (e.g. different lines forming at different depths in the atmosphere)?

Burki: The Baade-Wesselink method makes the assumption that the radial velocity and photometric measurements are sampling the same parts of the star. At minimum radius there seem to be some problems with the method. For RR Lyrae stars we exclude the phase of minimum radius.

Evans: If you have radial velocities and V,R photometry for Magellanic Cloud Cepheids you have a gold mine.

Cameron - Starspot imaging using VRI photometry

Mochnacki: Two groups (Penrod and Vogt at Lick and Rucinski and Horne at Cambridge) have tried this technique and failed. The use of VRI photometry, giving colour information, is an important step forward.

Rucinski: I think that the main point of the technique presented by Collier and Horne is that they defined a proper function "f" to be optimized, so that they avoided having bright spots adjacent to dark spots.

Warner: The doppler imaging technique has recently been put forward with comments that suggest that it is a new invention. But Otto Struve suggested the principle at least 30 years ago - a point that I have included in my own spectroscopy lectures for about 20 years.

Evans: Only looking at photometry doesn't constrain the spot. One needs to have very high signal to noise spectra as well.

Innis: I'd like to draw attention to the earlier observations of this star which show minimum light to be brighter than maximum light in the present series. This may mean that the star is more heavily spotted than the current model indicates.

Cameron: It is difficult to tell from a single dataset whether the long-term variations in mean flux from this star are caused by variations in the total emergent flux or geometric re-arrangements of the starspot distribution. We need several datasets obtained at intervals of a few weeks in order to see just how the star-spot distribution evolves with time, and whether the apparent variations in mean luminosity can be modelled purely in terms of the starspot distribution.

Genet: A 0.4 m automatic telescope on Mt Hopkins operated by Vanderbilt University (D.S. Hall) will soon begin UBVRI monitoring of all known (~60) RS CVn stars brighter than ~11 mag in the northern sky.

Evans: I would hope that you would add M dwarfs with traces of hydrogen emission since these are almost certain to show spots.

Djorgovski and Ebneter - A search for features in early type galaxies

A'Hearn: Have you used both your model-subtraction method and the Sekanina-Larson method on the same object to compare performance?

Djorgovski: Yes, it turns out that model galaxy removal works better for galaxies because you put in real isophotal information. For comets, the other technique would work better.

Scarfe: A very similar technique has been used by Gillman and Pritchet in Victoria to detect globular clusters all the way to the centre of M87.

Djorgovski et al - The structure of globular clusters

Warner: For what fraction of galactic globular clusters do you have surface density distributions?

Djorgovski: Approximately 90%. Combining with data from the literature there is some sort of profile for 135-140 globular clusters. We intend to completely reassess cluster morphology. Some of the clusters have no previous data, except photographs.

Duncan - Stellar activity

Gaustad: How small a telescope could one use for this work?

Duncan: On the 60" our magnitude range is $m_V \sim 3$ to 6; with a 24" telescope one should probably work on the brighter red giants. We can keep only ~ 50 giants under surveillance at any one time.

Warner: There is also the southern hemisphere for which none of this sort of work has been done.

Duncan: That's right! For instance, α Cen has never been monitored for any type of variability.

Scarfe: It would be very interesting to find out the variation of rotation period along the subgiant branch. Are there enough subgiants whose H & K flux is adequate to do this?

Duncan: Subgiants seem to be the least active chromospherically, one needs to be more intensive and accurate in ones observations. There are a couple which have shown rotational variation. They are certainly worth further investigation.

Cameron: Modulation of CaII in known active subgiants (RS CVn type) is often not observed. This is due to the wide distribution of active regions over the stellar surface which means that the rotational modulation is not able to be extracted from the data. The strongest emitters appear to be the most heavily spotted and the most reliable way of measuring the rotation period is broadband photometry.

Duncan: Rule of thumb: Younger, heavily spotted stars are analysed more efficiently with braodband photometry. Older stars, use K-line photometry.

Garrison: I'd like to underline the fundamental nature and essential importance of this kind of work, which can only be done by hard work over a very long timescale with a small or under-subscribed telescope. An example of its use is found in a paper by Bruce Campbell and me last year, which showed that it is possible to determine the inclination of the polar axis by combining the activity data with Campbell's hydrogen fluoride radial velocity measurements.

Evans: I am astonished that there is no section on flare stars in this meeting where one needs small telescope monitoring for starspots, flares and other evidence of activity, often with more than one telescope. There are important programs of co-operation with spacecraft observations and for extended longitude coverage.

Gaustad: Detection of rotation in red giants seems to me very important, for it may potentially be a way of distinguishing pre-main-sequence from post-main-sequence stars. How does the angular momentum of your stars compare with that of main-sequence stars of the same mass?

Duncan: All the red giants for which we have rotation periods are rotating extremely slowly. They are slower than one would expect by extrapolating from the rotational velocities for A stars on the main sequence to the giant stars region, under any of the commonly discussed schemes of angular momentum transport. David Gray has suggested that a phase of magnetic braking affects red giants at a critical B-V $\sim$ 0.8 - 0.9. One problem that we have is that the odds are very high (9 out of 10) that when looking at the random giant star it is a "clump" giant which may have already lost substantial mass (and also angular momentum) during its first ascent of the red giant branch.

Gray: I want to comment on this very point. There is no ambiguity regarding the evolution, especially for subgiants. Their evolution, if you believe the evolutionary calculations, does not show any "blue loop", but is nearly horizontal across the HR diagram from the main sequence to the K giant region and then upward.

Therefore the rotational discontinuity is likely due to a brake which rapidly decelerates the rotation velocities, and it happens near G0 for luminosity class IV stars. The rotational break is at G5 for the giants of class III (ref. Gray 1982 Ap. J. 262, 682; Gray and Nagar 1985 Ap. J. 298, Nov. 15th issue).

Gehrels - CCD scanning with a small telescope

Wisniewski: A very successful program and a very good case for a completely dedicated telescope for specific projects: dark time - comet and minor planet work; bright time - radial velocity work.

Hazen-Liller - Search for variable stars in globular clusters

Graham: Are you planning to continue this important work to other globular clusters where variable star characteristics are now known well?

Hazen-Liller: I am planning to finish studying all unstudied southern clusters that can be done with the Yale telescope, and have already gotten plates for most of them. I hope very much to be able to get the remaining necessary plates in the coming semester, although the 1-m Yale telescope is technically no longer available for photography. It seems sad that such a superb wide-angle instrument as the 1-m is now only to be used to observe a single star at a time, while the 0.9-m intended to be used on-axis, is being used for wide-angle work, i.e. the CCD.

Ryan: How are you comparing your plates?

Hazen-Liller: I use a blink microscope to find the variables, and an iris photometer to measure the magnitudes.

Hill et al - EX Hydrae timing data

Warner: This paper illustrates the spin-up of the white dwarf by the shortening of the period.

Ianna - Recent trigonometric parallaxes from two hemispheres.

Gliese: Some of the parallaxes of southern stars show very small observational errors, which are comparable with those from the north. Which other southern refractors are doing parallaxes:- The Cape, but it has a small output. Any other?

Ianna: No, but there is a proposal to do CCD parallaxes in Chile.

Warner: No observational work is currently being done with the Cape refractor. A new list of parallax stars will soon be published.

Gliese: We will soon know everything about photometry, variability and the velocity of objects in the southern skies, but we will not know their distance.

Gaustad: Is the parallax listed for NGC7293 the first parallax measured for a planetary nebula?

Ianna: No, this has a long history, starting with van Maanen who got $\pi = 0.04 (d=25pc)$. Our parallax puts it more than 100 pc away.

Innis et al - The Monash University Observatory: Equipment and research

No discussion of this paper.

Kennedy - Long-term period behaviour of contact binaries

Belserene: The details in the posted O-C diagrams show - as often pointed out at this meeting - the value of a small telescope which can accumulate a continuous data set. Without this the details would not have shown up.

Batten: I would like to underline the importance of the work discussed both in Kennedy's and Kreiner's (see later in this poster session) papers. Astronomers of the future will blame us if we do not determine times of minima. The task is ideally suited to small telescopes, as other papers in this session show. We should also remember that future generations may have different ideas from us about which systems are interesting. Therefore we should not confine our efforts to systems that interest us.

Kviz - Phase-night diagram for periodic variable stars.

Warner: Has this program been used to coordinate observations at different sites around the world?

Kviz: No, not yet.

Knee et al - Photometry of RZ Ophiuchi - an international collaboration

Scarfe: This needed careful phasing and observations around the world to cover the interesting phases.

Warner: An interesting result that one can have circumstellar matter drifting around, without a current exchange of matter.

Scarfe: It is possible that RZ Ophiuchi has a stellar wind, which may feed the circumstellar disk.

Kreiner - Times of minima and period changes of eclipsing binaries

Scarfe: Could you comment on the recent decrease in activity from the previously very active Namberg-Izmir group?

Kreiner: The decrease is largely due to the death of Dr Ibanoglu.

Wood: Observing times of minima and publishing them is an extremely important use of small telescopes. Period changes inferred from these observations can be used to study the evolution of these systems.

Sterken: It is indeed very important to systematically observe times of minimum, but one should also consider that with a little extra effort one should try to cover the whole eclipse phase, or even more.

Malasan et al - Photometric research at Bosscha Observatory

Osborn: I was impressed by the fact that here we have a case where useful work is being done in spite of serious problems: a small telescope, a refractor with an unstable objective, a limited sky for the telescope, and poor weather. Often we hear that a telescope is not being used because it is old and not in good condition. Given a suitably selected program and perseverence these instruments can be used profitably, as shown in the present case.

Menzies and Marang - A new B subdwarf

Scarfe: This system appears to have an enormous reflection effect. Is this in fact the case, and what sort of object must the secondary be?

Menzies: Yes, the largest I've ever seen! It makes it difficult to get a light curve program to fit the data. The primary is about 26,000 K and the face of the secondary is heated to about 13,000 K - 2 or 3 times its effective temperature.

Garrison: How close are the two stars?

Menzies: About 0.4 $R_{\odot}$ and their radii are about 0.2 $R_{\odot}$.

Warner: It looks as though your star should be the nucleus of a planetary nebula, but at a temperature of 26,000 K the system is probably old enough that the nebula will have expanded to a low surface brightness. Can you put any limits on the existence of such a nebula?

Menzies: If our star is the nucleus of a planetary nebula - the nebula probably has been blown away. There is no evidence for it in the CCD frame (2 arcmin x 2 arcmin) or in the spectrum (no emission lines). This object is cooler (and more evolved) than a similar star, AA Doradus ($T_{eff} \sim 40{,}000$ K).

Sullivan: Which programs have you used to model the light curve?

Menzies: Wilson-Devinney and Graham Hill's 'Light' program. The latter has problems with the large reflection effect. The system seems to be a single-line spectroscopic binary.

Moffett - Photometry of selected Cepheids

Moffett: Two initial comments: (1) the list published is not complete, and (2) it is not intended just for professional observers but also for amateurs with photoelectric photometers.

Scarfe: Thirty years ago Olin Eggen observed these stars and interpreted the variation as a bump, whose location in the light curve varied with period. How are these dips different?

Moffett: The bump Cepheids show an increase in luminosity during the bump portion of their light curve. The dip Cepheids show a clear decrease in light output during the dip phase. The theoretical models of Davis show a decrease in luminosity during the dip phase in the light curve.

Genet: Moffett asked for small telescope photometric observations by others including amateurs. A good way to obtain such observations is to place a written request in the IAPPP Communication. This journal goes to 600 professional and amateur astronomers in 40 countries. Advice on how to make such requests and follow up on them is given by R. Lines et al in IAPPP Communications issue No. 21. Typically amateurs with 8" - 12" telescopes are limited to the magnitude range 8 - 10.

Mohin et al - Photometry of the RS CVn binary II Peg

No discussion of this paper.

Osborn et al - Standard stars for comet photometry

A'Hearn: IHW comet filters have been distributed to ~ 70 observatories around the world. Standard stars have been observed so that all observers would have the same set of reference stars for the comet observations.

Djorgovski: Can you comment on the use of spectroscopic flux standards as an alternative to your standards? Then one can interpolate the standard's magnitude at the filter's effective wavelength.

A'Hearn: It's OK for very early type (<B2) spectroscopic standards. Other standards will produce different results because of strong absorption features, in particular the Balmer decrement.

Penhallow - Astrometry with small telescopes

Millis: What astrometric accuracy can be achieved with this system?

Penhallow: On comets, e.g. Giacobini-Zinner, our residuals using the Schmidt are slightly less than 1 arcsec, which compares favourably with other larger (and more expensive) telescopes. The 0.4m has residuals of ~ 0.6 arcsec. On asteroids, we can do better than 0.1 arcsec.

Perry and Crawford - A description of a catalogue of bright uvby....

Warren: Do you distinguish between primary and secondary standards in your catalogue? Does the catalogue contain positions, proper motions and other parameters, which are so useful to the observer at the telescope?

Crawford: They are all primary standards. They are bright stars, so the catalogue does not fulfil the faint star standard need. However the catalogue is intended primarily for use with small telescopes. The catalogue does contain much useful information but we would be glad of your comments so that we can make additions.

Warren: Davis Phillip in IAU Symposium No 111 published standard star data for various photometric systems. I have been thinking for some time of compiling such a comprehensive machine readable catalogue.

Crawford: The list we are publishing is not a synthesis of all published uvby observations. It only includes the Tucson and Danish data.

Scarfe - Photometry of apsidal motion stars - a progress report

Kennedy: How many stars show apsidal motion?

Scarfe: I have about a dozen under observation, but there are many more.

Kreiner: About 50.

Warren - The use of machine readable astronomical catalogues at small...

Mochnacki: What thinking has been done about the use of Data Base Management Systems and the standardisation of data formats? What about access to large data bases via networks?

Warren: Too many different formats make standardisation impracticable. However, CDS at Strasbourg are developing an on-line networking system.

A'Hearn: What networks are available for accessing data from either Goddard or Strasbourg?

Warren: Our data base of bibliographical information can be accessed directly by dial-up. It is not quite yet available on any of the standard networks. There is also work being done on networking Goddard to Europe.

Djorgovski: It would be nice but handling another persons software (and lots of it) could be a nightmare if bugs were found in the code.

Caton: The special interest group for Microcomputer use in Astronomy (SIGMUA), a subgroup of the AAS's Working Group on Astronomical Software, is setting up a telecommunications node at Goddard Space Flight Center. This will be used for exchange of programs and other ASCII text. It will operate under Kermit.

Mochnacki: Could I put in a plug for the standardisation of Kermit, since it is available on both mainframe and microcomputers?

Caton: Kermit has already been poorly implemented on some systems, not allowing the data from two machines to be transferred.

<u>Wisniewski</u> - Small telescopes and research from space

Garrison: The solution to the decline in small telescopes at national facilities could be simple in principle. Satellites should include ground-based support in their budgets. That would help small telescopes and provide the necessary data.

A'Hearn: Quite often it is a combination of space (e.g. IUE) <u>and</u> ground-based observations that give the best interpretation. I would support Bob Garrison's comment.

Warner: I would like the EXOSAT people to realize the limitations of optical observatories. We cannot observe in the daytime!

Garrison: They should also provide lists of objects requiring optical observations well ahead of time so that telescope allocation committees can accommodate their requests.

Warner: In defence of the X-ray astronomers some of their requests are when they have received "target of opportunity" time.

Redfern: EXOSAT has quite restrictive pointing constraints. However, a given set of X-ray data can be improved by many orders of magnitude by having optical data (e.g. optical counterpart of an X-ray source in globular cluster M15). In addition, EXOSAT can get long continuous runs on a given object which can be coordinated with a number of ground-based stations separated in longitude.

Innis: Perhaps we have made the case to include small optical telescopes on future X-ray and UV satellites.

Woodward et al - Arcsecond IR imaging of the BN-KL star formation... .

Latham: The improvement in IR arrays is even more dramatic than in CCD's. They are going from 1 to > 3000 InSb elements. (A 58 x 62 array from the Santa Barbara Research Centre is currently the best.) Not to mention the improvement in registration and photometry by having simultaneous channels. An IR CID for direct imaging out to 10 μm has been built by Aerojet which has been used by Goddard, The University of Arizona and the Smithsonian.

A'Hearn: There is also a 10-20 μm array, with many fewer pixels, which has produced some very interesting results in the last six months.

Woodward: We have just installed a continuously variable filter wheel, with 1% resolution, to do spectral line imaging in the IR. This will enable extinction to be measured to some of these regions using the hydrogen recombination lines. Small aperture telescopes are essential for use with these extended regions. Accurate astrometry can also be done and follow up photometry and polarisation work.

Crawford: I would like to add that much of this exciting new technology, on IR arrays and other items, will be discussed in depth at the next SPIE metting on Instrumentation in Astronomy, to be held next March (1986) in Tucson.

Zeilik - The RS CVn project at Capilla Peak Observatory

Zeilik: One could duplicate our hardware system for ~ $US85,000, and I would be happy to provide the software for those using an Apple Computer.

III. SPECTROSCOPIC RESEARCH PROGRAMMES

a) Invited and oral contributed papers

INSTRUMENTATION FOR LOW-RESOLUTION SPECTROSCOPY

M. M. Phillips
Cerro Tololo Inter-American Observatory
National Optical Astronomy Observatories
Casilla 603
La Serena
Chile

ABSTRACT. Instrumentation presently in use for low resolution, digital spectroscopy on 1 meter class telescopes at three observatories - Lick, Kitt Peak, and Cerro Tololo - is reviewed. A detailed comparison is made between the performance of a typical two-dimensional photon counter, the "2D-Frutti", and currently-available CCDs. Prospects for further detector advances center on large-format, low-noise, uv-sensitive CCDs.

1. INTRODUCTION

Two articles which are still the standard references on optical band K-corrections for giant elliptical galaxies appeared back-to-back in the 1971 October 15 issue of the Astrophysical Journal. These papers, by Schild and Oke (1971) and Whitford (1971), are especially relevant to the subject matter of this Symposium since they were based entirely on low-resolution spectrophotometry carried out with "small" telescopes. In fact, the spectra published by Oke and Schild were made with a 4-inch telescope, which even for this meeting must be considered extreme! Those of Whitford were obtained with the 36-inch Crossley telescope at Lick Observatory.

In drawing attention to these papers, there are at least two points to be made. First of all, it is clear that there are many important contributions to be made in modern astronomy with spectroscopic observations obtained on 1 meter class telescopes or smaller. This is especially true in the case of projects requiring frequent monitoring over several weeks or months, which are virtually impossible to carry out on larger, more oversubscribed telescopes. Equally relevant, but perhaps not so obvious, is the realization that there exists a certain class of problems which must almost by necessity be tackled with small telescopes. The problem facing Oke and Schild and Whitford was to calibrate the magnitudes of distant galaxies from the energy distributions of nearby, bright giant ellipticals. Since a large fraction of the total light of distant ellipticals is usually observed in photometric studies, and the colors of giant ellipticals are known to be a function of radius, it was imperative that very large apertures be used for the

J. B. Hearnshaw and P. L. Cottrell (eds.), Instrumentation and Research Programmes for Small Telescopes, 345–357.

nearby galaxies. Satisfying this requirement while maintaining adequate (≦100 Å) wavelength resolution ruled out all but the shortest telescope focal lengths, thereby effectively restricting the problem to small telescopes only. Instrumentation for low-resolution spectroscopy on small telescopes has changed considerably since 1971, but the lesson offorded by this example is still very much a valid one.

In this talk, I will purposefully avoid giving an all-encompassing review of modern instrumentation employed for low-resolution spectroscopy. Rather, I wish to concentrate attention on three observatories - Lick, Kitt Peak, and Cerro Tololo - where spectroscopic facilities for 1 meter class telescopes have been developed which are illustrative of the current "state of the art" with conventional Cassegrain spectrographs and digital detectors. I shall attempt to provide both a brief summary of the instrumental configurations employed at each site, and to mention a few of the research programs carried out recently with the same facilities as a way of underscoring the scientific potential of such instrumentation on small telescopes. The relative merits of two-dimensional photon counters and CCDs shall be examined in somewhat more detail, and brief consideration will be made of detector and spectrograph options for the near future.

2. LICK 1 METER NICKEL TELESCOPE

The 1 meter Anna L. Nickel Telescope went into operation at Lick Observatory in early-1980. This instrument was dedicated exclusively for the first few years to low-resolution spectroscopy, but is now also employed for digital imaging. The standard spectroscopic configuration consists of a copy of the image dissector scanner (IDS), developed by Robinson and Wampler (1972, 1973; see also Miller, Robinson, and Wampler, 1976), mounted on a conventional Cassegrain spectrograph. This combination is soon to be augmented by a new spectrograph designed especially for use with CCD detectors. In this talk, however, I shall concentrate on the older, dual-channel IDS spectrometer.

The IDS detector was designed at Lick over 15 years ago for the specific application of low- and moderate-resolution spectroscopy. The "front end" consists of a chain of three 40 mm, S-20 photocathode, electrostatically-focussed image tubes, coupled with fiber optics bundles. The phosphor of the third tube is repeatedly scanned by an image dissector at a rate roughly comparable to the characteristic decay time of the phosphor. The phosphor screen effectively serves as a short-term storage element, thereby enabling the system to approach true photon-counting performance. The spectrograph is equipped with dual entrance apertures so that separate spectra of object and sky are simultaneously imaged on the photocathode of the first image tube. The image dissector sweeps are preset so that the object and sky spectra are scanned alternately, with the data recorded as two 2048-pixel, one-dimensional arrays. During observations, the object is normally switched back and forth every 4 or 8 minutes between the two spectrograph apertures in order to optimize sky subtraction.

The general properties of the IDS spectrometer on the Nickel

Telescope are summarized in Table I. The IDS may be considered by some a relatively old-fashioned detector, but as shown in Table I, it is still quite competitive in many ways. Perhaps its chief disadvantage is the one-dimensional format. Another limitation results from the persistence of the phosphor screen which, while vital to the functioning of the instrument, can produce an annoying residual glow that lasts for several minutes, or even hours, after exposure to a very bright object. Hence, the observing program must be carefully planned to avoid scheduling faint objects immediately after bright standards.

In the more than 5 years that the IDS spectrometer has been available on the Nickel Telescope, several noteworthy papers have appeared based on data obtained with this facility. Antonucci and Cohen (1983) were able to monitor optical spectral variations of the nucleus of the Seyfert galaxy NGC 4151 over a 12-month period with a time resolution and accuracy rarely before achieved in such studies. Stauffer (1982) and Keel (1983a,b) both completed large spectroscopic surveys examining the nature of the ionized gas in spiral galaxy nuclei. In the area of stellar astronomy, Clarke, Capel, and Bowyer

TABLE I

	Lick 1 m	KPNO 0.9 m	CTIO 1 m		
	IDS	IRS	2D-Frutti	GEC CCD	RCA CCD
Detector Dimensions	40 mm (diam)	25 mm (diam)	40 mm (diam)	8.5 x 12.7 mm	9.6 x 15.4 mm
Peak Quantum Efficiency	~20%	~20%	~20%	~40%	~80%
Number of Independent Resolution Elements	~350	~300	~1000	288	256
Useful Wavelength Sensitivity	3800-7500 Å	3200-7500 Å	3200-7500 Å	5000-10000 Å	3700-10000 Å
Range of Wavelength Resolutions	3.5-20 Å	2.6-12.4 Å	1.4-7.5 Å	2.1-17 Å	2.9-23 Å

(1984) carried out a unique spectrophotometric study, covering one full outburst cycle of the dwarf nova SS Cygni, and chronicling for the first time ever the full sequence of changes in the hydrogen and helium emission lines with respect to the continuum variations. And, lest it be forgotten, the key spectroscopic observations that revealed the truly bizarre nature of SS433 were obtained with the same IDS and spectrograph when it was dedicated to the Lick 0.6 meter telescope (Margon et al. 1979).

3. KITT PEAK 0.9 METER TELESCOPE

Since early-1980, one of the Kitt Peak National Observatory (KPNO) 0.9 meter telescopes has been equipped with the so-called "Intensified Reticon Scanner" (IRS) dual-beam spectrometer. Like the Lick IDS system, the IRS records one-dimensional spectra produced from separate object and sky apertures. The detector chain consists of an S-20 photocathode, ITT proximity-focussed image intensifier followed by a Varo microchannel-inverter image intensifier and a Reticon dual 936-element array with fiber optic faceplate. Only 820 elements in each Reticon array are useable. The pixels are 30 microns wide (i.e., along the dispersion), and are long enough to allow the use of large (45 arcsec) apertures.

The dual Reticon array is normally read every 10 seconds. Electronic noise equivalent to ~4 photons per read per pixel is generated with each frame, so that readout noise is a problem only for faint objects. The Reticon also suffers from non-negligible dark current, fixed pattern noise, and pixel-to-pixel base line and sensitivity variations, all of which must be carefully calibrated via bias and quartz lamp scans. The same beam-switching observing technique described for the IDS is normally employed at the telescope.

The general properties of the IRS may be compared in Table I. This instrument is well suited for spectrophotometric investigations, with a precision of 3-4% achievable under favorable observing conditions (e.g., see Massey 1984). The system is stable enough to allow reasonably accurate (3-5 km sec^{-1}) radial velocity work as well. The disadvantages of the IRS are the same mentioned for the IDS, namely the one-dimensional character of the data and problems introduced by the persistence of the phosphor glow.

In a paper to be given at this symposium, George Jacoby will review in more detail the important scientific contributions which have come from instrumentation on small telescopes at KPNO, including the IRS. Certainly among the most prominent would figure the study by Hunter, Gallagher, and Rautenkranz (1982) of star formation rates in irregular galaxies, Barker's combined optical and ultraviolet observations of planetary nebulae (1985; and references therein), and Massey's spectrophotometry of northern Wolf-Rayet stars (1984). Special mention should also be made of the comprehensive spectral atlas based on IRS data put together by Jacoby, Hunter, and Christian (1984).

4. CERRO TOLOLO 1 METER YALE TELESCOPE

4.1 2D-Frutti

A copy of Steve Shectman's (1984) "2D-Frutti" two-dimensional photon counting detector is now in the final stages of commissioning on the Boller & Chivens spectrograph of the 1 meter Yale telescope at Cerro Tololo Inter-American Observatory (CTIO). This is the second copy of the 2D-Frutti to be built by CTIO, the first having been in regular service on the 4 meter telescope spectrographs for nearly a year. Unlike any of the other detectors mentioned in this talk, the 2D-Frutti is a genuine photon counting system. The front end of the detector is a 40 mm, S-20 photocathode, two-stage RCA "Advanced Carnegie" image tube. The output phosphor of this tube is reimaged with a transfer lens onto the photocathode of a 40 mm Varo electro-statically-focussed image tube, which is itself fiber optically coupled to 50/40 mm Varo microchannel-inverter image intensifier. To complete the chain, the output phosphor of the second Varo tube is imaged via a fiber optic minifier onto a 380 x 244 pixel Fairchild CCD with special fiber optic faceplate.

The size of the CCD is 11.3 x 8.8 mm, onto which is projected an area of approximately 35 x 28 mm from the spectrograph focal plane. The CCD is clocked at 15 MHz (2x standard TV rate), with full read out requiring about 7 msec. However, it is possible to read out only 1/2, 1/4, 1/8, or even 1/16 of the full verticle format, depending on the application. For the f/10 focal ratio of the 1 meter telescope, the full spectrograph slit length of 6.2 arcmin fits entirely within the 1/8th format (i.e., 32 horizontal lines of the CCD).

As the Fairchild chip is read out, individual photon events are identified and their locations tagged to a precision of 1/8th of a pixel through a digital centroiding technique. Double counts of photon events due to phosphor persistence are eliminated by subtraction of the previous frame before centroiding is carried out. In 32-line mode with 1/8th of a pixel centering both horizontally and vertically, the 2D-Frutti data format translates to 3040 x 256 pixels, with a resolution of 3 such pixels in either direction.

Representative performance figures for the 2D-Frutti on the 1 meter telescope are given in the third column of Table I. Dark current for the 2D-Frutti is so low that the detector performance is truly that of a photon counting system, even at very faint light levels. The wavelength stability of the 2D-Frutti is also quite good. However, as with any serial-readout photon-counting detector, coincidence losses set in and the detector becomes nonlinear at high count rates. Not surprisingly, the maximum permissible count rates for stellar objects are higher than those for pure continuum sources. The saturation limit for stars works out to be about 9 counts/resolution element/sec. On the 1 meter telescope spectrograph, this bright limit translates in the blue to B ≈ 12 for a wavelength resolution of 4 Å. Another somewhat bothersome property of the 2D-Frutti detector is the image distortion produced by the combination of image tubes, but this can be dealt with successfully (albeit somewhat painfully!)

through software at the reduction stage. Examples of a spectra taken with the 2D-Frutti on the 1 meter telescope are shown in Figure 1. These data were obtained at a resolution of ~2 Å.

4.2 CCDs

In November 1983, a CCD detector was used for the first time on the CTIO 1 meter telescope. The particular device was a GEC P8603 epitaxial chip, which has a 385 x 576 format with 22 micron square pixels. The rms readout noise of this first chip was ~20 electrons, but during 1984 a newer GEC CCD was commissioned having only 6-7 electrons noise. These chips are not thinned, but rather are front-illuminated, and thus have little response below 4000 Å. They have about 30-40% quantum efficiency from 5500-9000 Å, falling close to zero beyond 1 micron.

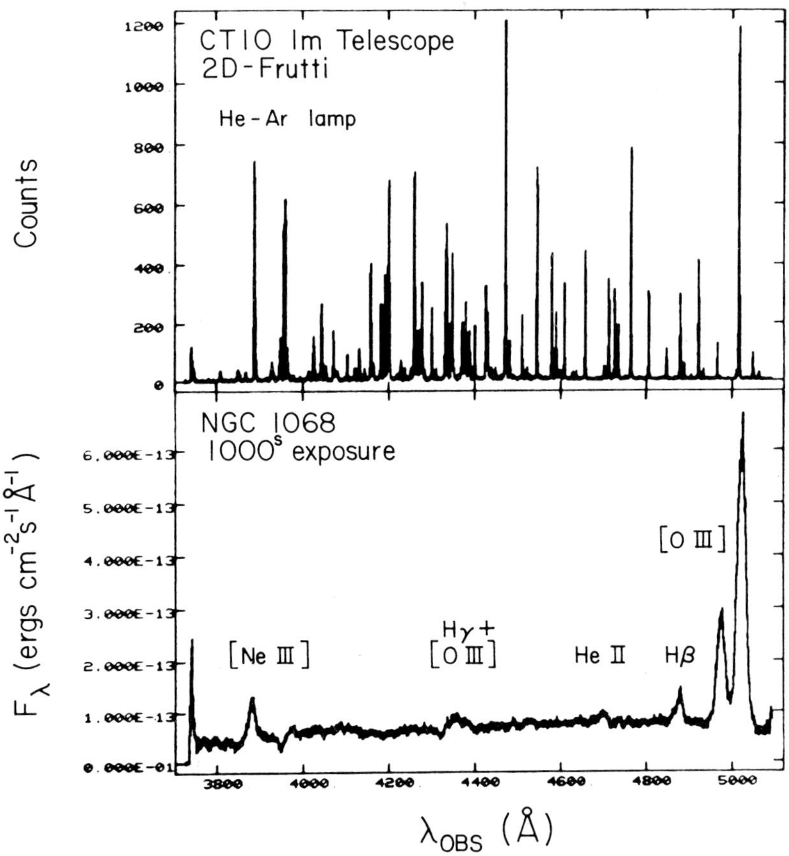

Figure 1. Examples of 2D-Frutti spectra taken on the CTIO 1 meter telescope.

The GEC CCDs suffer from several problems. Radiation events (including cosmic rays) are detected at a rate of about 2 per minute over the whole chip, and can be a considerable nuisance for long exposures. Some columns also show "traps" which are dead at low light levels. Even more bothersome are charge transfer problems which can lead to poor sky subtraction of the OH emission lines redward of 7000 Å. This problem is minimized by preflashing the chip - i.e., adding a flat background of roughly 150 photons per pixel to each exposure. Note, however, that this increases the effective readout noise to approximately 14 electrons.

An RCA SID501 thinned 320 x 512 pixel CCD with 30 micron square pixels was recently tested on the CTIO 1 meter telescope as a possible alternative to the GEC chips. These detectors offer significantly higher quantum efficiency than the GEC CCDs at most wavelengths, and are useful in the blue down to roughly 3700 Å. Charge transfer is also quite excellent, with no preflashing necessary. Unfortunately, the ~40 electrons readout noise of the RCA severely limits its usefulness on fainter objects (see next section). These devices also produce interference fringes with an amplitude of a few percent which require careful flat field calibration. Finally, the new generation of RCA CCDs suffers a high radiation event rate, amounting to ~6 events per minute over the whole chip for the particular device tested on the 1 meter telescope.

Basic properties of the GEC and RCA CCDs on the CTIO 1 meter telescope spectrograph are included in Table I. A few examples of scientific projects carried out with these detectors deserve mention. Vilas (1985) used the GEC CCD to obtain reflectance spectra of differing terrain on Mercury in search of absorption features due to crystalline Fe^{2+}. Another interesting project begun by Kemper with this detector (and now being continued on the 1.5m telescope) is the calibration of a new abundance parameter for RR Lyrae stars using Hα and the Ca II triplet lines near 8550 Å. The RCA was recently employed by Caldwell in a long-term project to obtain integrated-light spectra of nearby galaxies. Examples of data taken for the latter investigation are given in Figure 2.

5. 2D-FRUTTI VS. CCD

The 2D-Frutti and CCD systems represent different approaches to the problem of providing two-dimensional digital detectors for spectroscopy (on any size telescope). In Table II, I have listed what I feel to be the most relevant properties of a spectroscopic detector and have graded the 2D-Frutti, RCA CCD, and GEC CCD accordingly (+++ = good, ++ = fair, and + = poor). For the purpose of later discussion, I have included in this table similar grades for a Texas Instruments (TI) 800 x 800 pixel 3-phase CCD (with 15 micron square pixels), and a Tektronix 2048 x 2048 pixel CCD (with 27 micron square pixels). The performance of the latter device can only be surmised from preliminary data sheets, since Tektronix has not yet begun to deliver such CCDs. For both the TI and Tektronix CCDs, it is assumed that a uv-flooding technique (e.g., see Hlivak, Henry, and Pilcher 1984) has been used to increase the ultraviolet quantum efficiency to ~50%.

In comparing grades for the 2D-Frutti, RCA CCD, and GEC CCD, it is clear that the detector of choice is a strong function of the specific application. Obviously the 2D-Frutti performs best in the ultraviolet, whereas the GEC and RCA chips are superior at red wavelengths. Similarly, for certain spectrophotometric or radial velocity applications, and for very high signal-to-noise observations, the larger dynamic ranges of the GEC and RCA CCDs are a definite advantage.

A slightly different comparison of the relative performances of the same detectors (including the hypothetical cases of the TI and Tektronix chips) on the CTIO 1 meter telescope is given in Figure 3. At wavelengths of 4000 Å and 5500 Å, respectively, and for a range of stellar brightnesses, I have calculated the maximum signal-to-noise

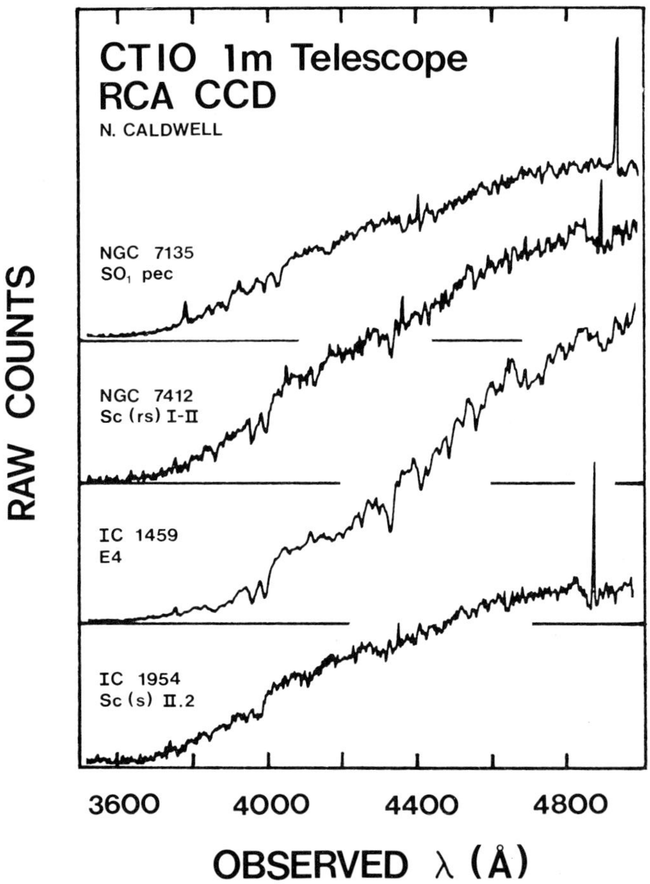

Figure 2. Integrated light spectra of nearby galaxies taken with an RCA CCD on the CTIO 1 meter telescope.

TABLE II

	Detector				
	2D-Frutti	GEC CCD	RCA CCD	TI CCD	Tek CCD
Format:					
Dimensions	+++	+	+	+	+++
Pixel Size	++	++	++	+++	++
Q.E.:					
3500 Å	++	+	+	+++	+++
5000 Å	++	++	+++	+++	+++
8000 Å	+	++	++	++	++
Geometric:					
Flatness	+++	++	++	+	++
Distortion	+	+++	+++	+++	+++
Noise	+++	++	+	++	++
Dynamic Range	+	+++	++	+++	+++
Wavelength Stability	++	+++	+++	+++	+++
Cosmetics	+++	+	++	++	++

ratio per resolution element achieved in a total exposure time of 1 hour. These curves are valid for a wavelength resolution of 6 Å, a slit width of 6 arcsec, and typical Cerro Tololo seeing (~1.5 arcsec). Sky subtraction is assumed to introduce no extra noise, since a long slit and two-dimensional detector allow a virtually error-free determination of the sky spectrum. A far more questionable assumption is that pixel-to-pixel flat fielding is carried out to infinitely high precision. In fact, because of the relatively low maximum count rate limit for the 2D-Frutti, a flat fielding accuracy of 1% is virtually never realized for this device. Hence, I have included in Figure 2 a curve showing the effect on 2D-Frutti data of no flat field correction, in which case the maximum signal-to-noise obtained is limited by the graininess of the front photcathode of the Carnegie image tube (~3% rms). A final assumption made is that the 1 hour integration time must be broken into three separate 20-minute exposures in the case of the CCD observations, so as to be able to eliminate radiation events and bad pixels during reductions.

As may be seen from Figure 3, the TI CCD would hold a clear

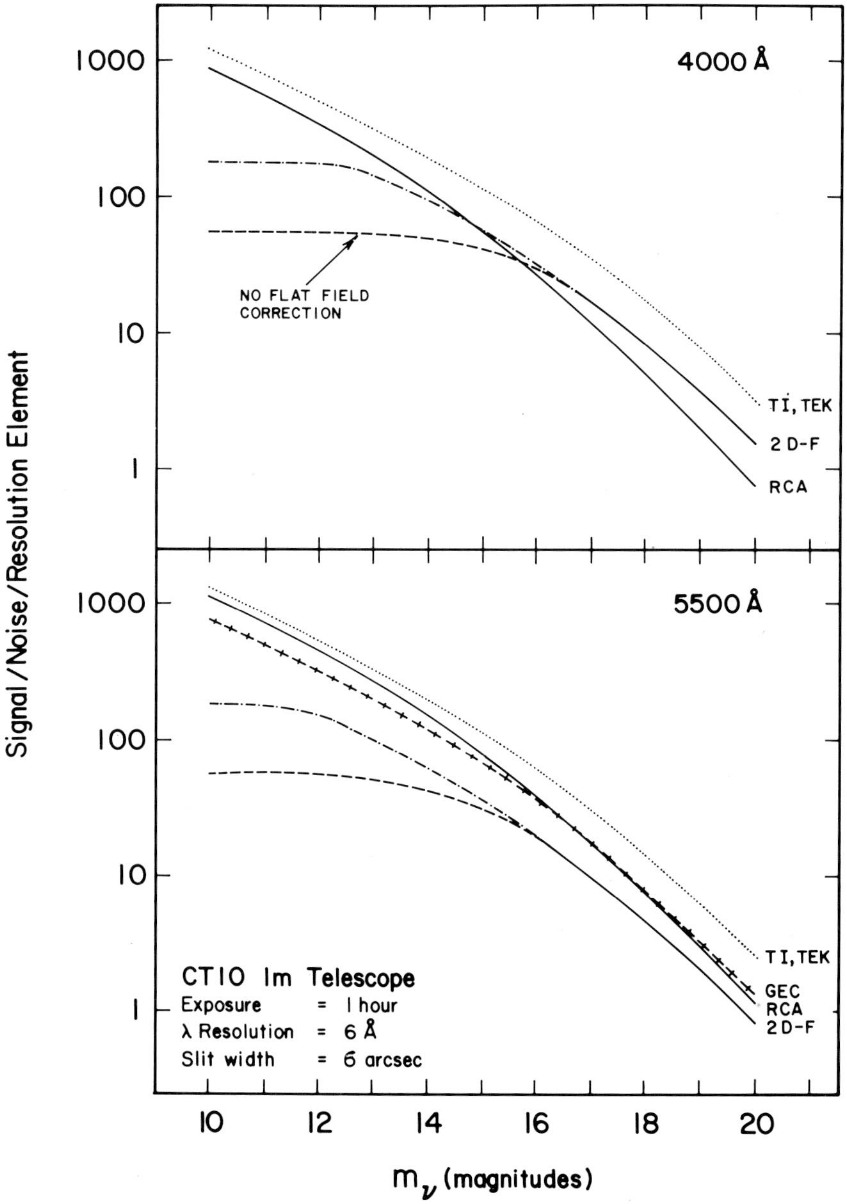

Figure 3. Calculation of the signal-to-noise per resolution element as a function of monochromatic magnitude obtained for various detectors on the CTIO 1 meter telescope. Assumptions behind the numbers are detailed in text. Note that the curves for the TI and Tektronix CCDs are purely hypothetical, as these detectors have never been used for spectroscopy on the 1 meter telescope.

advantage over the other detectors at both wavelengths. This is even more true in the ultraviolet (due to the high quantum efficiency resulting from uv-flooding), while in the red and near-infrared its performance is only rivaled by the GEC chips. Unfortunately, the physical package of the TI chips, and the large deviations from geometric flatness of the sensitive area, make the use of this type of detector with a fast spectrograph camera virtually impossible. Moreover, the small physical dimensions of the device would certainly not suit all projects. Thus, while the performance of each of these detectors is impressive for certain applications, it would seem that as yet no single detector provides <u>all</u> of the characteristics that are ideal for low-resolution spectroscopic applications.

6. FUTURE DEVELOPMENTS

What prospects are there in the near future for further detector improvements in the field of low resolution spectroscopy? Certainly the most promising developments are being made in CCD technology. As shown by the discussion above, currently-available CCDs have already become the detectors of choice for many applications. The recent announcement by the Tektronix Corporation of the development of large-format, low-noise CCDs offers even more promise for such devices. Current specifications call for two chip sizes - 512 x 512 and 2048 x 2048 - each with a 27 micron square pixel size. The format of the 2048 x 2048 chips is truly gargantuan (55 x 55 mm!) by electronic detector standards. The performance of these CCDs is expected to be comparable to that of the TI 800 x 800 devices (see Table II and Figure 3). If so, then the goal of a single optimum detector for both bright <u>and</u> faint object spectroscopy, from the ultraviolet to the near-infrared, may be finally realized. Unfortunately, the announced prices of the 2048 x 2048 devices may restrict there acquisition to larger telescopes, at least initially. Also, the 27 micron pixel size is a factor of ~2 larger than would be ideal for small telescope spectroscopy.

Fortunately, prospects for uv spectroscopy with CCDs do not rest entirely on the Tektronix chips. Over the years, considerable progress has been made with the use of fluorescent coatings on front-illuminated chips such as the GEC epitaxial CCDs. In a recent paper, Cullum <u>et al</u>. measured a peak quantum efficiency of more than 25% at a wavelength of 3400 A for a GEC device with a 3-4 micron thick fluorescent plastic coating. This particular coating causes no significant loss of resolution and is expected to be quite durable. Moreover, the coating seems to even produce a slight enhancement in quantum efficiency in the 4000-6000 A range.

In this talk, I have restricted myself to low-resolution spectroscopy with conventional Cassegrain spectrographs. However, advances in detector technology can be expected to spark interest in alternative optical configurations. For example, large-format CCDs with high quantum efficiency over a wide spectral range should be ideal for an echelette configuration on a small telescope. The same detectors may also eventually replace photographic plates for much grism and objective prism spectroscopy.

7. CONCLUSIONS

As astronomers, we are certainly witnessing exciting times with the 10 meter Keck Telescope project underway and other large telescopes such as the NNTT being proposed. However, a glance at the curves in Figure 3 shows that tremendous advances have also been made in the area of low-resolution digital spectroscopy on small telescopes. The potential of existing small telescopes for making fundamental contributions to astronomy, particularly for projects requiring frequent access to the telescope, is greater than ever, thanks largely to the development of modern electronic detectors. In the rush to build a new generation of giant telescopes, it is all the more important that we not lose sight of this simple fact.

REFERENCES

Antonucci, R., and Cohen, R. 1983, Ap. J., **271**, 564.
Barker, T. 1985, Ap. J., **294**, 193.
Clarke, J. T., Capel, D., and Bowyer, S. 1984, Ap. J., **287**, 845.
Cullum, M., Deiries, S., D'Odorico, S., Reiss, R. 1985, Astr. Ap., in press.
Ellis, R. S., Gondhalekar, P. M., and Efstathiou, G. 1982, M.N.R.A.S., **201**, 223.
Hlivak, R. J., Henry, J. P., and Pilcher, C. B. 1984, SPIE Proc., **445**, 122.
Hunter, D. A., Gallagher, J. S., and Rautenkranz, D. 1982, Ap. J. Suppl., **49**, 53.
Jacoby, G. H., Hunter, D. A., and Christian, C. A. 1984, Ap. J. Suppl., **56**, 257.
Keel, W. C. 1983a, Ap. J., **269**, 466.
Keel, W. C. 1983b, Ap. J. Suppl., **52**, 229.
Margon, B., Ford, H. C., Katz, J. I., Kwitter, K. B., Ulrich, R. G., Stone, R. P. S., and Klemola, A. 1979, Ap. J. (Letters), 230, L41.
Massey, P. 1984, Ap. J., **281**, 789.
Miller, J. S., Robinson, L. B., and Wampler, E. J. 1976, Adv. Electronics Electron Phys., **40B**, 693.
Robinson, L. B., and Wampler, E. J. 1972, Pub. A.S.P., **84**, 161.
Robinson, L. B., and Wampler, E. J. 1973, in Astronomical Obsevations with Television-Type Sensors, ed. J. W. Glaspey and G. A. H. Walker (Vancouver: University of British Columbia Press), p. 69.
Schild, R., and Oke, J. B. 1971, Ap. J., **169**, 209.
Shectman, S. A. 1984, SPIE Proc., **445**, 128.
Stauffer, R. 1982, Ap. J. Suppl., **50**, 517.
Vilas, F. 1985, Icarus, in press.
Whitford, A. E. 1971, Ap. J., **169**, 215.

DISCUSSION

Garrison: I have a somewhat controversial question. Noting that most of the spectra you showed were emission-line spectra, and considering that I have heard bad things about internal scattering in image tubes, what is your opinion of the use of IDS or Shectograph-type instruments for absorption-line work, especially that involving quantitative line strength or profile work? I'm thinking in particular of the recent work of Burstein et al. using IDS spectra to comment on the so-called "Super-Metal-Rich" stars. My impression is that image tube spectra always have a "washed-out" appearance.

Phillips: Absorption-line work isn't the strong point for a detector with lots of image-tubes in front of it.

Finkenzeller: I would like to draw your attention to how low resolution spectroscopy can help high resolution spectroscopy. If one is interested in calibrated, high resolution line profiles one requires a small entrance slit which is in conflict with spectrophotometric needs. Thus, we have chosen to use a coudé spectrograph concurrently with an IDS at a much smaller telescope. The flux level is established reliably there, and after some work with an image processing system the response of the high resolution spectrograph can be determined easily. In addition, this technique relieves the observer spending valuable time with the large telescope on mostly faint flux standards.

White: Spectrophotometry can be accomplished with extremely small aperture telescopes. Tüg, Lockwood, and myself have completed a spectrophotometric comparison of Vega and the Sun using a two inch and 20 micron aperture telescopes respectively.

LOW DISPERSION SPECTRAL CLASSIFICATION WITH SMALL TELESCOPES

Hugo Levato
Complejo Astronómico El Leoncito
Casilla de Correo 467
5400 San Juan
Argentina

ABSTRACT. We review the spectral classification work with low dispersion. It is mainly concentrated on the MK-System. First we summarize the original properties of this system, its evolution and its present properties. We describe the MK Process and emphasize its importance as a tool for future research on spectral classification. In the third section we comment briefly about the instrumentation and the techniques used in spectral classification work and finally we mention a list of research programs that can be undertaken with a 1m-telescope and a good spectrograph.

1. INTRODUCTION

The title of this paper implies a very broad subject, so I would like to define more precisely which are the points that I will touch on in this review. First there is a redundancy in the sense that spectral classification has been historically a subject for small telescopes and in the second place it will be necessary to define what is *low dispersion*.

Let me mention first that spectral classification is to arrange stars into a certain number of groups formed of similar objects, through the appearance of their spectra (lines, blends, continuum features). The morphologist does not know a priori about physics, envelopes, temperatures, abundances, gravity, stellar rotation, turbulence, etc.

Several different systems have been devised to classify stars using photographic plates or tracings of these plates where some quantitative measures were done. Among the range of very low dispersion we must include the *natural groups* illustrated by Morgan, Meinel and Johnson (1954).

I am not going to deal with natural groups; for a review of them it is useful to read McCarthy (1979, 1984).

An example of work with dispersions ranging from 150 Å/mm to 300 Å/mm was done by Nassau at Warner and Swasey Observatory. Those interested in this kind of work can read details in Nassau and van Albada (1947). Attempts to classify from continuum measures have been done by Barbier and Chalonge (1939, 1941). I will not deal with systems such

J. B. Hearnshaw and P. L. Cottrell (eds.), Instrumentation and Research Programmes for Small Telescopes, 359–370.

as this one or with dispersions lower than 130 Å/mm. In this review low dispersion will be defined to be between 130 Å/mm and 40 Å/mm, and I will introduce the ingredient of my personal experience which concerns primarily the MK system.

I consider the MK system to be the most powerful tool to classify stars. It has produced very important astrophysical results, it is widely used (although not very honestly sometimes) and it will be used in the future. So my review can more properly be entitled MK Spectral Classification with Small Telescopes.

I want to mention two very important meetings which are fundamental for every stellar astronomer. They are *Spectral Classification of the Future* (McCarthy et al. 1979) and *The MK Process and Stellar Classification* (Garrison 1984). The latter one is a masterpiece in the field and should be read by every stellar astronomer. The discussions I have found to be especially enlightening. Let us now describe *the system*, then we will refer very briefly to the technical requirements and finally in the last section I will describe some research that can be done with a spectrograph and a small telescope.

2. THE MK SYSTEM

2.1 The original frame

The original MK system was introduced by Morgan, Keenan and Kellman (1943) in one of the most famous papers ever written.

In the Introduction the authors wrote:

The plan of the Atlas can be stated as follows:

To set up a classification system as precise as possible which can be extended to stars of the eighth to twelfth magnitude with good systematic accuracy.

The actual process of classification is carried out in the following manner: (1) an approximate spectral type is determined; (2) the luminosity class is determined; (3) by comparison with stars of similar luminosity an accurate spectral type is found.

The instrument used to obtain the spectra was a small one-prism spectrograph attached to the 40-inch refractor at Yerkes Observatory. The reduction of collimator to camera was about 7 so one can open the slit on the telescope focal plane without losing resolution. The usable spectral region of the plates was λ 3920-4900 Å. The dispersion was 120 Å/mm at Hγ.

The Atlas was composed of 55 prints which represent the standard stars that defined the original MK system. The reproductions of the Atlas show the quality of classification spectra taken at the time.

The spectral types and luminosities are defined by the standard stars, several ratios between pairs of lines can be used but these ratios can not be used safely at different dispersions. Each dispersion has its own problems that must be studied independently.

Morgan and Keenan wrote in 1943:

There appears to be, in a sense, a sort of indefiniteness connected with the determination of spectral type and luminosity from a simple inspection of a spectrogram. Nothing is measured; no quantitative value is put on any spectral feature. This indefiniteness is however, only apparent. The observer makes his classification from a variety of considerations, the relative intensity of certain pairs of lines, the extension of the wings of the hydrogen lines, the intensity of a band, even a characteristic irregularity of a number of blended features in a certain spectral region. To make a quantitative measure of these diverse criteria is a difficult and unnecessary undertaking. In essence the process of classification is in recognizing similarities in the spectrogram being classified to certain standard spectra.

It is not necessary to make cephalic measures to identify a human face with certainty or to establish the race to which it belongs; a careful inspection integrates all features in a manner difficult to analyze by measures. The observer himself is not always conscious of all the bases for his conclusion. The operation of spectral classification is similar.

It is useful to recall that in the original frame each type and luminosity was defined by several standards for example α Leo, β Lib and β CMi defined the B8 V box.

2.2 The evolution of the MK system

As with any frame of reference in active use, the MK system has evolved to deal with the many interesting problems involved in astrophysical research. In 1978 Morgan, Abt and Tapscott (1978) published *The Revised MK Spectral Atlas for Stars Earlier than the Sun* while in 1976 Keenan and McNeil had already published *An Atlas of Spectra of the Cooler Stars: Types G, K, M, S and C*. The revision by Keenan and McNeil of stars later than the sun had two main purposes.

(1) The assignment of types and luminosities can be strongly influenced by abundance effects even when the overall appearance of the spectra does not show conspicuous peculiarities.

(2) For the later classes it is not sufficient to speak of *the metal abundance* and to assume that the concentrations of all the elements heavier than helium vary in step.

So the Atlas for the Cooler stars has introduced a few modifications in the subdivisions of types G, K and M especially for interpolation, as K2, K2+, K2.5 and K3-.

In luminosity class the change with respect to the original system was the recognition that there are some stars in the Large Magellanic Cloud that are more luminous than the Ia class in our Galaxy so they introduced the luminosity class 0. Also in some parts of the HR diagram it was possible to subdivide the luminosity classes from II to V, using the subscripts *a* or *b*.

Perhaps the most significant change was the introduction of a compact notation that signal differences in the features defining the abundance groups. These groups are labelled in the Atlas as: Hδ, Ca or Na, Fe, CN, CH, Spectral Type C, C^{13}, Ba, Zr and Li. The symbol is followed by an index from 1 to 5 to indicate the amount of departure of the feature from the intensity in the spectrum of a star with solar composition. The index is positive or negative if the intensity is higher or lower respectively. The plate material for this revision was taken with the 72-inch Perkins reflector at Flagstaff, with the 60-inch telescope at Cerro Tololo and with the 32-inch Schotland telescope at Perkins Observatory. The dispersions ranged between 75 and 80 Å/mm.

In the Morgan, Abt and Tapscott Atlas the revision had three purposes, to present an improved version of the MK system defined in 1943, to decrease the *noise* of the classification defining only one standard star in each box and to demonstrate how rich the prospects can be for classification in the future, using several conceptual improvements that were incorporated into the MK-78 System as the authors named it. The consequence was a finer structure of the MK Diagram for stars earlier than the Sun. Some parts of the written booklet that accompanies the reproductions are also masterpieces because a lot of important concepts are stated in a couple of lines.

A major improvement of the Morgan and Abt work is the extension of the spectral range used to λ 3500 Å and also that the MK-78 system permits perfectly well the inclusion of localized third dimensions.

The Atlas is clearly not definitive because there is a large body of stellar spectra that have not been studied yet with the MK system.

The reproductions of this Atlas show the quality of the spectrograms that can be taken at the present time for spectral classification. The spectra were taken with the spectrographs at Cerro Tololo and Kitt Peak, and they have a demagnification slit-plate of almost 10. They are very efficient and have good definition.

2.3 The Present System

We can summarize the current properties of the MK System as follows:

(a) It is an empirical system. No physics is involved a priori.

(b) Types and luminosity classes are defined by the patterns and intensities of spectral lines, bands and blends which exist in the spectrum of a standard star.

(c) It is defined in the spectral range (λλ 3500-4900 Å) for earlier types than G2 and (λλ 3900-4900 Å) for later types, using as a detector the photographic plate (Kodak IIa-0 emulsion). The dispersion is 125 Å/mm for earlier types, 75-80 Å/mm for later types and the resolution ranges from 1.5 to 3 Å.

(d) A third dimension is included for the stars later than G2 to take account of the abundance differences.

(e) A constant spectral type is assumed for all the standard stars.

More than 130000 stars have been classifed in the MK System (for a review about the compilation of data see Jaschek, 1984).

2.4 The MK Process

One of the most important results of the meeting *The MK Process and Stellar Classification* was the definition of a procedure for the classification of specimens which do not fit the MK System. The methodology that makes possible the construction of systems of classification based on the characteristics of stellar spectra is called the MK Process. Morgan (1984) gave details and examples about its usage.

The application of the MK Process to classify certain types of objects in our Galaxy and in other galaxies opens a lot of research possibilities with small telescopes equipped with good (but not necessarily very expensive) spectrographs and different kinds of detectors including the photographic plate.

Before going to the technical needs and the research possibilities I would like to show in Figure 1 what is wrong and what is good in the application of the MK System today from my personal point of view. I

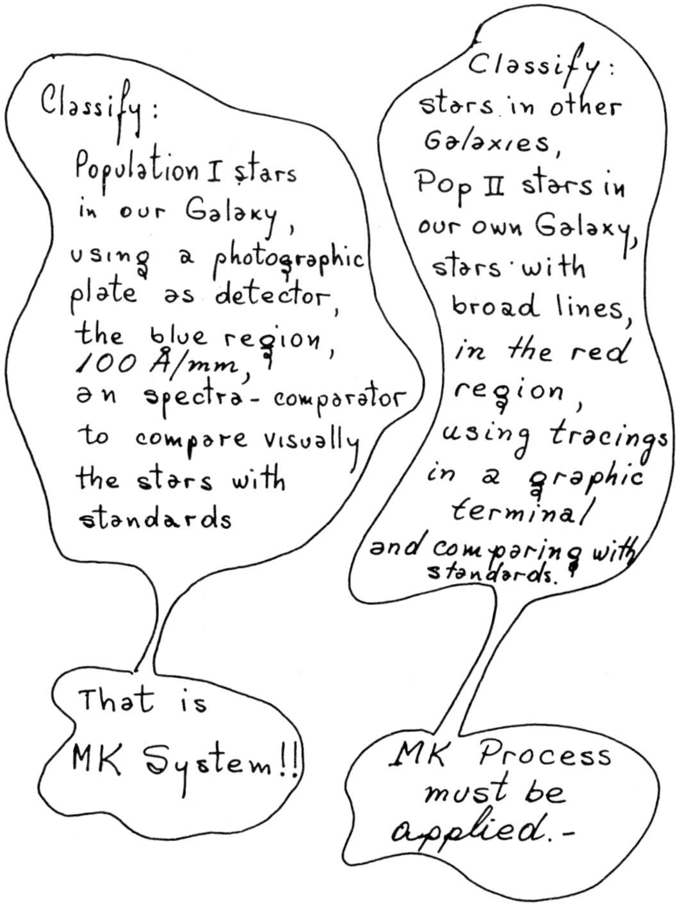

Figure 1. The MK System and the MK Process

would like to emphasize that this is what I consider wrong and good because I perfectly agree with the principle introduced by David Crawford (1984) about *live and let live*, so it is not my intention to advise other people what they must do.

From the figure it is clear that for me it is not possible to use the MK System as defined by Morgan, Abt and Tapscott (1978), for stars in other galaxies, for stars of different population than I in our own Galaxy (i.e. globular cluster stars). It is not possible to apply it to broad-line stars; it is not possible to classify, in the MK System in a rigorous way, spectra which are not recorded on a photographic plate or outside the wavelength range λ 3500-4900 Å. Also to classify in the MK System it is necessary to compare the stars visually with the proper standards.

For working at other wavelengths, with other detectors, in other galaxies or on stars for which the MK System is not designed, an extremely powerful tool exists in the *MK Process*. I believe it must be used when the MK System can not be applied and then the results can be studied and discussed in the framework of the MK System.

3. INSTRUMENTATION AND TECHNIQUES

What you need for research in spectral classification is a telescope with a smooth tracking and a fast spectrograph with a good optical quality. The telescope is supposed to be of the order of 1m, a small telescope, in order to pursue the aim of this meeting. The spectrograph must have optics which transmit in the UV and the demagnification from the telescope to camera focal planes should be between 7 and 10 so you can use the spectrograph with advantage even in places where the seeing is not very good. The detector may be the photographic plate or a CCD or a Shectograph-type instrument. The cheapest is a photographic plate followed by the combination with an image tube. There are some tubes that give acceptable results concerning resolution. Then you have a CCD or Reticon in the middle of the price range and finally the Shectman system. Perhaps the latter is not yet in the price range of small institutions.

After the observations have been secured you can use an eyepiece, a comparator or a microscope (it depends on your budget and on your visual memory) in the case that your detector is the photographic plate, intensified or not. If you like tracings, you can trace your plate and classify from it; or from tracings on a graphics terminal if you obtain the data with a CCD, Reticon or Shectograph. But also you can convert your tracing to resemble a photographic plate through modern plotter devices. So the combination is probably unlimited and the observer can do what he likes, but the important point is to keep oneself honest with the frame of reference he is using: the MK System or another one built through the MK Process. Figure 2 summarizes the technical possibilities.

A word about detectors. There are several very good reviews about them. Let me mention two excellent ones: Eccles et al. (1983) and Gray (1984). The first one is a book very complete and especially written for astronomers. I would like to take the final paragraph of Gray's paper.

1. Don't think in terms of building an electronic detector system - do you make your own photographic plates?!

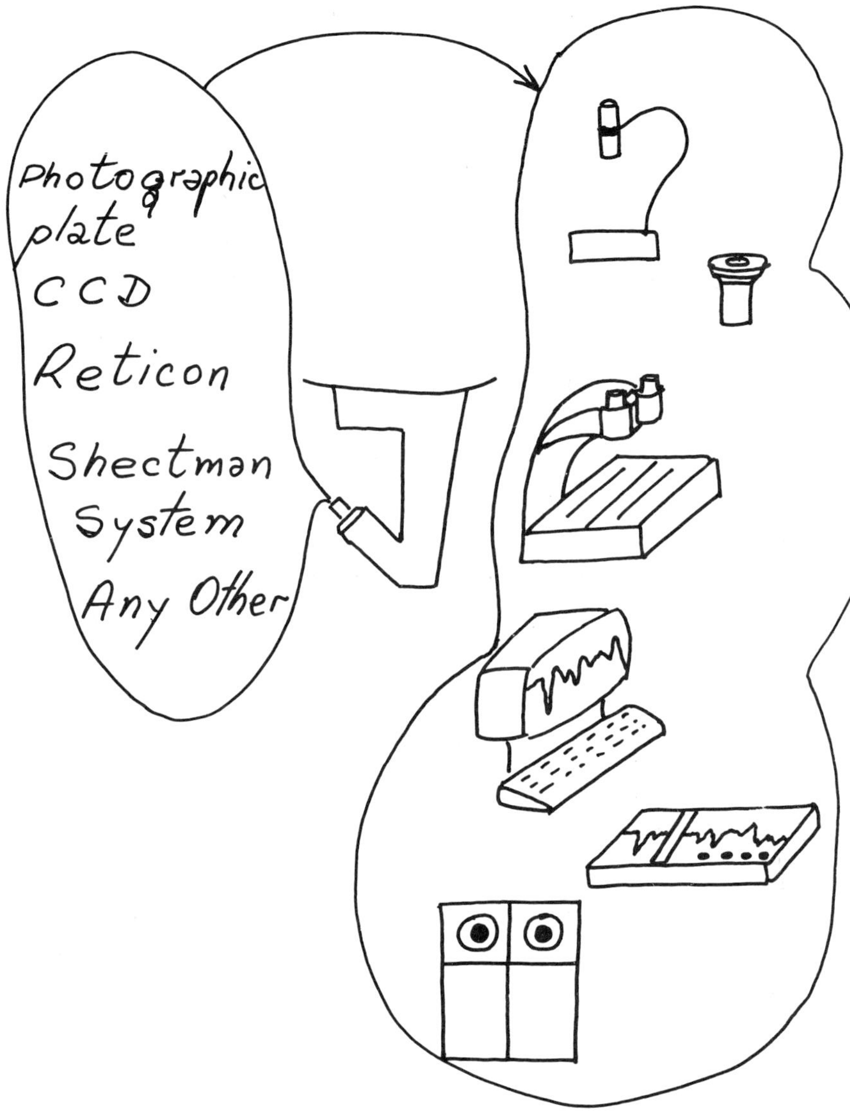

Figure 2. Different techniques for spectral classification.

2. Ask yourself. Can the electronic system give me the spectral-classification data that I need but can't get photographically?

Experience indicates that it is necessary to go to S/N ratios of 500 or more to see on a spectrum record from a CCD, the same details that one sees on a photographic plate. My impression is that for spectral classification work it would be justified to use CCD frames if you can combine them in the future with automatic spectral classification. I would like not to be misunderstood on this point; 'modern detectors' are wonderful tools that permit us to reach, with small telescopes, objects unthinkable 10 years ago, but we are trying here to get an accurate spectral type so we need to select a system and the prescriptions of this system must be followed.

For me there is another point of great concern with the modern detectors. In general, research programs on spectral classification are long range projects which are fundamental work and one would like to have an instrumental configuration to remain the same for a long period of time.

It is very common today that new detec tors especially at large observatories but even on small telescopes, are changed very often with the additional fact that some of these observatories are discontinuing the usage of the most traditional configurations. These are severe drawbacks for systematic projects.

4. RESEARCH PROGRAMS

A lot of work remains to be done in spectral classification using small telescopes. I will review only some of them very briefly.

(a) Research on the MK System itself
It is necessary to explore certain localized regions of the MK Diagram to introduce some detailed third dimension.
Morgan (1984) gives two examples namely the Hg-Mn stars and the Sr II stars. Also the λ Bootis stars whose discovery was reviewed by Abt (1984) deserve special attention.

(b) Standards
It is necessary to include one standard in each box of the MK Diagram which is not full yet.
Concerning faint standards it is possible to define new ones with the procedure, mentioned in the Toronto meeting, of boot-strapping from bright magnitudes to fainter ones using different telescopes. I believe that with a 1m telescope it is possible to work reasonably well as far as magnitude 12. Garrison is working on the problem of establishing faint standards.

(c) It is necessary to use the MK Process to explore the yellow-green and the red region of the spectra of the stars, or to establish new systems in these spectral regions completely autonomous from the MK. This is perfectly possible with a 1m telescope.

(d) It is necessary to explore spectral classification with other detectors than the photographic plate. During the meeting at Toronto many people expressed the desire that an Atlas showing an array of frames in some *natural* order would be very important. You can classify from the tracings using the MK Process if you reach high S/N values.

(e) The devising of a homogeneous system for Pop. II stars can be done with a small telescope and this is very necessary to complement the MK System.

(f) It is necessary to use the MK Process to explore a system for broad-line stars for which the MK System is not designed.

(g) Eclipsing Binaries
It is a difficult task but there are hundreds of eclipsing binaries that can be observed with a fast spectrograph on a 1m telescope. Morgan (1984) pointed out some of the difficulties that one is going to encounter when classifying eclipsing binaries but I believe that a project on eclipsing binaries is worth undertaking.

(h) Spectroscopic and Visual Binaries
Some work has been done recently on spectroscopic and visual binaries. In particular Abt (1981, 1985a) published 1865 MK types of members of visual systems. He found new Ap stars λ Boo stars, Am stars and SB2s.
Corbally and Garrison (1984) determined MK classification of 170 visual binaries with separations between 1 and 5 arcseconds.
But there is still a lot of place for research projects in this field. There are still spectroscopic binaries with orbits computed in Batten et al. (1978) which do not have MK types.

(i) Trapezium Systems
Some work has been done recently by Abt (1985b) and by Levato still unpublished. Candidates selected from the Catalogue of Trapezium Systems compiled by Allen et al. (1977) were observed in both cases. Abt observed 31 systems and found only 11 Trapezium systems, Levato observed around a 100 systems from Las Campanas with Garrison's spectrograph and found the same proportion of true Trapezium systems.
Continuation of programs on Trapezia can be done with small telescopes.

(j) Variable Stars
There is a lot to be done in this field. Some kinds of variables have been studied recently in two important contributions: Miras by R. Crowe (1984) and Cepheids by Gauthier and Garrison (1984) but other kinds of variables are practically untouched.

(k) Open Clusters
A lot has been done on these objects but let me point out that MK spectral classification is lacking even for some clusters nearer than 500 pc. Recent compilations of data by Mermilliod (1984) indicate that there are only 4000 stars in 225 open clusters with MK spectral classifications. This is a dramatic situation. I have worked with Helmut Abt on several clusters in the mid 70's. The results that we have got with respect to the occurrence of peculiar stars have been summarized by Abt (1984). From CTIO we continued the work on several

interesting southern objects like the Trumplers in the Carina region, Cr228, IC2391, NGC2287. We have almost finished the classification of the 550 stars in the Orion Association which were included by Warren and Hesser (1977) in their photometric study. These investigations produced dozens of peculiar objects, composite spectra, double-line binaries, Be stars, and other kinds of exotic objects. If we are going to understand completely these specimens the spectroscopic observations of open cluster members must be pursued.

(1) It is necessary to take advantage of the strong effort of Nancy Houk's (1975, 1978, 1982) classification of the HD stars in the MK System. The new catalogues are a gold mine for picking up interesting objects whose spectra for several reasons could not be precisely defined. The excitement of astronomy compared with other sciences is that astronomy does not know yet all the specimens under study.

Summarizing, down to magnitude 12 there are more than 3 million stars that can be reached with a modest telescope + spectrograph combination from a good astronomical site; less than 4% have MK types so there is a lot of work to do; photographic plates, CCDs, Reticons, Shectographs, microscopes, comparators, tracings and computers are welomed; keep honest and go ahead.

I thank very much Bob Garrison and Helmut Abt for sending me details of their current projects on spectral classification.

REFERENCES

Abt, H.A. 1981, *Astrophys, J. Suppl.* 45, 437.

Abt, H.A. 1984 in *The MK Process and Stellar Classification,* ed. R.F. Garrison (Toronto: David Dunlap Observatory, University of Toronto), p.369.

Abt, H.A. 1985a, *Astrophys. J. Suppl.* 59, 95.

Abt, H.A. 1985b, *Astrophys. J. Suppl.* (in press).

Allen, C., Tapia, M. and Parrao, L. 1977, (unpublished).[2]

Barbier, D. and Chalonge, D. 1939, *Ann. d'Astrophys. 2, 479.*

Barbier, D. and Chalonge, D. 1941, *Ann. d'Astrophys. 4, 30.*

Batten, A., Fletcher, J. and Mann, P. 1978, *Seventh Catalogue of Orbital Elements for Spectroscopic Binary Systems,* Publ. *Dom. Astrophys. Obs.* 15.

Corbally, C.J. and Garrison, R.F. 1984, in *The MK Process and Stellar Classification,* ed. R.F. Garrison (Toronto: David Dunlap Observatory, University of Toronto), p. 367.

Crawford, D.L. 1984, in *The MK Process and Stellar Classification,* ed. R.F. Garrison (Toronto: David Dunlap Observatory, University of Toronto), p.191.

Crowe, R. 1984, in *The MK Process and Stellar Classification,* ed. R.F. Garrison (Toronto: David Dunlap Observatory, University of Toronto), p.348.

Divan, L. 1979, in *IAU Colloquium 47: Spectral Classification of the Future, Ricerche Astronomische* 9, ed. M.F. McCarthy, A.G.D.

Phillips and G.V. Coyne (Vatican City: Vatican Observatory), p. 247.

Eccles, M.J., Sim, M.E. and Tritton, K.P. 1983, *Low Light Level Detectors in Astronomy,* (Cambridge: Cambridge University Press).

Garrison, R.F. 1984, *The MK Process and Stellar Classification,* ed. R.F. Garrison (Toronto: David Dunlap Observatory, University of Toronto).

Gauthier, R.P. and Garrison, R.F. 1984, in *The MK Process and Stellar Classification,* ed. R.F. Garrison (Toronto: David Dunlap Observatory, University of Toronto), p.362.

Gray, D.F. 1984, in *The MK Process and Stellar Classification,* ed.R.F. Garrison (Toronto: David Dunlap Observatory, University of Toronto), p.112.

Houk, N. 1978, *Michigan Catalogue of Two-Dimensional Spectral Types for the HD Stars,* Vol. 2, (Dept.of Astronomy, University of Michigan, Ann Arbor).

Houk, N. 1982, *Michigan Catalogue of Two-Dimensional Spectral Types for the HD Stars,* Vol. 3, (Dept. of Astronomy, University of Michigan, Ann Arbor).

Houk, N. and Cowley, A.P. 1975, *Michigan Catalogue of Two-Dimensional Spectral Types for the HD Stars,* Vol. 1, (Dept. of Astronomy, University of Michigan, Ann Arbor).

Jaschek, C. 1984, in *The MK Process and Stellar Classification,* ed. R.F. Garrison (Toronto: David Dunlap Observatory, University of Toronto), p.94.

Keenan, P.C. and McNeil, R.C. *An Atlas of Spectra of the Cooler Stars: Types G, K, M, S and C* (Columbus: Ohio State University Press).

McCarthy, M.F. 1979, in *IAU Colloquium 47: Spectral Classification of the Future,* p.103.

McCarthy, M.F. 1984, in *The MK Process and Stellar Classification,* ed. R.F.Garrison (Toronto: David Dunlap Observatory, University of Toronto), p.55.

Mermilliod, J.C. 1984, in *The International Course on Data Handling in Astronomy and Astrophysics,* in press.

Morgan, W.W. 1984, in *The MK Process and Stellar Classification,* ed. R.F. Garrison (Toronto: David Dunlap Observatory, University of Toronto), p.18.

Morgan, W.W., Abt, H.A. and Tapscott, J.W. 1978, *Revised MK Spectral Atlas for Stars Earlier than the Sun,* (Chicago and Tucson: Yerkes and Kitt Peak National Observatories).

Morgan, W.W., Keenan, P.C. and Kellman, E. 1943, *An Atlas of Stellar Spectral* (Chicago: The University of Chicago Press).

Morgan, W.W., Meinel, A.B. and Johnson, H.M. 1954, *Astrophys. J.* 120, 506.

Nassau, J.J. and van Albada, G.B. 1947, *Astrophys. J.* 106, 20.

Warren, W. and Hesser, J. 1977, *Astrophys. J. Suppl.* 34, 115.

DISCUSSION

Gliese: You have mentioned the MK classification on objective prism plates by Nancy Houk. I have used MK types as second order distance indicators and I became aware that, at least in the region K2V to M0V, these objective prism MK types have a somewhat different spectral type-luminosity relation than the slit MK types (differences up to $\Delta M = 0\overset{m}{.}5$ occur).

Mochnacki: Both the previous speakers (Phillips and Levato) referred to image tubes. As you know it has been difficult to obtain UV-sensitive intensifiers in recent years. There is good news: VARO are now producing UV-sensitive proximity-focussed microchannel plate intensifiers, with gains of up to 15,000 and sizes up to 40 mm. This is particularly useful for photon-counting and classification applications.

Garrison: Two comments. First, Morgan and Abt are working on a definition of a classification system for pop II stars. Second, it is possible to build an MK spectrograph for about $US2000 worth of materials. I have drawings and will supply them freely.

Scarfe: The lack of MK classifications for spectroscopic binaries could well be worse in a future orbit catalogue. For example the majority of the objects for which Griffin has published orbits in the Observatory do not have MK types.

CASSEGRAIN ECHELLE SPECTROSCOPY WITH SMALL TELESCOPES

J. B. Hearnshaw
Mt John University Observatory
Department of Physics
University of Canterbury
Christchurch, New Zealand

ABSTRACT. This paper comments on the growing use of cassegrain échelle spectrographs for high resolving power astronomical spectroscopy over the last two decades. Some of the theoretical and practical advantages of échelle spectrographs are outlined, as well as some of the problems that arise in their use. Design parameters of some recent échelle spectrographs in various observatories are summarised. A novel use of échelle spectrographs without a cross-dispersing element for radial velocity spectrometry is proposed. The design parameters of a fibre-fed échelle spectrograph that will achieve a resolving power of about 7×10^4 with a 2 arc second input from a 1m telescope are outlined.

1. INTRODUCTION

In 1956 the distinguished American stellar spectroscopist, Dean McLaughlin, wrote: 'Detailed analysis of stellar spectra with high dispersion is strictly the province of the largest telescopes with coudé spectrographs. It seems hopeless to dream of doing such work with lesser instruments' (McLaughlin, 1958).

He was wrong, though in hindsight it was a reasonable error to make at that time.

George Harrison's (Harrison, 1949) classic paper on échelle gratings had appeared a few years earlier, but the échelle was not applied to stellar spectroscopy until 1965. Harrison pointed out that for any grating used in the Littrow configuration ($\alpha \simeq \beta$), the angular dispersion is given by $d\beta/d\lambda = 2\tan\beta/\lambda$. Here α is the angle of incidence and β the angle of diffraction. At a *given wavelength* the angular dispersion is independent of the groove spacing d of the ruling, but increases monotonically with β, and hence α. A blazed échelle grating should have facets of width s at a blaze angle $\theta_B \simeq \alpha$. If $t = d\sin\theta_B$ is the step depth (see Fig.1), then the free spectral range, defined as the wavelength interval between successive orders at a fixed value of β, is $\lambda/n = \lambda^2/2t$ (or in terms of wave number it is $\frac{1}{2t}$, which is the same for all orders), while the full angular width of one order is $2\lambda/s$ radians. A coarsely ruled échelle thus produces many orders at

J. B. Hearnshaw and P. L. Cottrell (eds.), Instrumentation and Research Programmes for Small Telescopes, 371–384.

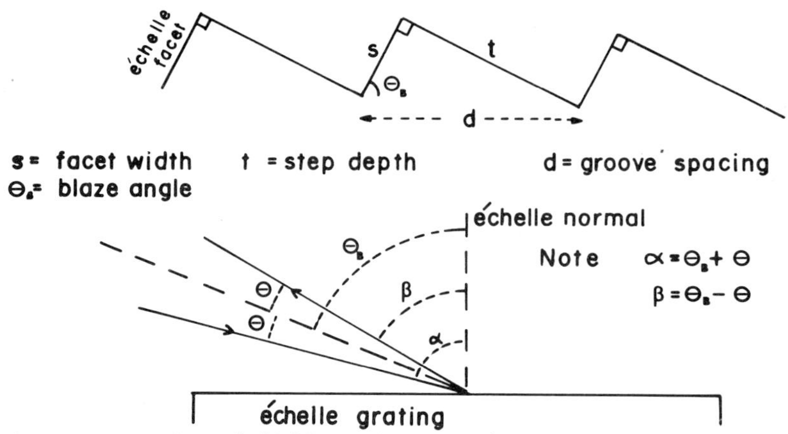

Fig. 1 Schematic diagram of échelle angles and dimensions

the same angle of diffraction, each with small angular spread and high dispersion. Harrison also described the use of a cross-dispersing element, either a prism or a conventional grating, to achieve order separation. Because of the λ^2 - dependence of the free-spectral range, a prism will give more uniform order separation than a grating.

Although échelle gratings were occasionally used for solar spectroscopy during the 1950s and 1960s from the ground (Pierce et al., 1951) or from rockets (Tousey et al., 1967), the first reported use for stellar spectroscopy is from the Crimean Astrophysical Observatory by Kopylov and Stechenko (1965). Soon afterwards Schroeder (1967) and Schroeder and Anderson (1971) described the échelle spectrographs built at Wisconsin for the f/13.5 cassegrain focus of the 0.91m Pine Bluff Observatory telescope. The latter instrument achieved a resolving power $R(=\lambda/\delta\lambda)$ of about 27000 using a slit of angular width $\theta_S = 1.7$ arc seconds, and was therefore able to compete with coudé spectrographs with larger telescopes. At 650 nm the reciprocal dispersion was 2.5Å/mm.

2. SOME NOTES ON THE THEORETICAL ADVANTAGES OF ECHELLE SPECTROGRAPHS

The basic equation for the resolving power of any slit-limited Littrow grating spectrograph, including échelle instruments is

$$R = \frac{2L \sin \theta_B \cos \theta}{\theta_S D} \qquad (1)$$

Here
L is the width of the grating, normal to the grooves, which is illuminated
θ_B is the blaze angle
θ_S is the angular size of the slit in radians
D is the aperture of the telescope
and θ is a small angle such that

$$\alpha = \theta_B + \theta$$
and $$\beta = \theta_B - \theta$$ in the centre of an order.

Most échelle gratings are blazed at 63°26' = $\tan^{-1}2$, though a few have $\theta_B = \tan^{-1}4$. The angle θ is small (i.e. $\alpha \simeq \beta$) and chosen to be positive ($\alpha > \beta$) as this gives higher R than $\alpha < \beta$ (Chaffee and Schroeder,

1976) for a given value of collimator focal length f_{coll}, which, because of flexure, is one of the limiting parameters (generally $f_{coll} \lesssim 1m$) in a cassegrain spectrograph. However $\alpha < \beta$ was adopted in the Johns Hopkins échelle spectrograph (McClintock, 1979) because of mechanical constraints.

Equation (1) for the resolving power can usefully be written in its equivalent forms

$$R = \frac{2B \tan \theta_B}{\theta_s D(1 - \tan \theta_B . \tan \theta)} \simeq \frac{2B \tan \theta_B}{\theta_s D} \qquad (2)$$

where B is the collimator beam diameter, or

$$R = \frac{2f_{coll} \tan \theta_B}{W(1 - \tan \theta_B . \tan \theta)} \simeq \frac{2f_{coll} \tan \theta_B}{W} \qquad (3)$$

where f_{coll} is the collimator focal length and W the slit width. The last equation thus shows that a smaller value of f_{coll} can be employed for a given R if θ is positive, while equation (1) shows the advantage of the Littrow condition of small θ (typically a few degrees) for a given échelle grating. In addition, the Littrow condition ensures that photons of a given wavelength are distributed in no more than two adjacent orders, and not three or more (for larger θ). The total wavelength range of an order in the Littrow condition is just twice the free spectral range, and the intensity maximum of one order coincides with the first minimum at the ends of the adjacent orders, thus ensuring the best throughput efficiency.

Hence in the Littrow condition (small θ) equation (1) becomes

$$R \simeq \frac{2L \sin \theta_B}{\theta_s D}$$

This relationship illustrates the advantages of large gratings operating at large blaze angles and fed by light from a narrow slit to achieve high resolving power. Since $\sin \theta_B \simeq 0.9$ for $\theta_B = \tan^{-1} 2$, there is little advantage in increasing θ_B yet further. On the other hand doubling $\tan \theta_B$ to 4 for a given groove spacing also doubles the free spectral range and angular spread of the orders which may no longer fit onto a given detector such as an image tube or CCD. Thus $\tan \theta_B = 2$ is about the optimum value.

The resolving power formula also emphasises that small aperture telescopes achieve higher resolving power, other factors being equal. This at first sight paradoxical result simply arises from the fact that large telescopes usually have a smaller focal plane scale (arc s. mm^{-1}), which necessitates opening up the slit to pass a given fraction of the seeing disk.

The high angle of incidence of the light in an échelle instrument has two advantages : (i) a high value of $\sin \theta_B$ and (ii) the illuminated width of grating L is considerably greater than the beam size $B = L \cos \alpha$ ($\simeq 0.4L$, for $\tan \theta_B = 2$). In principle, resolving powers as high as $R = 10^5$ can thus be achieved for an échelle on a 1m telescope with a 1 arc s slit provided an échelle grating with L = 28cm or a beam size of about B = 11cm can be employed. This would be about the practical limit for a cassegrain échelle spectrograph on a 1m telescope, and resolving powers of half this are now routine on a number of instruments in operation.

For a given telescope and slit size, the resolving power advantage of the échelle over a coudé spectrograph is simply the ratio of the $L \sin \theta_B$ terms in the first formula for R above. Writing subscript e for cassegrain échelle, c for conventional coudé, then

$$\frac{R_e}{R_c} = \frac{L_e \sin \theta_{Be}}{L_c \sin \theta_{Bc}} \simeq \frac{B_e \tan \theta_{Be}}{L_c \sin \theta_{Bc}}$$

The cassegrain échelle is limited through flexure by its beam size Be of about 10cm, while the coudé is limited by the availability of very large gratings, say about L_c up to 30cm. Taking $\theta_{Be} = \tan^{-1}2$, $\theta_{Bc} = 8\frac{1}{2}°$ (a typical value) then Re/Rc = 4.4. A more conservative estimate with Be = 7.5cm gives Re/Rc = 3.3, or an advantage of 1.3 magnitudes for the cassegrain échelle over the coudé. A similar result has been demonstrated by Schroeder (1974).

This is the main reason why cassegrain échelle spectroscopy with small telescopes has been increasing in popularity. Another factor is that the cost of a cassegrain échelle spectrograph is probably about one order of magnitude less than that of a coudé spectrograph. We installed our échelle instrument at Mt John Observatory in 1977 after an expenditure of about $US 12,000 and $2\frac{1}{2}$ man-years of workshop time for spectrograph construction.

3. PRACTICAL PROBLEMS OF REDUCING ÉCHELLE SPECTRA

The échelle spectrograph has made high resolving power stellar spectroscopy with relatively small aperture cassegrain telescopes an attainable goal. However the format of échelle spectra presents a number of problems which have been extensively discussed in the literature (e.g. Chaffee and Schroeder, 1976; Hearnshaw, 1981; Spite, 1980). The four most often cited complications are tilted lines, non-uniform widening, non-linear dispersion and the curved continuum over a given diffraction order. The first three of these problems can all be largely overcome. Thus tilted lines arise from the combination of échelle and cross dispersions. The loss of resolution may not be serious unless the spectrum is widened considerably. In this case a microdensitometer with a rotatable slit assembly overcomes the resolution loss. Alternatively the line tilt can be largely eliminated for a given échelle order by illumination of the échelle at a small angle γ to the plane normal to the grooves. For a $\tan \theta_B = 2$ échelle, $\tan \gamma$ is a quarter the ratio of cross grating to échelle dispersions.

The non-uniform widening arises simply because it is not necessary to trail the star along the slit in our f/10 cassegrain instrument to produce a reasonably wide spectrum, as both collimator astigmatism and the seeing produce a spectrum typically around 200 μm wide in the spectrograph focal plane. For solid state detectors this problem in any case vanishes; for photographic work it can be treated either by tracing the spectrum with a tall microdensitometer slit, spanning the full width of the spectrum (Hearnshaw, 1981), or by tracing the spectrum in a number of narrow parallel strips along the length of the spectrum (Peterson and Title, 1975). The former technique entails some loss of signal-to-noise and the necessity of measuring the cross density profile encompassed by the measuring slit; the latter method invokes complex software to control the microdensitometer with precision in two dimensions.

The non-linear dispersion comes from the change in the angular dispersion (proportional to sec β) along the length of an order. For an échelle with $\tan\theta_B = 2$ and 79 grooves/mm, a diffraction order in the yellow spectral region will have an angular width in the useful region (over one free spectral range) of about $\beta = \pm 2.5°$ from the order centre. There is a change of sec β of about 20 per cent over this spectral region. However such non-linearity is relatively easy to handle using a cubic least squares polynomial fit to the stellar lines, or to the lines of a thorium comparison lamp.

The curved continuum of the short échelle orders is their most serious drawback. The short free spectral range (λ/n, with n typically in the range 25 to 75) is often comparable to the intrinsic width of spectral lines in early type stars. Observation of such broad lines may result in orders without any continuum being present, which may make subsequent profile analysis very difficult or impossible. In Belfast, McKeith et al (1978) have used a Bausch and Lomb 316 grooves/mm échelle operating in relatively low diffraction orders (n = 5 to 11) giving a large free spectral range. This minimises order curvature and makes the study of broad lines in early type stars possible. Most other échelles have 79 or 31.6 grooves/mm which are therefore best restricted to observations of the spectra of late type stars, or to narrow stellar or interstellar lines in early type stars. On the other hand their shorter orders fit more readily onto two-dimensional detectors.

4. DESIGN CHARACTERISTICS OF SOME ÉCHELLE SPECTROGRAPHS

Table I lists some cassegrain échelle spectrographs that have been used on small telescopes. This table gives basic design details of the majority of such instruments although it is certainly not a complete list. Echelle spectrographs used primarily on telescopes of greater than 1.5m aperture, or installed at the coudé focus are not included. Of the fourteen instruments referenced in this table, all but two were completed in the last ten years, and seven of them since 1980. All use $\tan\theta_B = 2$ échelle gratings except for the Lunar and Planetary laboratory instrument where $\tan\theta_B = 3.2$ (Brown et al, 1982). Most of the échelles have 79 grooves/mm; but several employ 31.6 grooves/mm to shorten the free spectral range, so as to fit the orders onto small aperture detectors. Only the Queen's University Belfast échelle uses a 316 groove/mm échelle, where the large free spectral range is suitable for broad lines in early-type stars (McKeith et al, 1978). The cross

TABLE I Some cassegrain échelle spectrographs used on small telescopes

Instrument	Year completed	Echelle $g.mm^{-1}$; $\tan\theta_B$	Cross disperser	Reciprocal at 5000 Å (n/mm)	Typical highest resolving power	Telescopes employed	Reference
1. Uni of Wisconsin	1971	73.5 2	grating after echelle	1.9	1.6×10^4 for 1"slit	0.9m Pine Bluff Obs.	(1)
2. Smithsonian Astrophys. Obs.	1973	31.6 2	grating after echelle	2.0	3.4×10^4 for 1"slit	1.5m Mt Hopkins Whipple Obs.	(2)
3. Harvard Coll. Observatory	1975	79 2	grating after echelle	2.0	3.4×10^4 for 1"slit	1.5m Oak Ridge Obs.	(3)
4. Johns Hopkins	1976	79 2	flint glass 45° prism before echelle	1.4	4×10^4 for 1"slit	0.91m Goddard Space Flight Center	(4)
5. replicas of the HCO spectrograph							
a)	1977	79 2	grating after echelle	2.0	4.3×10^4 for 2"slit	0.61m Mt John University Observatory, N.Z.	(5)
b)	1978	"	"	"	5.2×10^4	1.0m Ritter Observatory Toledo, Ohio	
c)	1980	31.6 2	"	"	5.2×10^4 for 1"slit on 1m tel	1.0m Lowell Observatory 1.8m Perkins telescope at Lowell	
d)	1982	"	"	"	2.7×10^4 for 1"slit on 1.88m tel	1.88m David Dunlap 0.61m Las Campanas	

6. Queen's Uni Belfast	1978	31.6 2 NB n ~ 5 to 11	double pass flint prism	0.6	4×10^4 for 1" slit on 0.9m tel	0.9m Royal Greenwich Obs. 1.0m NHO, La Palma 0.75m SAAO, South Africa 1.88m Kottamia, Egypt	(6)
7. Australian National University	1980	79 2	grating after échelle	5.0	3×10^4	1.0m Siding Spring Obs. 2.3m SSO Alt-Az tel. 4.0m AAT	(7)
8. University of Vienna	1982	79 2	grating after échelle	7.7	5×10^4	1.5m Figl Observatory for Astrophysics (three copies of this spectrograph are in use in Mexico and Italy.	(8)
9. University of Arizona, Lunar and Planetary lab.	1982	79 3.2	filter between slit and collimator	1.23	1.5×10^5	0.61m Mt Hopkins Whipple Obs., Arizona 1.55m Catalina Obs., Mt Bigelow, Arizona	(9)
10. Pennsylvania State University	1985	79 2	a) double pass prism after échelle with silvered back surface or b) grating after échelle	6.5 1.7	1.2×10^4	1.6m Penn State Observatory " " (fibre optics feed)	(10)
11. Osmania Uni., Hyderabad	1986	79 2	grating after échelle	5	5×10^4	1.22m Japal-Rangapur Observatory	(11)

References

(1) Schroeder and Anderson (1971)
(2) Chaffee and Schroeder (1976)
(3) Latham (1977)
(4) McLintock (1979)
(5) Hearnshaw (1977a,1978)
(6) McKeith et al (1978); Bates et al (1985)
(7) Rodgers (1985)
(8) Weiss et al. (1981)
(9) Schneider et al. (1985)
(10) Ramsey et al. (1985)
(11) Bhatia et al. (1984)

dispersing element is usually a reflection grating after the échelle in the optical train, though some instruments use prisms. If the cross disperser is before the échelle as in the Johns Hopkins design (McLintock, 1979) this introduces a wavelength-dependent line tilt which further complicates the reduction procedure.

The Mt John design is typical of those with a cross-dispersion grating after the échelle, and is shown in Fig 2. This spectrograph has been used photographically for detailed abundance studies of bright stars (see for example Desikachary and Hearnshaw (1982)), with a Varo single stage electrostatic image tube to about magnitude V = 6.5 in an hour's exposure and with a Reticon diode array. Fig 3 is an example of a diode array Hα spectrum of the active chromosphere binary HR4492 (V = 5.3) with a signal-to-noise of about 200 to one and R = 5 x 10^4. The exposure was obtained in 75 min. on the Mt John 0.61m telescope by P.J. MacQueen, who also designed and built the diode array system.

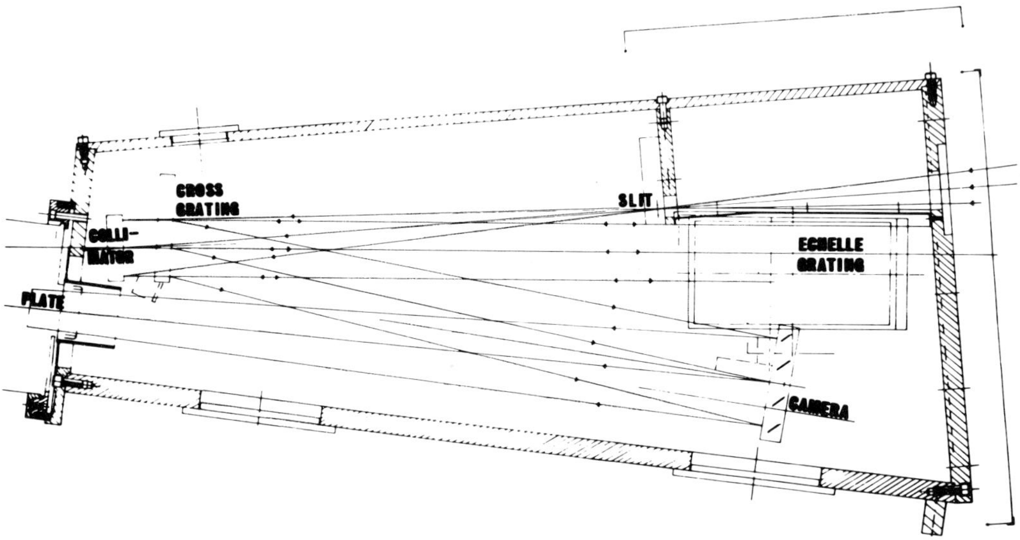

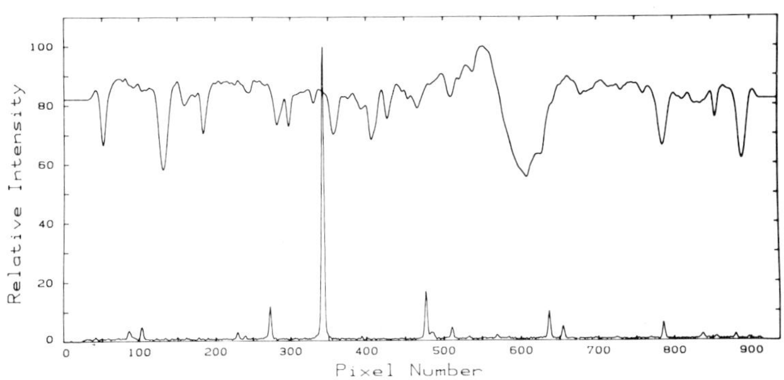

Fig. 2 (above) MJUO échelle spectrograph

Fig. 3 (left) HR 4492 at Hα showing chromospheric emission, MJUO Reticon spectrum 1985 Oct 29.

The reciprocal dispersions in Table I are typical of those normally associated with high dispersion coudé spectrographs and are mostly in the range 1 to 5 Å/mm. Spectrographs designed for smaller telescopes are generally limited to a single camera and hence there is no choice of values of the dispersion available. More meaningful is the resolving power which has either been quoted from published data or from private communications, or calculated from design parameters in the references. In most cases the resolving power has been scaled to a 1 or 2 arc second slit; if a slit size is not stated then the figure refers to the highest values typically attained. Since most of these spectrographs use Bausch and Lomb or PTR Optics échelle gratings whose ruled width is 206 mm (this is the value of L if all this width is illuminated) the resolving powers are all comparable for use on telescopes around a metre in aperture with say a one arc second slit. A resolving power of 5×10^4 is quite typical in most cases, allowing a wide range of line profile or abundance work to be undertaken. Values of 10^5 or more have been achieved with the Queen's University Belfast instrument for studies of narrow interstellar spectral lines (Bates et al, 1985) and by Schneider et al (1985) for studies of planetary atmospheres.

The Pennsylvania State University spectrograph built by Ramsey and his colleagues (Ramsey et al, 1985) deserves special mention because it has been designed from the start for fibre-optic coupling to the 1.6m Penn State telescope. This is in fact the third generation spectrograph developed from the first fibre-coupled échelle in 1979. This fact emphasises the flexibility of fibre-fed spectrographs for design upgrading, which is one of the attractive features of a system off the telescope comprising bench mounted optical elements and permitting routine access.

5. ECHELLE GRATINGS USED FOR RADIAL VELOCITY SPECTROMETRY WITHOUT CROSS-DISPERSION.

The plate factor or reciprocal dispersion (Å/mm) of an échelle spectrograph is inversely proportional to the order number and hence proportional to the central wavelength for each order of diffraction. This just balances the linear increase of $\Delta\lambda$ with λ for a Doppler-shifted spectrum, resulting in a constant shift in the focal plane of

$$\Delta s = 2 \tan \theta_B f_{cam} V_R/C$$

for the lines in all the échelle orders when a star of radial velocity V_R is observed (Hearnshaw, 1976b). This is of course the principle of the Coravel spectrometer.

Let us suppose that the radial velocity of a star is to be determined by cross correlating the spectrum with the spectrum of a standard velocity star, using a linear diode array as detector. This is the technique used by Latham (1985) at the Center for Astrophysics. In the Mt John Reticon diode array, about 40Å spectral coverage in a single order is recorded on the 936-pixel detector in one exposure. If on the other hand the cross-dispersion grating is replaced by a plane mirror, then many diffraction orders fall on the detector together. The result is a jumble of short 40Å intervals from different orders in the spectrum, but still preserving the unique value of Δs which is common to all the different orders for a given star. A given pixel can receive photons from say 30 different orders and hence as many different wavelengths, yet the noise is still primarily the read-out noise of each pixel rather than that from photon statistics.

This concept is about to be tested at Mt John where we have recently installed a plane mirror in our spectrograph to replace the cross-dispersion grating. Using a 1mm thick Schott BG 18 filter limits the spectral coverage to about thirty orders between 3400 to 6300 Å and eliminates longer wavelengths contaminated by telluric lines. Theoretically the advantage arises because on average about thirty times as many lines will be used for the radial velocity solution in the cross-correlation procedure, which might be expected to give a speed gain of about square root thirty or nearly two magnitudes.

6. DESIGN PARAMETERS FOR A HIGH RESOLVING POWER FIBRE-OPTICS ECHELLE SPECTROGRAPH

The formula for spectrograph resolving power in section 2 shows that the limitation to achieving high resolving power with a cassegrain instrument on a small telescope is the focal length of collimator and camera mirrors due to flexure. A fibre-fed spectrograph however avoids these problems and allows the use of a wide diameter collimated beam sending light to a large échelle. The largest échelle produced by the Milton Roy Company (formerly Bausch and Lomb) has a ruled area of 204mm x 408mm. The purpose of this section is to evaluate the possible performance of a spectrograph using such a large grating. The pioneering work of Ramsey et al (1985) (see also Barden et al (1981)) at Penn State on fibre-fed échelle spectrographs has already demonstrated the viability of this technique.

First, I assume the detector will be a Tektronix 56mm x 56mm CCD chip. Using a 31.6 groove/mm échelle with $\theta_B = \tan^{-1}2$ (catalogue no 35-13-43-411) gives the smallest free spectral range and allows maximum coverage of the spectral orders in one exposure with this detector. The large échelle will accept a beam diameter of B = 180mm. There are numerous advantages to operating a spectrograph off the telescope with a fibre; Ramsey et al (1985) have discussed the pros and cons. Beam degradation, due to spreading of the output cone from the fibre, is one of the disadvantages. This modifies the resolving power formula to

$$R = \frac{2L \sin \theta_B}{r\, \theta_f\, D} = \frac{2B \tan \theta_B}{r\, \theta_f\, D}$$

where r is the fibre degradation factor (r = ratio of input to output focal ratios and is always greater than unity), and θ_f is the angular diameter of the fibre input in the focal plane of the telescope. In this calculation it is assumed the fibre output acts as the slit of the spectrograph without an image-slicer.

Beam degradation is minimised for fast input beams so the best results are obtained with a fibre input from the prime focus. The resolving power is then in general

$$R = \frac{1.48 \times 10^5}{r\, \theta_f\, D}$$

with the large échelle (D in m. and θ_f in arc seconds)

$$\text{or} \qquad R = \frac{7.4 \times 10^4}{rD}$$

for a 2 arc second fibre. The degradation can be as small as 10 per cent for f/3.5 input (Schiffer, 1981; Barden et al, 1981), corresponding to r = 1.10. Hence a 1m telescope will achieve a resolving power of about 67000 and send nearly all the light into the spectrograph under only average seeing conditions. A 34μ diameter fibre corresponds to 2 arc seconds at the prime focus of such a telescope. The f/3.2 collimator would require a 580mm focal length. Although this is short, the large size of the components makes a mounting on the telescope impractical. The camera focusses the fibre's output onto the detector. The diameter of the circular monochromatic image should be about 51μ as this spans the width of two of the CCD pixels for optimum performance. Allowing for an anamorphic de-magnification factor of $\cos\beta/\cos\alpha = 4/3$ (for $\theta = 4^o$) shows that a camera with a focal length of 1150mm is required. Finally one free spectral range is contained in a length of spectrum of $\delta\ell = \frac{\lambda}{s} f_{cam}$, where s = 14.15μm is the groove facet width for this échelle grating (see Fig 1). At λ = 600nm (in the 94th diffraction order) this length is 49mm, so complete spectral coverage of the orders on the CCD is nicely achieved.

An alternative design in which the light is taken from the f/7.9 Cassegrain focus of a 1m telescope will achieve a resolving power of 60000 with a 1.65 arc second fibre (this corresponds to 63μm fibre diameter, e.g. Galite 5000/1 ST). The efficiency is lower due to increased beam degradation by the fibre. The collimator is then f/5.2 (based on a degradation factor of 1.5) and the optimum camera focal length is 1010mm. This gives a 43mm long spectrum for one free spectral range in the 94th order, which again permits continuous spectral coverage on the 56mm wide detector. Table II gives proposed details of these two fibre-fed échelle spectrograph designs.

Such high resolving power with relatively wide slits should revolutionise high resolving power stellar spectroscopy in the near future. Contrary to the predictions of McLaughlin thirty years ago, the best results will come from smaller telescopes (D ~ 1m) coupled to spectrographs with large components by a narrow fibre. Increased light gathering power can then be achieved by using an array of several small fibre-linked telescopes, with the separate fibres brought together to form a linear spectrograph slit (see Angel et al, 1977). The cost of small light-weight alt-az telescopes under remote control with fast prime focus fibre feeds and no capacity for instrument load or direct observer access could be substantially less than the conventional load-carrying equatorial cassegrain telescopes of today.

TABLE II Design parameters for a high resolving power wide beam fibre-fed échelle spectrograph

---

a) Prime focus system with 10 per cent beam degradation
Echelle grating : 204mm x 408mm (L = 408mm)
31.6 grooves/mm
$\tan\theta_B = 2$ ($\theta_B = 63°26'$)
Beam size : B = 180mm
Fibre input : f/3.5 prime focus
(For 1m telescope 2 arc s corresponds to a 34µm core diameter)
Fibre output : f/3.2 ; degradation factor 1.10
Collimator : f_{coll} = 580mm (f/3.2)
Camera : f_{cam} = 1150mm
Order length (at 600nm; n=94) : $\delta\ell$ = 49mm for one free spectral range giving continuous spectral coverage on a Tektronix TK 2048 56mm x 56mm CCD chip.
Diameter of fibre image on detector : 51µm ≡ 2.0 pixels (for TK 2048 CCD)

Resolving power : $R = \dfrac{1.35 \times 10^5}{\theta_f D}$

For a fibre with θ_f = 2 arc s and a D = 1m telescope : R = 67000

b) f/7.9 cassegrain focus system with 50 per cent beam degradation
Echelle grating and beam size as for (a) above.
Fibre input : f/7.9 cassegrain focus (as on Mount John 1m telescope)
Fibre core diameter : 63µm Galite 5000/1 ST corresponding to θ_f = 1.65 arc s. for a 1m telescope
Fibre output : f/5.2; degradation factor 1.50
Collimator : f_{coll} = 936mm (f/5.2)
Camera : f_{cam} = 1010mm
Order length (at 600nm; n = 94) : $\delta\ell$ = 43mm for one free spectral range giving continuous spectral coverage on a Tektronix TK 2048 56mm x 56mm CCD chip.
Diameter of fibre image on detector : 51µm ≡ 2 pixels (for TK 2048 CCD)

Resolving power : $R = \dfrac{9.9 \times 10^4}{\theta_f D}$

For a fibre with θ_f = 1.65 arc. s and a D = 1m telescope : R = 60,000

REFERENCES

Angel, J.R.P., Adams, M.T., Boroson, T.A. and Moore, R.L. 1977, *Astrophys. J.* 218, 776.
Barden, S.C., Ramsey, L.W. and Truax, R.J. 1981, *Publ. Astron. Soc. Pacific* 93, 154

Bates, B., McKeith, C.D., Jorden, P.R. and van Breda, I.G. 1985, *Astron. Astrophys.* 145, 321.
Bhatia, R.K., Swaminathan, R. and Vijas, M.L. 1984, *Bull. Astron. Soc. India* 12, 79.
Brown, R.A., Hilliard, R.L. and Phillips, A.L. 1982, *Appl. Optics* 21, 167.
Chaffee, F.H. and Schroeder, D.J. 1976, *Ann. Rev. Astron. and Astrophys.* 14, 23.
Desikachary, K. and Hearnshaw, J.B. 1982, *Mon. Not. R. Astron. Soc.* 201, 707.
Harrison, G.R. 1949, *J. Optical Soc. America* 39, 522.
Hearnshaw, J.B. 1977a, *Proc. Astron. Soc. Australia* 3, 102.
Hearnshaw, J.B. 1977b, *Observatory* 97, 5.
Hearnshaw, J.B. 1978, *Sky and Telescope* 56, 6.
Hearnshaw, J.B.1981, *AAS Photo-Bulletin* 26, 9.
Kopylov, I.M. and Stechenko, N.V. 1965, *Proc. Crimean Astrophys. Observ.* 33, 308.
Latham,D.W. 1977, *I.A.U. Coll.* 40, 45.
Latham,D.W. 1985, *I.A.U. Coll.* 88
McClintock, W. 1979, *Publ. Astron. Soc. Pacific* 91, 712.
McKeith, C.D., Dufton, P.L. and Kane, L. 1978, *Observatory* 98, 263.
McLaughlin, D.B. 1958, in *The Present and Future of the Telescope of Moderate Size*, chapter 14, p.205 ed F.B. Wood, *Univ. of Pennsylvania Press*, from the Univ. of Pennsylvania Symposium of June 1956.
Peterson, R.C. and Title, A.M. 1975, *Appl. Optics* 14, 2527.
Pierce, A.K., McMath, R.R. and Mohler, O. 1951, *Astron. J.* 56, 137.
Ramsey, L.W., Brungardt, C., Huenemoerder, D.P. and Rosenthal, S. 1985 *Bull. Amer. Astron. Soc.*, in press.
Rodgers, A.W. 1985, private communication.
Schiffer, J.G. 1981, *Mitt Astron. Gesellscheft* 54, 182.
Schneider, N.M., Brown, R.A. and Hunten, D.M. 1985, private communication.
Schroeder, D.J. 1967, *Appl. Optics* 6, 1976.
Schroeder, D.J. 1974, in *Methods of Experimental Physics* vol. 12(part A), chapter 10, p. 463.
Schroeder, D.J. and Anderson, C.M. 1971, *Publ. Astron. Soc. Pacific* 83, 438.
Spite, M. 1980, *Astron, Astrophys.* 81, 365.
Tousey, R., Purcell, J.D., Garrett, D.L. 1967, *Appl. Optics* 6, 365.
Weiss, W.W., Barylak, M., Hron, J. and Schmiedmayer, J. 1981, *Publ. Astron. Soc. Pacific* 93, 787.

DISCUSSION

Evans: If you put all the orders on top of each other, aren't you going to wipe out all the detail such that you won't have anything left to fix a radial velocity?

Hearnshaw: I don't think so. Although the lines will get weaker in percentage terms, you will still have the same number of detected photons between the continuum and a given line centre.

Hiltner: Your argument for the use of échelle spectrographs on small telescopes is true for all spectrographs as Bowen pointed out 40 years ago.

Hearnshaw: Yes, this is true.

Andrews: What will be the limiting magnitude of the échelle with the new detector system built by Phillip MacQueen.

Hearnshaw: On our 0.6m telescope the parameters are: 5.5 mag star at a signal to noise of 250 : 1 in one hour in 2 arcsec seeing.

Rowe: Your formula for spatial resolution is equivalent to the Rayleigh criterion for spatial resolution.

Ryan: John commented on the problem of the continuum for each échelle order being curved, and mentioned that the width of Balmer lines in early stars may be so large that the continuum is not apparent. I would like to add that the same applies to late type stars where there may be no continuum window over the length of an order. However, this problem is readily overcome by also exposing a lower dispersion spectrogram, say 60 Å/mm. The continuum can be determined on this, and by measuring the intensities of several lines relative to the continuum, a continuum can be drawn for the échelle spectrogram.

Levato: A Richardson-type coudé spectrograph can be as fast as an échelle spectrograph.

Hearnshaw: This is because they are using special coatings. The same can be done in an échelle system, so one still maintains the factor of 3 or 4 advantage.

Saxena: There is another échelle spectrograph which is being used at our institute on a 1 m telescope.

Hearnshaw: The list in the paper is only 80-90%. There are a few others for which I don't have all the information.

Garrison: I still think I would like a coudé given no monetary limitations. One doesn't have to worry about stability problems.

RADIAL VELOCITIES WITH CORAVEL: RESULTS ON STELLAR VARIABILITY AND DUPLICITY

G. Burki and M. Mayor
Geneva Observatory
1290 Sauverny, Switzerland

ABSTRACT. The complete radial velocity curve has been determined with CORAVEL for many pulsating stars of various classes: cepheid stars in the Galaxy and in the Magellanic Clouds, RR Lyrae, δ Scuti and SX Phoenicis stars. These measurements allow the determination of the radius variation and of the surface acceleration of these stars. In addition, the mean stellar radius of many of these stars has been determined by applying the Baade-Wesselink method.

Systematic surveys of definite groups of binary or multiple stars are in progress with CORAVEL in order to determine the distribution functions of the orbital parameters. The eccentricity distributions for the binaries in the open clusters Pleiades, Praesepe, Coma Ber and Hyades are presented and their dependence on the physical processes (star formation mechanisms, mass exchange, tidal circularization, dynamical evolution) is briefly discussed.

1. INTRODUCTION

The very impressive advantages of the cross-correlation technique for the determination of radial velocities was shown some twenty years ago by Roger Griffin (1967). The efficiency gain defined by $Q = (t\ \varepsilon^2)^{-1}$, where t is the integration time and ε the radial velocity uncertainty, is more than 1000 times larger than in the photographic technique. Some fields of research which used to be unaccessible with traditional spectroscopy and large telescopes are now within the capabilities of 1 to 1.5 m telescopes owing to the cross-correlation spectroscopy (see Mayor, 1985). As examples we may cite the measurement of radial velocities in globular clusters with an uncertainty smaller than 1 kms^{-1} or the measurement of 10th magnitude RR Lyrae stars with an integration time shorter than 5 minutes. This last condition is necessary for a correct description of the phase of minimum radius, where the radial velocity variation is of the order of 1 $(kms^{-1})min^{-1}$.

The first CORAVEL scanner (Baranne et al., 1979), built jointly by

J. B. Hearnshaw and P. L. Cottrell (eds.), Instrumentation and Research Programmes for Small Telescopes, 385–400.

Geneva and Marseille observatories, has been in full-time operation since 1977 at the Geneva 1 m telescope at the Haute-Provence Observatory (France). In the southern hemisphere measurements have been carried out since 1981 with a second CORAVEL on the Danish 1.5 m telescope at the European Southern Observatory in La Silla (Chile).

The CORAVEL scanner is particularly well suited for small telescopes due to its compactness (focal length 57 cm) and due to the fact that it could be mounted to the Cassegrain focus. An echelle grating with a cross disperser allows an efficient use of the stellar light (λλ 3600-5200 Å) combined with a large dispersion (2 Å/mm). The actual masks in both CORAVELs are based on Arcturus and allow the measuring of stars later than spectral type about FO.

The performances of the cross-correlation spectroscopy are not limited to radial velocity measurements. Such a technique has been successfully applied to stellar rotation measurements (Benz and Mayor, 1981, 1984; Benz et al., 1984), for the determination of stellar metallicities (Mayor, 1980), for the study of the turbulence in pulsating stars (Benz and Mayor, 1982; Benz and Stellingwerf, 1985) and for the detection of magnetic fields (Borra et al., 1984).

We shall not discuss here the number of important investigations done with the same instruments because it is not the aim of this review. Some of them are:

- The study of more than 400 supergiants in the LMC (Prévot et al., 1985).
- A similar investigation on 232 supergiants in the SMC (Maurice et al., 1985).
- The determination of the rotation velocity field in ω Cen and 47 Tuc and the detection of the heavy remnants in their nuclei (Meylan and Mayor, 1986), based on more than 500 radial velocities in each cluster (Mayor et al., 1983).
- A catalogue of 790 southern stars in order to achieve the completeness of the Bright Star Catalogue (Andersen et al., 1984).

Among the numerous investigations published or in progress, the study of pulsating stars (sect. 2) and of stellar duplicity (sect. 3) are good examples of the possibilities offered by an efficient instrumentation, combined with the continuity of measurements allowed by a small telescope dedicated exclusively to this task. Also, the completion of the various programs has frequently been made possible thanks to the kind contribution of several colleagues.

2. PULSATING STARS

2.1. Yields of the Radial Velocity Measurements

2.1.1. Pulsational velocity, radius variation and acceleration. From the radial velocity curve $V_r(t)$ of a pulsating star, several interesting curves can be derived. First, the pulsational velocity

$$\dot{R}(t) = -\beta(V_r(t) - \overline{V}_r) = -\beta \Delta V_r$$

where β is the conversion factor from radial to pulsational velocity. Second, the curve of the radius variation

$$\Delta R(t) = \int \dot{R}(t)dt = R(t) - R_o$$

where R_o is the mean stellar radius. Third, the acceleration of the stellar surface $\ddot{R}(t)$. This quantity is related to the gravity and to the pressure gradient by the expression

$$\ddot{R} = -(GM/R^2) - (1/\rho)(\partial P/\partial R) = -g + g_{eff}$$

Figure 1 shows two sets of such curves derived from CORAVEL observations of the small amplitude cepheid V636 Cas (Burki and Benz, 1982) and of the RR Lyrae star RR Cet (Burki and Meylan, 1986a). For the calculations of g, the values $M = 5\ M_\odot$ and $0.6\ M_\odot$, $R_o = 76\ R_\odot$ and $6.7\ R_\odot$ have been adopted for V636 Cas and RR Cet respectively. It clearly appears on Fig. 1 that the pulsational behaviour of the two stars exhibits important differences. In particular, the movement of the stellar atmosphere near the minimum radius is much more violent in the case of the RR Lyrae star.

2.1.2. Mean radius determination. The Baade (1928) - Wesselink (1946) method allows the determination of the mean radius R_o of a pulsating star from the knowledge of the curves of radial velocity, magnitude (generally V) and one color index related to T_{eff}. The simultaneity of the photometric and spectroscopic observations is required if either the period or the shape of the various curves change with time. Many versions of the Baade-Wesselink method exist, we only mention some of the recent ones: Balona and Stobie (1979a,b), Imbert (1981), Burki and Benz (1982).

An important variant of the method is due to Barnes et al. (1977). They use the remarkable relation between the surface brightness and color index V-R (Barnes-Evans relation) to derive, from the radial velocity, V and V-R curves, the mean radius and the distance of the star. A similar method, based on photometric calibrations in the Geneva system allowing the determination of T_{eff} for each photometric measurement, was applied by Burki and Meylan (1986a) to derive the mean radius and the distance of RR Lyrae stars. In recent years, many determina-

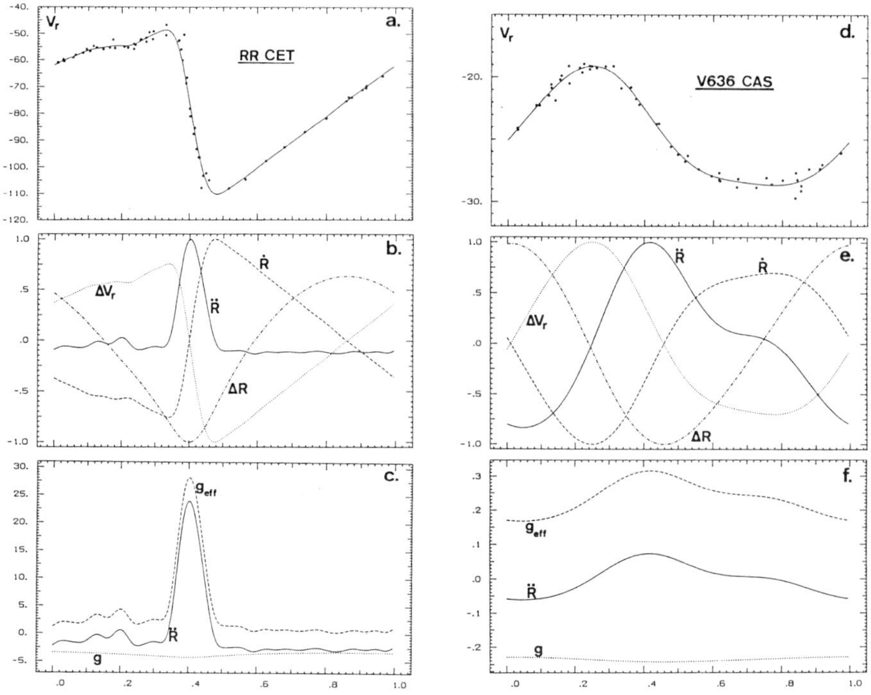

Figure 1. a. V_r curve of RR Cet (RR Lyrae).
b. Various curves derived from V_r. The value +1.0 of the ordinates is +36 kms^{-1} for ΔV_r, +49 kms^{-1} for $\dot{R}$, +0.51 $R_\odot$ for ΔR and +24 kms^{-2} for $\ddot{R}$.
c. $\ddot{R}$, $-g$ and g_{eff} curves in ms^{-2}.
d.-f. Same as a-c, but for V636 Cas (cepheid). The value +1.0 (e.) is +5.6 kms^{-1} for ΔV_r, +7.7 kms^{-1} for $\dot{R}$, +1.11 $R_\odot$ for ΔR and 0.073 ms^{-2} for $\ddot{R}$.

tions of the mean radius of pulsating stars have been performed, allowing a considerable improvement of the period-radius relation. This relation was essentially examined in the case of classical cepheids (see e.g. Fernie (1984) and Burki (1985a,b) for recent studies) and in the case of pop. II cepheids (Woolley and Carter, 1973). A complete log P - log R diagram, grouping all the classes of pulsating stars for which Wesselink radii have been determined, i.e. classical cepheids, W Virginis, BL Herculis, RR Lyrae, δ Scuti and SX Phoenicis stars, has been established by Burki and Meylan (1986c) and is summarized in another paper in this volume.

2.1.3. Turbulence variations (with CORAVEL). In the case of non-pulsating stars, the correlation function (dip) of CORAVEL is very well represented by a gaussian curve

$$c(v) = 1 - H \cdot \exp\left(-(v-V_r)^2 / 2\sigma^2_{gauss}\right)$$

where H is the depth and σ_{gauss} the width of the dip. The minimum value of σ_{gauss} is observed in the case of non-rotating and non-pulsating F and G stars of classes V to III: 6.88 kms^{-1} for the CORAVEL at the O.H.P.. The increase of the dip width by stellar rotation was described by Benz and Mayor (1981) and the effect of stellar pulsation was studied by Benz and Mayor (1982). They showed that the variation of the dip width with the pulsation phase is much too large in the cases of the cepheids SV Vul, X Cyg and δ Cep to be explained only by the movement of the stellar surface (geometrical effect). Following the spectroscopic study by van Paradijs (1971), Benz and Mayor (1982) suggested that the increase of turbulence during the phase of large pulsation velocity is the cause of this excess of line broadening. In fact, as shown by Fig. 2 (X Cyg), the maxima of the broadening correspond exactly to the maxima of $\ddot{R}(t)$.

This variation of the CORAVEL dip (or photospheric lines) broadening during the pulsation cycle is observed in all the pulsating stars, even in very small amplitude cepheids like V440 Per or V636 Cas (Burki and Benz, 1982). As to the peculiar cepheid HR 7308 (see section 2.2), the amplitude of dip width variation is strongly correlated to the amplitude of pulsation (Burki et al., 1982).

2.2 Cepheid Stars

M. Imbert (Marseille Observatory) has undertaken an extensive research to determine the mean radius of about 25 galactic cepheids. Three cases are already published (Imbert, 1981, 1983, 1984): XY Cas, AD Gem and SU Cyg. This last cepheid is a member of a binary system and the orbital elements were calculated simultaneously with the Fourier coefficients of the pulsation curve.

Two very small amplitude cepheids have been discovered, V636 Cas (see Fig. 1) and V440 Per (Burki et al., 1980). The latter is the classical cepheid with the smallest amplitude known so far, 5.6 kms^{-1} in radial velocity and $0\overset{m}{.}09$ in V (total amplitudes). The mean radii of these stars were determined by Burki and Benz (1982) using simultaneous photometry. Radius determination has also been performed for nine short period cepheids (Burki, 1985a): V465 Mon and possibly DX Gem are spectroscopic binaries, SW Tau could be a pop. II cepheid and EU Tau could be an overtone pulsator.

One of the most typical examples of fruitful research with a small telescope was the CORAVEL monitoring of the unique cepheid HR 7308 (V473 Lyr). The variability of this star was discovered by Breger (1969) and the cepheid classification is due independently to Percy et al. (1979), Burki and Mayor (1980) and Henriksson (1980). The remarkable characteristics of this short period (P = 1.49 d.) cepheid is the very large variation of the pulsation amplitude in a time of about 1400 d.:

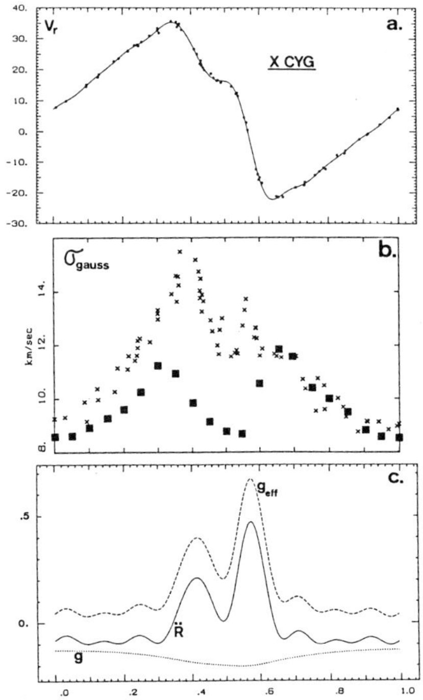

Figure 2.

a. V_r curve of X Cyg (cepheid).

b. Variation of the CORAVEL dip width (crosses) and predicted geometrical effect (squares).

c. $\ddot{R}$, $-g$ and g_{eff} curves in ms^{-2}.

the last maximum and minimum total amplitudes were 35 kms^{-1} (summer 1985) and 2.3 kms^{-1} (summer 1982). Two estimations of R_o are based on CORAVEL data, 34 ±5 $R_\odot$ (Burki et al., 1982) and 44 ± 9 $R_\odot$ (Burki et al., 1986). Note that the mean radius does not vary during the 1400 d. period, since the pulsation period of 1.49 d. shows very little, if any, variation with time. These large values of R_o are in favor of an overtone pulsation (at least the 2nd one). The radial velocity curve, and probably also the light curve, is almost sinusoidal at minimum amplitude and becomes asymmetrical with increasing amplitude. Thus, HR 7308 offers the unique possibility for studying, in the same star, the variation of the shape of velocity (and light) curve with the intensity of the non-linear effects acting on the driving mechanism of the pulsation. Unfortunately, the physical cause for the amplitude variation in HR 7308 has not yet been found. The various possible mechanisms are: i) a star entering or leaving the instability strip (Burki and Mayor, 1980), ii) an interaction between pulsation and convection (Stellingwerf, 1984), iii) instability of the limit cycle in non-linear dynamical calculations (Auvergne, 1985).

A monitoring of about twenty cepheids in the Magellanic Clouds by M. Imbert is in progress. The radial velocity curves are published for six of them (Imbert et al., 1985) and the case of HV909 is shown in Fig. 3. The resulting σ_{O-C} of the fitted curves are from 1.1 to 2.7 kms^{-1}, depending on the stellar magnitude (from 14.1 to 15.5 in B at

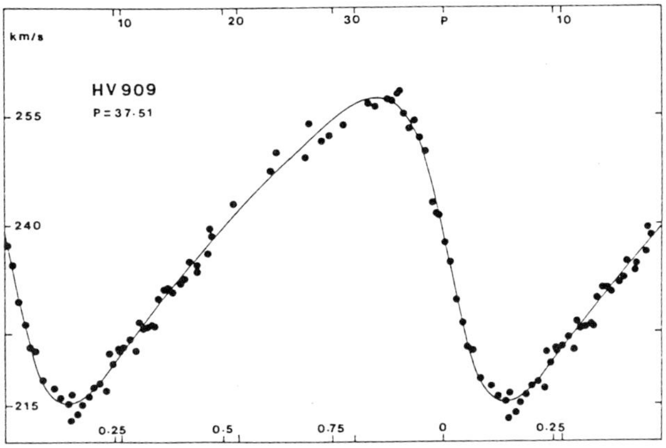

Figure 3.

Radial velocity of a cepheid in the LMC (Imbert et al., 1985).

minimum light). The accuracy on the centre-of-mass velocities is from 0.2 to 0.4 kms^{-1}. A Baade-Wesselink analysis will be published for these stars.

2.3. RR Lyrae Stars

L. Prévot, Marseille, and Burki and Meylan, Geneva, have undertaken independent but similar programs of radial velocity measurements on RR Lyrae stars in order to determine the radii, distances and absolute luminosities and to study the possible dependence of the luminosity on the physical parameters, in particular the metal content. Concerning the Geneva program, partial or complete radial velocity curves and quasi-simultaneous Geneva photometry were obtained for 10 stars and the entire analysis is completed for two cases: RR Cet (see Fig. 1) and DX Del (Meylan et al., 1986; Burki and Meylan, 1986a,b). The essential results are given in Table 1. The uncertainties on the parameters are only of mathematical nature and result from the least squares procedure. The total uncertainties are roughly twice as large as the values given.

	RR Cet	DX Del
Period	0.55 d.	0.47 d.
[Fe/H]	-1.3	-0.1
R_o	6.7 ± 0.2 $R_\odot$	5.5 ± 0.2 $R_\odot$
$\overline{M}_V$	0.28 ± 0.14	0.49 ± 0.16
distance	760 ± 40 pc	750 ± 45 pc

TABLE 1. Physical parameters of two RR Lyrae stars

At present, the Baade-Wesselink method has been successfully applied by various authors to 7 RR Lyrae stars (see Burki and Meylan, 1986b, for references). By adopting the unique value $0^m.30$ for the individual $\overline{M}_V$ uncertainty a mean value $<\overline{M}_V> = 0.57 \pm 0.09$ is obtained for these RR Lyrae. This value is brighter, but compatible at the 1 σ level, than the recent value 0.76 ± 0.14 derived by Hawley et al. (1986) by means of a new statistical parallaxes analysis.

2.4. δ Scuti stars

A large number of measurements has been performed on δ Scuti stars (including the pop. I dwarf cepheids). The aims of the Marseille and Geneva programs are the study of the pulsation mode(s), the analysis of the multiperiodicity and the determination of the mean radii.

Imbert (1980) obtained the first velocity curve for V1719 Cyg and suggested that the observed amplitude variation is due to a second pulsation mode, a hypothesis recently confirmed by Mantegazza and Poretti (1986).

A new δ Scuti star, V356 Aur, was discovered by Burki and Mayor (1981). Despite a simultaneous survey in photometry and radial velocity it was not possible to determine the mean radius, because the difference in phase between the light and colour curves is almost zero: $\Delta\phi = -0.003 \pm 0.010$. This could be the signature of a non-radial oscillation mode with odd spherical harmonic order ℓ, but the large uncertainty on $\Delta\phi$ does not completely exclude a radial mode.

Bardin and Imbert (1981, 1982, 1984) determined the orbital parameters of two binary δ Scuti stars: V644 Her ($P_{orb} = 11.86$ d) and SZ Lyn ($P_{orb} = 1181.5$ d, $P_{puls} = 0.121$ d).

For BS Aqr, Burki and Meylan (1986c) obtained the following parameters from a Baade-Wesselink analysis: $R_o = 3.2 \pm 0.4\ R_\odot$, $M_V = 1.4 \pm 0.3$, $d = 410 \pm 60$ Pc, $M = 1.8\ M_\odot$. These values are in good agreement with the classification of this dwarf cepheid as a pop. I δ Scuti star.

2.5. SX Phoenicis stars

This is the new classification for the short period ($0.03 \lesssim P \lesssim 0.08$ d) pulsating stars showing low metallicities, high space motions and low luminosities (e.g. Breger, 1980). This class of variable stars groups three blue stragglers in ω Cen (Jorgensen and Hansen, 1984) and five field stars (SX Phe, BL Cam, CY Aqr, DY Peg, KZ Hya) which are studied with CORAVEL. For DY Peg, the following values have been determined: $R_o = 1.4 \pm 0.4\ R_\odot$, $\overline{M}_V = 3.2 \pm 0.5$, $d = 250 \pm 80$ pc (Burki and Meylan, 1986c).

In the colour-magnitude diagram, DY Peg and the three variable blue stragglers in ω Cen form a surprisingly well-defined clump whose coordinates are $(B-V)_o \cong 0.20$ and $M_{V_o} \cong +3.0$. Then, DY Peg has probably the same evolutionary status as the three ω Cen blue stragglers. As to the possible physical interpretation of the blue stragglers as the result of a binary evolution, it must be noted that no binary motion has been detected in DY Peg on a time base of two years.

2.6. FG Sge

This is the nucleus of a planetary nebula ejected about 6000 years ago. Since 1950 its T_{eff} has been decreasing with a gradient of about $-300^{o}K/y$ (Stone, 1979) and the corresponding annual increase of the pulsation period is about 5 d (Jurcsik and Szabados, 1979). The star was monitored with CORAVEL and the resulting radial velocity curve has the following essential parameters (Mayor and Acker, 1980): 110 d for the period and 2.1 ± 0.3 kms^{-1} for the amplitude in 1978, 115 d and 3.7 ± 0.5 kms^{-1} in 1979. In addition, by means of photometric data given by various authors, it was possible to estimate the mean radius for the period 1978-1979, $R_o = 140 \pm 30$ $R_\odot$, and the other parameters, $M = 1.2 \pm 0.6$ $M_\odot$ and $M_V = -4.3 \pm 0.7$.

3. STELLAR DUPLICITY

3.1 Generalities

In the solar neighbourhood, roughly 2/3 to 3/4 of the mass locked in stars belongs to binary or multiple systems. The distribution functions of the orbital parameters of these systems depend on the star formation mechanisms, the stellar evolution (mass exchange, tidal circularization etc.) and the dynamical evolution. Despite the high frequency of stellar systems, these distribution functions are still badly known due to various observational selection effects.

Among the recent efforts to present unbiased views of definite groups of stars, the study by Abt and Levy (1976) on G-type dwarfs can be cited. Also, the work carried out by Griffin and Gunn to study the duplicity in the Hyades is to be mentioned: the distribution N(log P) for the SB (spectroscopic binaries) is slowly increasing in the range $1 < \log P < 3$ (Griffin, 1985), in good agreement with the correspon-ding distribution derived by Abt and Levy. Such a result would not have been obtained from the study of the actual general catalogues of binary stars, which are affected by selection bias.

Radial velocities derived by cross-correlation devices will certainly reduce this bias and an important part of the measurements can be made with small telescopes. Photoelectric scanners have already made significant contributions in this domain. Let us mention the outstanding work done by R. Griffin at Cambridge, mainly on late-type field giants, the high percentage of binaries in barium stars discovered by McClure (1984) and the measurements made in M3 by Gunn and Griffin (1979) and later by Pryor et al. (1985).

A series of systematic surveys based on CORAVEL measurements and related to binary stars are in progress, in particular on the following objects:

- Late-type dwarfs in the open clusters Pleiades, Praesepe, Coma Ber, α Per, ζ Scl and several others in the southern hemisphere. This program is carried out by J.C. Mermilliod and one of the authors (M.M.).
- Red giants in open clusters (more than 800 stars).
- Low main sequence stars in the solar neighbourhood (a program of the Geneva observers with the collaboration of J.L. Halbwachs).
- Galactic supergiants of spectral types F to M (see Burki and Mayor, 1983).
- SB with known orbits in order to detect a third, distant companion by nodal precession (Mazeh and Mayor, 1983; Mayor and Mazeh, 1986).
- Visual binaries with known visual orbits (more than 100 systems, see paper by Duquennoy and Mayor in this volume).
- Halo stars in the southern hemisphere (A. Ardeberg, H. Lindgren and other CORAVEL observers).
- Giant stars (more than 100 are measurable) in the nucleus of ω Cen, in view of detecting the binaries.

These large surveys for which the detection level of binaries is about 1 to 2 kms^{-1} should give new constraints on the physical and dynamical properties of the binary stars.

3.2. Eccentricity-Period Relation and Distribution of Orbital Eccentricities

3.2.1 Late-type dwarfs in open clusters. The radial velocity survey of the main open clusters with CORAVEL has allowed the determination of 22 orbits, essentially in Coma Ber, Pleiades, Praesepe and a few in the Hyades. By adding the orbits determined by Griffin and his collaborators (see Griffin et al., 1985 and references therein) in the same clusters but mainly in the Hyades, we now dispose of a sample of 46 SB later than F5V. Recall that an advantage of the binaries in open clusters is the knowledge of their age and evolutionary status.

The e vs. log P diagram for these systems is shown in Fig. 4a. The discontinuity appearing at $P \sim 5.7$ d can be interpreted in terms of tidal circularization; this value of the limit period is in agreement with the expected one for main sequence stars with an outer convection zone and having the same age as these clusters (Mayor, Mermilliod, 1984). However, one cannot reject the possibility that the circularization process has operated during the stellar evolution prior to the main sequence. Indeed, for a given period, the time required for the circularization, t_{circ}, is considerably shorter during the pre-main sequence evolution than during the main sequence, since the radii of the protostars are larger, the convection zone is more developed and the gravity is lower. This clearly appears from the relation giving, for a specified mass ratio, the expression of t_{circ}:

$$t_{circ} \sim (a/R)^5 \; g \; P^2 \; / \; (\lambda \; \eta \; v_c)$$

where λ and η are the relative thickness and mass of the convective zone and v_c is the convective velocity (Lecar et al., 1976).

The normalized eccentricity distributions f(e) for our sample of SB with dwarf components are shown in Figs. 5a (P < 5.7 d) and 5b (P < 5.7 d). The distribution in Fig. 5b is certainly biased in the case of very eccentric orbits, but we estimate from the analysis of the data that this effect is only important for e larger than 0.8.

It is interesting to compare these distributions f(e) with that obtained from the study of Harrington and Miranian (1977) for the orbits of visual binaries. They have shown that selection effects diminish the probability of obtaining orbital elements in the case of very large values of e. Their "completeness factor", derived by numerical simulations, permits to correct the observational distribution f(e). This corrected distribution, established for visual binaries with log P $\gtrsim$ 3, is shown in Fig. 5c. Note that the study of dynamical decay of larger stellar systems predicts a similar distribution f(e) (Harrington and Miranian, 1977) and that these distributions are not far from Ambartsumian's (1937) distribution f(e) = 2e (dashed line in Fig. 5c), expected if the orbital distribution in phase space is a function of energy only.

In consequence, the eccentricity distribution f(e) for late-type SB of very short period is depending on the tidal circularization mechanisms. For very long periods, f(e) can be explained in terms of the dynamical decay of larger stellar systems. In the intermediate range of periods (0.76 $\lesssim$ log P $\lesssim$ 3), f(e) probably gives direct information on the initial conditions.

3.2.2. Late-type giants in open clusters. The e vs. log P diagram established from orbital elements of SB with a red giant as primary component (Fig. 4b) is very similar to the diagram for red dwarfs (Fig. 4a), except that the limit period for the circularization appears at a larger value: between 127 d and 179 d (Mayor and Mermilliod, 1983). It seems that the tidal dissipation due to low gravity, large convective zone and big radii in giant stars is sufficient to explain this observed limit period (Mayor and Mermilliod, 1984).

During stellar evolution, the radius of a component of mass M_1 in a binary star will fill its Roche lobe if the orbital period is shorter than a critical value P_{RL}. Such a limit depends on the mass M_2 of the other component. By using the Kopal (1978) value for this limiting period in the circular case (e = 0), we have estimated P_{RL} for non-circular orbits, under the hypothesis that the primary component just fills its Roche lobe at periastron during its evolution (see Table 2).

The masses of the three giants in SB with circular orbits (see Fig. 4b) deduced from the age and evolutionary tracks of the parent clusters are between 1.5 and 1.8 $M_\odot$. From Table 2 it appears that these three systems could have experienced a small mass transfer if their initial orbital eccentricities on the main sequence were slightly larger than 0.3.

3.2.3. Late-type supergiants. The study of the binary F-M supergiants with known orbital elements shows that all systems with a period

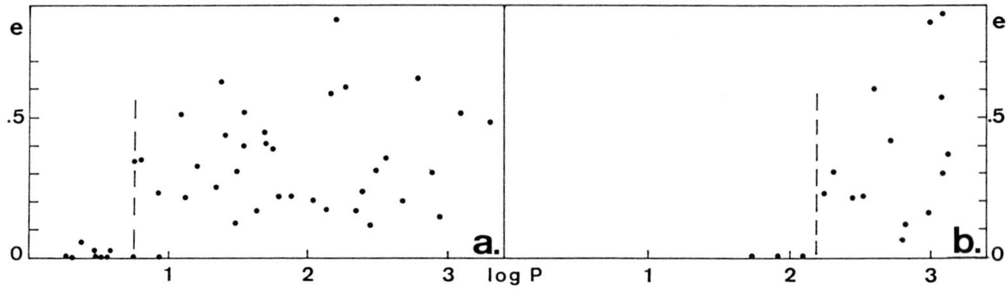

Figure 4. e vs. log P diagrams for the SB of late-types in open clusters. a. Dwarfs. b. Giants

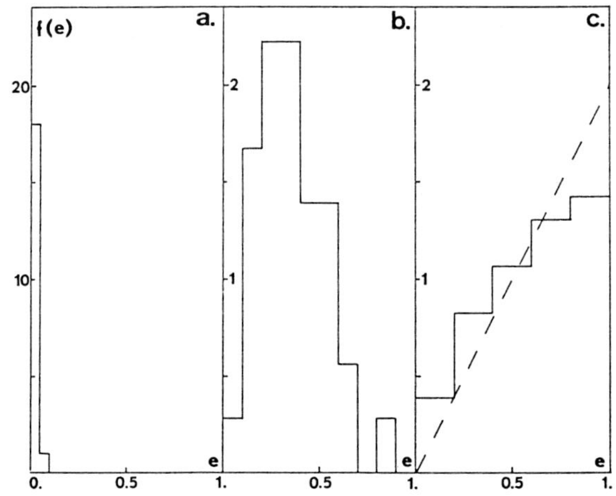

Figure 5. Distribution of the orbital eccentricities.

a. Short period SB (P < 5.7d) in open clusters with a late-type dwarf as primary component.

b. Same as a, but for P > 5.7 d.

c. Visual binaries (P ≳ 1000 d), with a correction for the observational bias (see sect. 3.2.1.).

M_2/M_1	$M_1 = 1\ M_\odot$			$M_1 = 3\ M_\odot$		
	e = 0.0	0.3	0.6	e = 0.0	0.3	0.6
1.0	20.8	35.6	82.2	39.2	67.0	155.0
0.6	19.4	33.2	76.6	36.5	62.0	144.2
0.2	15.9	27.2	62.8	30.0	51.3	118.5

TABLE 2. Limit period P_{RL} $(M_1/M_2, M_1, e)$ in days

$P_{circ} = P(1-e)^{1.5}$ smaller than a critical value P_{crit} have a nearly circular orbit (P_{circ} is the period of a hypothetical circular orbit which would have for radius the distance between components at the periastron of the real elliptical orbit). The value of P_{crit} depends on the luminosity class: 350 - 440 d for class Ib, 1400 - 3900 d for class Ia. This circularization of the orbits results probably from the transfer of angular momentum during the phase of binary mass exchange. In several systems with circularized orbits, the star actually observed as a supergiant was probably the secondary component at the origin of the system (see Burki and Mayor, 1983, 1984).

REFERENCES

Abt, H.A., Levy, S.G., 1976, Astrophys. J. Suppl. 30, 273
Ambartsumian, V.A., 1937, Sov. Astr. A.J. 14, 207
Andersen, J., Nordstroem, B., Ardeberg, A., Benz, W., Lindgren, H., Imbert, M., Martin, N., Mayor, M., Maurice, E., Prévot, L., 1985, Astron. Astrophys. Suppl. 59, 15
Auvergne, M., 1985, preprint
Baade, W., 1928, Astron. Nachr. 228, 359
Balona, L.A., Stobie, R.S., 1979a, Monthly Notices Roy. Astron. Soc. 187, 217
Balona, L.A., Stobie, R.S., 1979b, Monthly Not. Roy. Astron. Soc. 189, 649
Baranne, A., Mayor, M., Poncet, J.L., 1979, Vistas in Astronomy 23, 279
Bardin, C., Imbert, M., 1981, Astron. Astrophys. 98, 198
Bardin, C., Imbert, M., 1982, Astron. Astrophys. Suppl. 47, 319
Bardin, C., Imbert, M., 1984, Astron. Astrophys. Suppl. 57, 249
Barnes, T.G., Dominy, J.F., Evans, D.S., Kelton, P.W., Parsons, S.B., Stover, R.J., 1977, Mon. Not. Roy. Astr. Soc. 178, 661
Benz, W., Mayor, M., 1981, Astron. Astrophys. 93, 235
Benz, W., Mayor, M., 1982, Astron. Astrophys. 111, 224
Benz, W., Mayor, M., 1984, Astron. Astrophys. 138, 183
Benz, W., Stellingwerf, R.F., 1985, Astrophys. J., in press
Borra, E., Edwards, G., Mayor, M., 1984, Astrophys. J. 284, 211
Breger, M., 1969, Astrophys. J. Suppl. 169, 79
Breger, M., 1980, Astrophys. J. 235, 153
Burki, G., 1985a, IAU Coll. 82, Ed. B.F. Madore, Cambridge Univ. Press, Cambridge, p. 34
Burki, G., 1985b, IAU Coll. 82, Ed. B.F. Madore, Cambridge Univ. Press, Cambridge, p. 71
Burki, G., Benz, W., 1982, Astron. Astrophys. 115, 30
Burki, G., Mayor, M., 1980, Astron. Astrophys. 91, 115
Burki, G., Mayor, M., 1981, Astron. Astrophys. 97, 4
Burki, G., Mayor, M., 1983, Astron. Astrophys. 124, 256

Burki, G., Mayor, M., 1984, IAU Symp. 105, Eds. A. Maeder and A. Renzini, D. Reidel Publ., Dordrecht, p. 403
Burki, G., Mayor, M., Benz, W., 1982, Astron. Astrophys. 109, 258
Burki, G., Mayor, M., Waelkens, C., 1980, Astron. Astrophys. 91, 276
Burki, G., Meylan, G., 1986a, Astron. Astrophys., in press
Burki, G., Meylan, G., 1986b, Astron. Astrophys., in press
Burki, G., Meylan, G., 1986c, Astron. Astrophys., in press
Burki, G., Schmidt, E.G., Arellano Ferro, A., Fernie, J.D., Sasselov, D., Simon, N.R., Percy, J.R., Szabados, L., 1986, Astron. Astrophys., in preparation
Fernie, J.D., 1984, Astrophys. J. 282, 641
Griffin, R.F., 1967, Astrophys. J. 148, 465
Griffin, R.F., 1985, The distribution of periods and amplitudes of late-type spectroscopic binaries, in "Interacting Binaries", Eds. P.P. Eggleton and J.E. Pringle, Reidel Publ. Co., p. 1
Griffin, R.F., Gunn, J.E., Zimmermann, B.A., Griffin, R.E.M., 1985, Astron. J. 90, 609
Gunn, J.E., Griffin, R.F., 1979, Astron. J. 84, 752
Harrington, R.S., Miranian, M., 1977, Publ. Astr. Soc. Pac. 89, 400
Hawley, S.L., Jefferys, W.H., Barnes, T.G., Wan Lai, 1986, Astrophys. J., in press
Henriksson, G., 1980, 5th European Regional Meeting, Ed. P. Ledoux, Liège, p. B.1.7.
Imbert, M., 1980, Astron. Astrophys. 86, 259
Imbert, M., 1981, Astron. Astrophys. Suppl. 44, 319
Imbert, M., 1983, Astron. Astrophys. Suppl. 53, 85
Imbert, M., 1984, Astron. Astrophys. Suppl. 58, 529
Imbert, M., Andersen, J., Ardeberg, A., Bardin, C., Benz, W., Lindgren, H., Martin, N., Maurice, E., Mayor, M., Nordstroem, B., Prévot, L., 1985, Astron. Astrophys. Suppl. 61, 259
Jorgensen, H.E., Hansen, L., 1984, Astron. Astrophys. 133, 165
Jurcsik, J., Szabados, L., 1979, IBVS 1722
Kopal, Z., 1978, Dynamics of close binary systems, Reidel Publ. Co., Dordrecht, p. 414
Lecar, M., Wheeler, J.C., McKee, C.F., 1976, Astrophys. J. 205, 556
Mantegazza, L., Poretti, E., 1986, Astron. Astrophys., in press
Maurice, E., Andersen, J., Ardeberg, A., Bardin, C., Imbert, M., Lindgren, H., Martin, N., Mayor, M., Nordstroem, B., Prévot, L., Rebeirot, E., Rousseau, J., 1985, Astron. Astrophys. Suppl., subm.
Mayor, M., 1980, Astron. Astrophys. 87, L1
Mayor, M., 1985, IAU Coll. 88, Eds. A.G. Davis Philip and D.W. Latham, L. Davis Press, p. 35
Mayor, M., Acker, A., 1980, Astron. Astrophys. 92, 1
Mayor, M., Imbert, M., Andersen, J., Ardeberg, A., Baranne, A., Benz, W., Ischi, E., Lindgren, H., Martin, N., Maurice, E., Nordstroem, B., Prévot, L., 1983, Astron. Astrophys. Suppl. 54, 495
Mayor, M., Mazeh, T., 1986, Astron. Astrophys., in prep.

Mayor, M., Mermilliod, J.C., 1983, in "Les étoiles binaires dans le diagramme HR". Proceedings of the 5th meeting of Strasbourg Observatory, p. 45

Mayor, M., Mermilliod, J.C., 1984, IAU Symp. 105, Eds. A. Maeder and A. Renzini, D. Reidel Publ. Co., p. 411

Mazeh, T., Mayor, M., 1983, IAU Coll. 62, Lowell Obs. Bull. no. 167, Ed. R.S. Harrington and O. Franz, p. 120

McClure, R., 1984, IAU Symp. 105, Eds. A. Maeder and A. Renzini, D. Reidel Publ. Co., p. 187

Meylan, G., Burki, G., Rufener, F., Mayor, M., Burnet, M., Ischi, E., 1986, Astron. Astrophys. Suppl, in press

Meylan, G., Mayor, M., 1986, Astron. Astrophys., submitted

Percy, J.R., Baskerville, I., Trevorrow, D.W., 1979, Publ. Astron. Soc. Pacific 91, 368

Prévot, L., Andersen, J., Ardeberg, A., Benz, W., Imbert, M., Lindgren, H., Martin, N., Maurice, E., Mayor, M., Nordstroem, B., Rebeirot, E., Rousseau, J., 1985, Astron. Astrophys. Suppl. 62, 23

Pryor, C.P., Latham, D.W., Hazen-Liller, M.L., 1985, Center for Astrophys., preprint no. 2127

Stellingwerf, R.F., 1984, private communication

Stone, R., 1979, Publ. Astron. Soc. Pacific 91, 389

van Paradijs, J.A., 1971, Astron. Astrophys. 11, 299

Wesselink, A.J., 1946, Bull. Astron. Neth. 10, 91

Woolley, R., Carter, B., 1973, Mon. Not. Roy. Astron. Soc. 162, 379

DISCUSSION

Rucinski: You showed us a diagram showing the synchronisation of binaries in a few clusters differing in age by a factor of at least ten. Do you see differences in limiting synchronisation periods as a function of the cluster age?

Burki: Unfortunately the number of binary star orbits in each cluster is not sufficient to conclude anything at present.

Cameron: What duplicity rate do you find among the main sequence G-stars in these clusters?

Burki: The frequency of binaries does not require a complete orbit. Mermilliod and Mayor have already published this value : 34 ± 5% for the global rate of main-sequence binary stars in the Hyades, Pleiades, Praesepe, α Persei clusters.

Duncan: It is worthwhile doing something that someone has done before, since all the parameters may not have been determined. Marcy, a Mt Wilson post-doc, has just finished a study of the binary frequency of M-dwarfs. He finds a frequency of $< 10\%$, and he has a velocity accuracy of $0.2 kms^{-1}$. Thus the M-dwarfs behave quite differently to the G-dwarfs. Only 2 out of 40 M-dwarfs showed companions in this survey.

Latham: His selection was for single stars not visual binaries. I think he would find a higher frequency of (spectroscopic) binaries amongst the visual binaries.

Levato: If you change the mask in CORAVEL for an earlier spectral type in the same wavelength region, what is the precision that you get on an early type star.

Burki: The precision of our machine is proportional to the number of lines one considers. The cross-correlation technique will therefore realize lower precision for early-type stars, but will be better than by traditional methods. We are not proposing to make an early-type mask as we first want to finish our programs on late-type stars.

Moffett: What would be the cost of reproducing the CORAVEL instrument?

Burki: About $US 100,000.

MEASUREMENT OF LINE PROFILES

David F. Gray
Department of Astronomy
University of Western Ontario
London, Ontario, Canada   N6A 3K7

ABSTRACT. The basic requirements for high precision spectral line profile measurements are reviewed, with the observatory at the University of Western Ontario serving to illustrate several of the points.

1. ADVANTAGES OF SMALL TELESCOPES

While there is some risk of repeating what has already been said at this meeting, I list here some of the advantages of using a small telescope:

* More observing time, and on a continuous basis, is often available. This is particularly important for stars showing time variablity, e.g., H and K emission or Zeeman broadening, or for projects requiring a large sample of objects.

* You can afford to make repeated observations to thoroughly establish calibrations and equipment stability, and verify astrophysical results.

* Equipment in the testing-debugging stage can often be left in place (Need not be removed for other observers).

* Facilities are usually easier to get to (observing can be integrated into your normal life) and to operate (I usually observe alone).

* Less support staff is needed. This makes for less expense and simpler coordination of activities.

* A breakdown does not automatically mean the loss of your observing run and the end of your observing for another year. When a breakdown does occur, you are on home territory with all your usual tools and test equipment. Other users of the telescope are close at hand and informal switching of the observing schedule can salvage most of the observing time.

* Justification of your observing plans is often minimal

J. B. Hearnshaw and P. L. Cottrell (eds.), Instrumentation and Research Programmes for Small Telescopes, 401–412.

Also notice that the intrinsic accuracy of measurement is not decreased because of small telescope aperture; observations will simply be limited to brighter stars. In fact, greater accuracy may be possible because a person has more time at the telescope to tune his equipment.

2. LINE PROFILE MEASUREMENTS

In this hour, we concentrate on some aspects of measuring line profiles. The widths of spectral lines vary systematically across the HR diagram and are controlled mainly by rotation and turbulence (very strong lines such as the Balmer lines excepted). Since Doppler broadening is often what we deal with, resolution is best specified in velocity units. For many F, G, and K stars, resolution of 2-4 km/s is adequate for measuring line profiles. For late-G and K dwarfs, somewhat higher resolution is desirable because the lines are very narrow. In early-type stars, rotation often produces very broad lines, in which case even modest spectral resolution will resolve the lines. In what follows, I shall concentrate on line profiles in the spectra of the late-type stars.

3. THE SPECTROGRAPH

Attaining high spectral resolution basically hinges on the diffraction grating, and not the size of the telescope. Choose a grating large enough to satisfy two constraints,

1) the diffraction pattern of the grating, of characteristic width W/λ radians, where W is the width of the grating, must map into a wavelength interval less than the desired resolution, and

2) no light should be lost by overfilling the grating.

For a fixed spectral resolution, point 2) actually amounts to an upper limit on the entrance slit width, W', set by the size of the grating. Upper limits on W' are undesirable because it means light will be lost at the entrance slit, a particularly nasty loss when working with a small telescope. The situation is easily understood in terms of the following equation:

$$\Delta\lambda = \frac{-\cos\alpha}{f_{coll}}\,\frac{d}{n}\,W' \qquad (1)$$

where $\Delta\lambda$ is the projection of W' into the spectral plane (equal to the spectral resolution in most cases), α is the angle of incidence onto the grating, d is the spacing between rulings, n the order in which the grating is used, and f_{coll} is the focal length of the collimator. One makes f_{coll} as long as possible in order to make W' appear angularly small. "As long as possible" in this case is determined by the f-ratio of the divergent telescope beam as it passes from the slit to the collimator, coupled with the grating size.

In practice, one buys the largest grating available, makes f_{coll} just long enough to fully fill the grating, and then the slit width must be set to give the desired $\Delta\lambda$ in equation (1).

Obviously, from equation (1), we also choose the grating to have d/n as small as possible.

Choice of camera depends basically on the detector. One sets the camera focal length, f_{cam}, such that $\Delta\lambda$ from equation (1) corresponds to two pixels. If f_{cam} is shorter, the spectrum will not have the measured points close enough together to get all the information. If f_{cam} is longer, the light will be spread out unnecessarily and exposure times will be correspondingly longer. The wavelength interval recorded also varies inversely with f_{cam}, so too long a focal length causes a double loss.

4. DETECTORS

There is not much point in using photography for modern high resolution spectroscopy. Differences in line profile shapes of a fraction of one per cent need to be measured, and photography can reach this level of precision only with great difficulty, if at all.

Electrical detectors have a signal-to-noise ratio given approximately (assuming Gaussian error distributions) by

$$S/N = \frac{n}{\sqrt{n + r^2}} \tag{2}$$

where n is the number of detected photons and r is the electronic noise of the system. When $r \ll n$, we have $S/N \sim \sqrt{n}$, and in principle the precision is limited only by the time available for collecting photons. A number of currently available detectors, such as Reticons and CCD's, have a significantly large r. When $r \gg n$, the detector is usually operating in an integrating mode, and $S/N \sim n/r$. The size of r in this case is critical. Typical values of r range from ~ 500 photons for Reticons down to ~ 50 photons for CCD's. Apparently the CCD would gain a factor ~ 10 in exposure time over a Reticon, and even lower r's for CCD's are expected in the near future, but this gain is only for relatively low S/N. In the current context, we need high S/N, hence large n, and the transition is made to the case where S/N varies as $\sqrt{n}$. In effect, r determines where this transition occurs. The actual gain from reduced r is somewhat less than expected at first glance (Fig. 1). For example, from Fig.1, at a S/N of 40, the CCD gains a factor of 7, but at S/N of 500, the CCD gain is only a factor of 1.6. Furthermore, if the net quantum efficiency of the CCD is a factor of two or three less than for the Reticon, as is often the case, then n is reduced by this same factor with a corresponding reduction in S/N. For the example in Fig. 1, the Reticon has the advantage for S/N > 200. As a rule of thumb, r is the more important factor for low S/N, while quantum efficiency is the more important factor for high S/N.

There are other factors that limit the ultimate signal-to-noise ratio attainable, such as the total number of charges that can reside on a pixel (< 1 E 5, or S/N < 300, for currently available CCD's, while for Reticons it is ~ 2 E 7, or S/N ~ 4500) and the inability to fully remove pixel-to-pixel sensitivity differences.

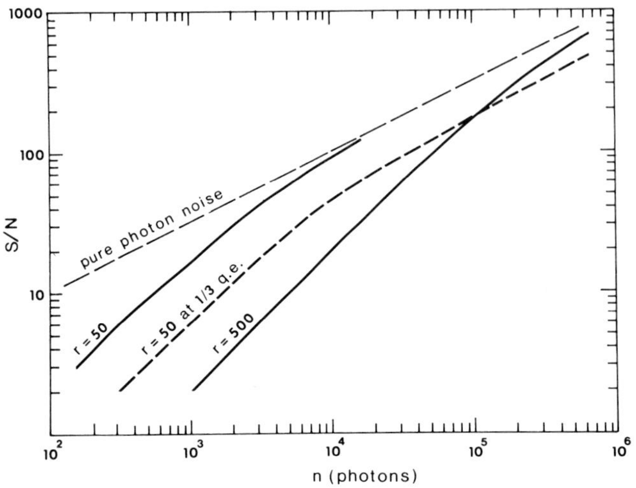

Fig. 1. Signal-to-noise ratio is shown vs. the number of photons incident on the detector, according to equation (2). A quantum efficiency of 1.0 was chosen for the two solid curves; a value of 1/3 for the shorter dashed curve, as labeled. Readout noise, r, is listed. At low S/N, r is the more important factor, but quantum efficiency is the more important factor for high S/N.

5. EXPOSURE TIME DETAILS

Exposure times for a fixed S/N can be reduced by increasing the optical throughput of the system. Frequent resurfacing of mirrors is advantageous. Better still is to use super-reflecting surfaces. An extra factor of two can easily be lost on a coude optical train if care is not exercised. Grating efficiency is also important in this regard. And high order gratings (echelles) can be annoying to use because the blazing has so many "off-blaze" portions across the spectrum. Image slicers should be used for cases where the stellar seeing disk is significantly larger than the entrance slit (and if it is not, beware of the stellar image itself acting as an entrance slit--and a moveable one at that). The image slicer can also result in more efficient photon delivery inside the spectrograph, as I will explain below.

There are problems associated with long exposure times, and since longer exposures might be correlated with smaller telescope aperture, they should be considered here. But an observer with even modest experience will be on the lookout for thermal leakage, slow electrical drifts inherent in silicon-diode detectors, and the presence of background skylight. I will, therefore, discuss only one item which I feel has not received sufficient attention, and that is the line broadening introduced by the diurnal rotation of the Earth. Simply stated, the radial component of the Earth's rotation changes during the exposure and "trails" the spectrum in the direction of dispersion. For a star of declination δ, at an hour angle t, being observed from a latitude ϕ, the elementary equation is $v = 0.465 \cos \delta \sin t \cos \phi$. Spurious broadening from diurnal radial velocity shifts is obviously limited to < 0.93 km/s, but since currently attainable measuring errors for line width parameters are ~ 0.1 km/s, even something much less than this upper bound can be an important systematic error.

Furthermore, it is easy to see that v changes most rapidly when the star is crossing the meridian, which is where all good observers strive

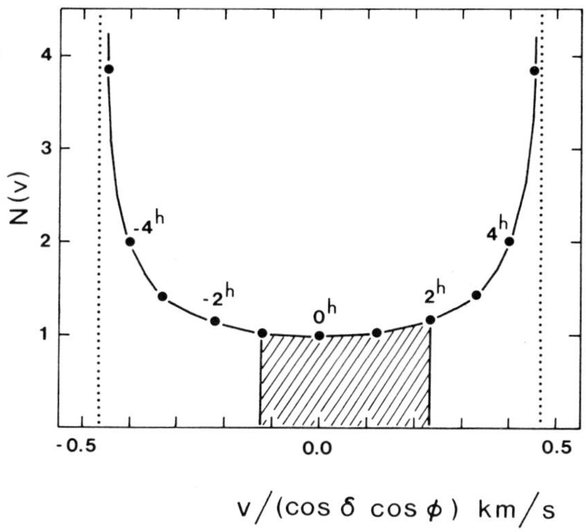

Fig. 2. The distribution of diurnal radial velocity is shown for 0° declination and 0° latitude. Integral hour angles are shown by the dots. A δ-function spectral line would have the shape of the solid curve (turned over if in absorption) for a star observed from horizon to horizon. The shaded region shows the broadening for a three hour exposure extending from -1^h to $+2^h$ in hour angle. The width amounts to 0.35 km/s.

to catch their program stars. The actual blurring introduced into the line profile will be the convolution of the instantaneously recorded profile with a segement of the distribution of v. Since dv/dt goes as cos t, the distribution of v, N(v), goes as 1/(cos t). For example, a star on the equator, observed from a low latitude observatory from 1 before meridian crossing to 2 after, will have v ranging from -0.12 km/s to +0.23 km/s, as shown in Fig. 2. A systematic error in line width of 0.35 km/s is certainly not negligible, and yet, I have never heard of anyone correcting for this important phenomenon.

6. LINE PROFILE MEASUREMENTS AT THE UWO OBSERVATORY

The 1.2 m telescope at the University of Western Ontario is about 25 km north of the campus (Fig. 3). Our horizontal coude room is vastly less burdensome on the taxpayer than one built to the 43° latitude slant (Fig. 4). The extra mirror needed to bring the light beam horizontal is supercoated for a broad spectral band and introduces a loss of 5%. The concrete floor acts as an optical table. It was laid on undisturbed subsoil and is physically isolated from the walls. The walls and ceiling are constructed of heavy concrete to provide a large thermal balast. Out side of the concrete is insulation and a double metal cladding. Typical temperature changes are as low as 1°C/week. Room lights and body heat introduce the largest thermal changes. The biggest air leakage occurs through the electrical outlets (on windy days).

The 200 µm wide entrance slit is part of a Richardson image slicer (Richardson 1966 P.A.S.P. 78, 436, Fig. 5). The effective entrance aperture of the slicer is 4".5 by 6".8. Four slices are used so that the rectangular shape of the grating is matched by the collimator beam. The collimator focal length is 6960 mm.

The Bausch and Lomb grating has 316 l/mm and a blaze angle of 63°26'. The visible spectrum is measured in the 15th through the 6th orders, and intermediate bandwidth interference filters are used to

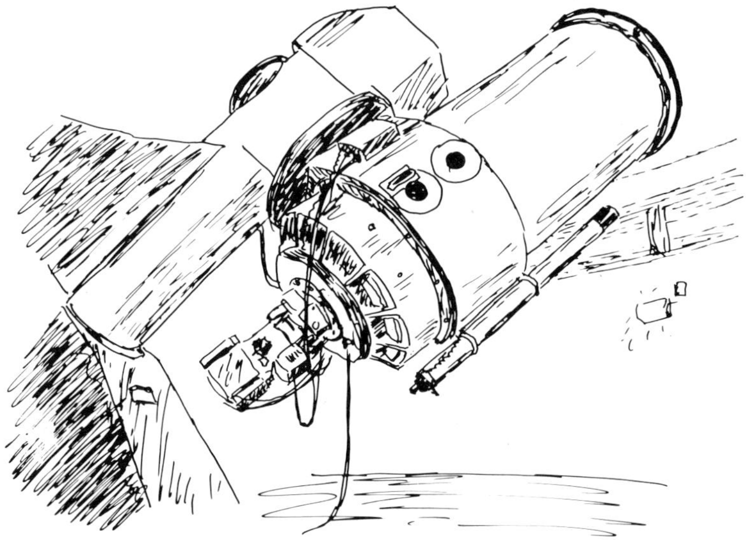

Fig. 3. The University of Western Ontario (U.W.O.) 1.2 m telescope was built by Boller and Chivens, Inc. and installed in 1969.

Fig. 4. The U.W.O. observatory is shown from the South. The horizontal coudé room can be seen as the extension of the building to the East.

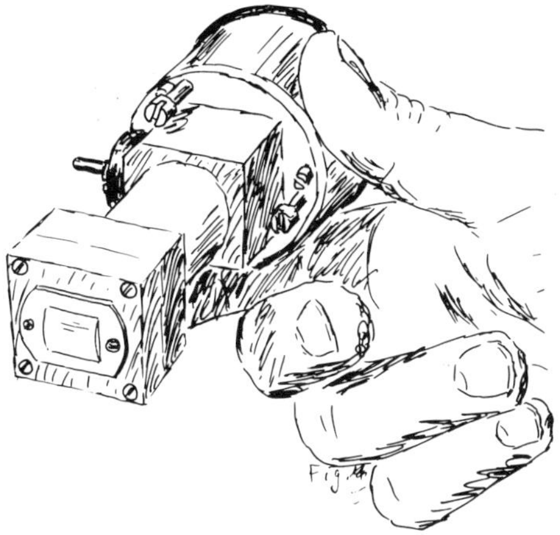

Fig. 5. This is a 4-slice Richardson image slicer. It comprises the entrance aperature to the coude spectrograph, and has an effective aperture of 4".5 by 6".8 on the sky. The optical throughput is more than double that of a simple slit.

isolate orders. These 25 mm diameter filters have peak transmission as high as 85%, and are somewhat more efficient than a cross-disperser for order sorting. Furthermore, the extra reflection from a cross-disperser is eliminated, so the orientation of the cameras need not be changed when switching between this quasi-echelle and conventional gratings working in lowest orders. The camera is a Schmidt, and a 45° pick-off mirror is used to move the focus out of the optical beam. Two slices from the image slicer pass above the pick-off mirror (Fig. 6) and two pass below, resulting in no light being lost by this centrally located mirror. The focal length of the camera is 559 mm. The detector I have used in recent years is an 1872 diode Reticon. The individual diodes are 15 μm wide by 750 μm tall. Again, the image slicer is arranged to make the spectrum 750 μm high.

At 6250 A, working in the 9th order, the dispersion is ~ 0.038

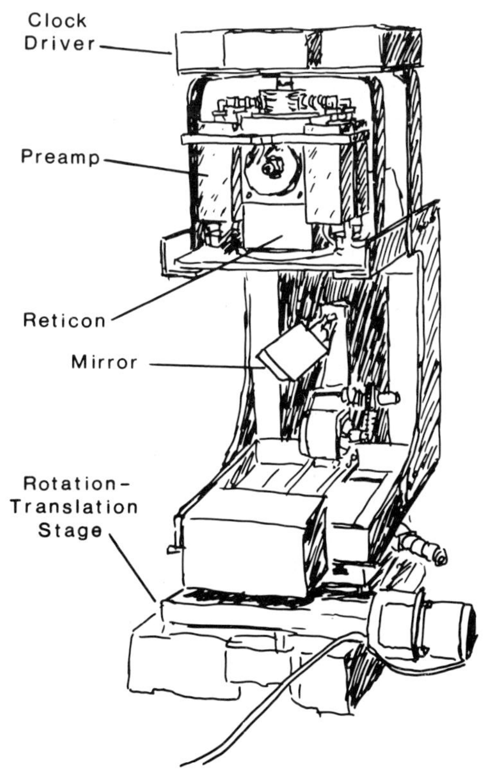

Fig. 6. This Reticon mount is inside the Schmidt camera. The light enters from the left, the spherical camera mirror is on the right (not shown). The focus is above the pick-off mirror on the Reticon inside the vacuum chamber. The whole assembly is mounted on a precision microscope stage used for optical alignment and microscanning.

A/diode and the resolution is ~ 3.5 km/s. This system will attain a signal-to-noise ratio of 100 on a 6th magnitude star in one hour. Signal-to-noise ratios of several hundred are routinely used.

An example of stellar data is seen in Fig. 7. A number of startlingly weak features can be seen. Line blending at the 1% level is

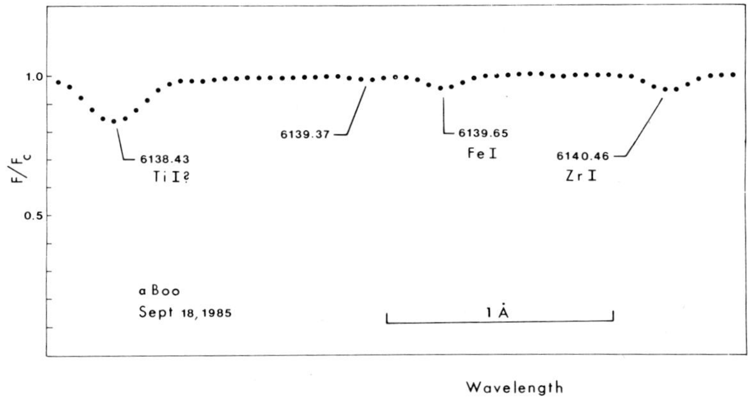

Fig. 7. A small portion of a typical exposure is shown here. A few lines are labeled; the two weakest ones are not.

obviously commonplace in cool stars. Only with high signal-to-noise data of this sort can we expect to measure line shapes and deduce interesting things about velocity fields in stellar atmospheres. The Fourier transform of a typical line shows that the "signal" information contained in the profile is adequately known, i.e., the Nyquist frequency, given by $\sigma_N = 1/(2\Delta\lambda)$, is high enough, or alternatively that $\Delta\lambda$ is small enough. In K dwarfs, the spectral lines become so narrow that finer sample spacing is desirable (Fig. 8). Microscanning is

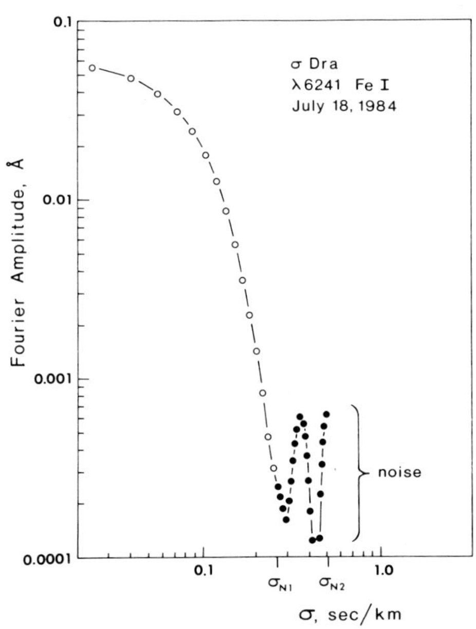

Fig. 8. The Fourier transform of λ6240.65 in σ Dra (K2 V) is shown for two data spacings. The first Nyquist frequency, σ_{N1}, corresponds to a single Reticon exposure with $\Delta\lambda$ = 0.037870 A between diodes. This data spacing is just adequate for line broadening analyses. The second Nyquist frequency, σ_{N2}, results when two exposures are interlaced.

introduced to alleviate this problem--two succesive exposures are taken with a half-diode shift introduced between them. The data numbers are then interlaced (Fig. 9). One example of the improvement from micro-scanning is seen in Fig. 10, where a line bisector is shown. A line bisector is a measure of a line's asymmetry, in this case caused by granulation on the stellar surface. (The shape of the bisector is not very sensitive to the spectral resolution, which remains unchanged.) I think you will agree that the bisector from the interlaced data is superior to those from the individual exposures alone.

You may wonder, since double the exposure time is needed to take the two successive exposures when microscanning, why the camera focal length is not simply made twice as long. That would give twice the density of measurements as required, and since the photons are spread out twice as much, the two exposures would be wrapped into a single one of twice the duration. One would save the (minor) effort of interlacing the two data sets. But microscanning has these advantages: 1) the field coverage is twice as large, and 2) microscanning is not used when it is not needed, so for most stars, the exposures are not doubled. (They would always be doubled if the longer camera focal length were used!) Another important detail is the double height spectrum that would result from doubling the camera focal length, i.e., half the light would be off the detector, and a cylindrical lens would have to be introduced in front of the Reticon to salvage part of it, or else the length of the entrance slit would have to be cut in half, reducing the optical throughput.

I have been please with the productivity of our coude installation at Western Ontario. Coude spectroscopy has many advantages over Cassegrain spectroscopy, especially when high resolution and high signal-to-noise data are basic requirements, but some Cassegrain work can also be done with a small telescope by using an echelle to get high angular dispersion. If you do not have a coude spectrograph, you might consider using a floor mounted echelle spectrograph fed by fiber optics from the Cassegrain focus.

7. EXAMPLES OF RESEARCH USING SPECTRAL LINE PROFILES

One can hardly guess at what the future might hold for high resolution spectroscopy, but here are some currently active research areas:

* velocity fields (turbulence) ⎤
* rotation (v sin i) ⎦ —— from line shape and broadening

* magnetic field measurments using Zeeman broadening

* starspot tracers--following bumps across line profiles

* stellar granulation measurements from asymmetries in lines

* radial oscillation in classical variable stars ⎤
* nonradial oscillations ⎦ —— from time series of profile or radial velocity changes

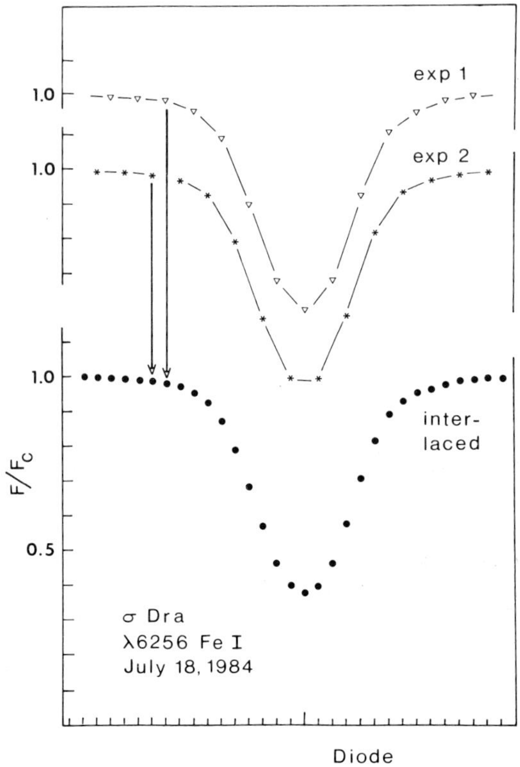

Fig. 9. Microscanning allows two succesive exposures to be interlaced to double the number of points on the profile. (The spectral resolution is not changed.) The topmost curve is from the first exposure; the next down from the second. The combined measurements are shown in the bottom-most profile. The array is moved half a diode width along the direction of dispersion by the precision microscope stage between exposures.

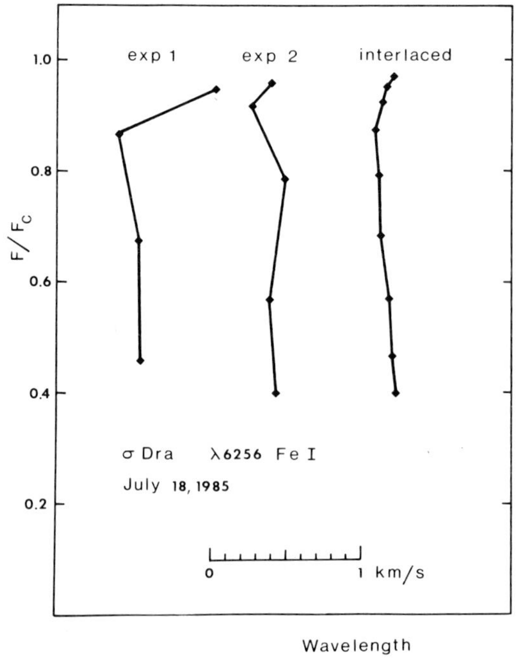

Fig. 10. Line bisectors, which show the asymmetry of the profile, are plotted for the three profiles of Fig. 9. Without microscanning, a narrow-lined star like this one does not give a sufficiently large number of points on the bisector causing them to be ill defined (left and center). The bisector on the right is for the same exposures interlaced. The "C" shape, so familiar in the solar lines, can be clearly seen.

* geometrical shapes of close binary system from the "rotation" profile

* Ca II H & K and H_α variability

Currently attainable signal-to-noise ratios of 100 to 1000 allow these phenomena to be studied. Spectral resolution better than ~ 5 km/s is needed, except for rotation measurements in earlier-type stars. Some wider lines, such as the Na D lines, the Ca II IR triplet, or Balmer lines could also be studied with somewhat lower spectral resolution. For the cooler dwarfs resolution of ~ 2 km/s is desirable.

The list above gives just some of possibilities, and you can expand it as needed to suit your own interests. At the same time, many of the listings are far from mature research topics, and there is still plenty of work to be done....

I am grateful to the Natural Sciences and Engineering Research Council of Canada for continuing financial support.

DISCUSSION

Gaustad: If the original spectrum is truly critically sampled, it should contain all the information possible, and a proper mathematical interpolation should provide any desired degree of fineness in the line profile. Therefore I don't understand the necessity for a repeated exposure to get interlaced data points.

Gray: First, you increase the signal-to-noise with the second exposure (by $\sqrt{2}$). Second, a simple 5-point interpolation will not give the true profile.

White: Have there been any advances beyond the Richardson and Bowen image slicers? I was thinking in terms of fibre optics.

Gray: The Richardson image slicer is impossible to improve upon. The Bowen one is not always in focus along the slit. In our system 2 slices go above the pick-off mirror and 2 go below it. Thus we don't lose any light on this mirror. The slicer changed the focal ratio from f/31 to f/41 which then fitted our grating.

Duncan: An appeal for additional observations. Some of the phenomena (e.g. rotation and chromospheric activity) associated with stars are time dependent. If one is trying to determine the ages of stars using these phenomena, there is a real lack of young stars of known age. Stromgren photometry is able to differentiate the older stars, 2-10 x 10^9 years, but there is a deficiency of stars with ages known to be $< 10^9$ years. One could find these stars as components of visual binaries. There are a lot of binaries known where the primary is an A or B star, which is young, but most of the secondaries don't have spectral types. We need some of these to be of later spectral type to calibrate with their time dependent phenomena.

ABUNDANCE ANALYSES : METHODS, RESULTS AND IMPLICATIONS

P. L. Cottrell
Mount John University Observatory
Department of Physics, University of Canterbury
Christchurch, New Zealand

ABSTRACT. The methods by which abundance analyses can be undertaken, especially with the growing use of high quantum efficiency digital detectors, are reviewed. The importance of the differential curve of growth technique is discussed, followed by a review of some recent results involving both late-type dwarf and giant stars. These include (1) the F and G dwarf abundance analyses of the elements sodium through nickel, which show an important differentiation between stars belonging to the young and old disk populations, (2) the possibility that departures from local thermodynamic equilibrium can effect sodium and aluminium abundances in late-type giants and supergiants and (3) that significant differences exist between the abundances of carbon, nitrogen and oxygen in old disk and young disk giant stars. Finally, the implications of these abundances in terms of stellar and galactic evolution are briefly addressed.

1. INTRODUCTION

Sometimes as astronomers we narrow our visions into a specific area, whereas we should be broadening our horizons. We hope that this symposium has a breadth which will/and has helped.

In spectroscopic abundance analyses we often get involved in very specific analyses in which we overlook the broader picture. We need to take a (large) step back and look at:

- more than an individual star;
- more than a group of stars of a given spectral type and luminosity class;
- and even more than the solar neighbourhood (although this is a useful region);
- and one should at least be aware of the implications of abundances to the Galaxy as a whole and consequently to extra-galactic systems.

One person who possibly applied the latter approaches better than anyone else was Beatrice (Hill) Tinsley (a graduate of the University of Canterbury), and reading her published works and various other

J. B. Hearnshaw and P. L. Cottrell (eds.), Instrumentation and Research Programmes for Small Telescopes, 413–424.

articles about her has been a refreshing experience. I'll return (briefly) to some aspects of her work and others in relation to the *implications* of abundance analyses, when I've talked a little about the *methods* and some new abundance *results*.

And so to begin my talk after the preaching and publicity, I'll deal with some of the more "specific" aspects which relate to the use of stellar abundances in understanding stellar nucleosynthesis, stellar evolution and galactic evolution using samples of "normal" late-type dwarf and giant stars. I must also bring my talk within the auspices of this symposium on small telescopes. I will take a little "LOC licence" and discuss observations from "slightly" larger telescopes but as has been/and will be shown at this symposium, the size of the telescope is not as important as the *nature* and *quality* of the *dispersing* and *detecting* elements. In particular here at Canterbury, a post-graduate student (Phillip MacQueen) has designed, built and commissioned a diode array system for our échelle spectrograph (which many of you would have seen at Mount John University Observatory yesterday), with which we will be able to address similar astrophysical problems to those which I will talk about here today. Charge-coupled devices have/and will continue to revolutionize data taking and therefore smaller telescopes will be able to undertake programmes to determine abundances that 20 years ago would have only been possible with *large* allocations of *large* telescope time.

2. METHODS

Historically, the curve of growth (COG) technique is the method used to deduce stellar atmospheric and abundance parameters from high resolution ($\Delta\lambda \leqslant 0.2$Å or $\lambda/\Delta\lambda \geqslant 30000$) data. It is still used extensively today although in slightly different forms.

The most useful is the differential COG technique developed in the 1940s by Wright (1947) and Greenstein (1948), and emphasised by Gustafsson (1981) especially in relation to the use of model stellar atmospheres. The differential aspect enables the astronomer to eliminate the need for many absolute parameters (e.g. oscillator strengths) and also to minimize the problems of departures from sphericity and local thermodynamic equilibrium when one is working differentially between objects of similar effective temperature (T_{eff}) and luminosity class. These and many other aspects of model atmosphere techniques have been excellently reviewed by Gustafsson (1981, 1983).

Gustafsson (1983) and Cayrel de Strobel (1983) both concluded that the primary method for abundance analyses must be for high resolution, high signal to noise data which can be used to calibrate low resolution spectroscopic (e.g. Laird 1985a,b; Zinn 1985; Langer, Kraft and Friel 1985) or photometric (e.g. Twarog 1980a,b; Hartwick and McClure 1980) data. Although the basic abundance analyses for this began with Wright (1947) and Greenstein (1948) they have been continued by a large body of people on a variety of objects (e.g. Wallerstein 1962; Spite 1968; Hearnshaw 1972; Lambert and Ries 1977, 1981; Peterson 1981; Clegg, Lambert and Tomkin 1981; Kjaergaard et al 1982; Edvardsson,

Gustafsson and Nissen 1984; Tomkin, Lambert and Balachandran 1985). One feature of the analyses listed above is that they were/are for large (> 10 stars) samples which enable other workers in the field to calibrate their derived properties and hence make significantly better use (in a statistical sense) of an even larger sample of data. However a list of standards (similar to those in the discussion of the paper by Cayrel de Strobel 1976), but including a few more stars (e.g. ϕ^2Ori) accessible from southern latitudes, especially New Zealand, would be useful.

Now I would like to outline the procedure which has been undertaken to improve the number of large samples for which stellar abundances are available. This will involve the recent work of Edvardsson et al (1984) [EGN], Tomkin et al (1985) [TLB] and Cottrell and Sneden (1986) [CS]. EGN and TLB described analyses of F and G dwarfs which relate directly to some of the problems involved in *galactic* evolution, whereas the third paper deals with G and K giants and involves implications in terms of *stellar* evolution. (These analyses involve "normal" stars, varying only in their principal parameters, T_{eff}, log g and metallicity, [M/H]. The peculiar giant stars (e.g. see Scalo 1981) also have to be fitted into the overall picture of stellar and galactic evolution.)

This procedure requires one to combine together the observational system(s) available to the user with the types of stars involved (e.g. a restricted spectral type and luminosity class). The system must be capable of a resolution of 30,000 and a signal to noise ratio of ~ 100 : 1 which will enable the all important weak spectral lines to be measured (see Gustafsson 1981). The wavelength regions required to proceed with the analysis should be carefully chosen to give you just the correct amount of data for the problem posed, and not a severe case of data poisoning. One also needs to look at the distribution of spectral lines for the species under investigation.

For example in our analysis (CS) we were endeavouring to extract atmospheric parameters and abundances with reasonable uncertainties (e.g. ± 100 K in T_{eff} and <0.2 dex in the abundances) using spectral data from a restricted number of wavelength intervals (Table I).

TABLE I

Wavelength regions, principle features and atmospheric parameters.

Central Wavelength (λ_c in Å)	Wavelength Range (Å)	Principle Features	Principal Atmospheric Parameters
5110	5060 - 5160	C_2	C abundance
5380	5330 - 5430	TiI/II, FeI/II	T_{eff}, log g, [M/H]
5950	5900 - 6000	TiI, FeI	T_{eff}, [M/H]
6325	6275 - 6375	[OI]	O abundance
8030	7980 - 8080	^{12}CN, ^{13}CN	N abundnace, $^{12}C/^{13}C$ ratio

This has been extended at Mount John University Observatory by Begley (1985) initially with image tube plates, but in the near future with the diode array to determine Na ($\lambda_c \sim 5680$Å and ~ 6160Å) and Aℓ ($\lambda_c \sim 6700$Å) abundances in a large sample of young and old disk giant stars.

Having gathered ones' data your problems have only just begun, especially if you don't have access to a computer on which you have the ability to massage your data (Fourier filtering, continuum placement and a wavelength dispersion solution) into a suitable format from which the strengths of your chosen spectral features can be measured. Chris Sneden and I chose to use an automatic measuring programme (by adopting a gaussian approximation to the line profile) to determine the equivalent widths of the 150 lines in the ~ 40 stars (that is more than 6,000 lines) in our giant star sample. As in EGN and TLB strict selection criteria were placed on the lines. We chose lines for which there was no known blend within 0.5Å and only used lines in the subsequent abundance analysis with log ($^{W}\lambda/_{\lambda}$) $\leqslant -4.8$.

Our differential analysis was conducted relative to β Gem and ε Vir, which have similar T_{eff} and gravities to our programme star sample. Our analysis required us to initially deduce atmospheric parameters from our observations followed by a detailed abundance study of, in particular, CNO. By working differentially with respect to stars of similar spectral type and luminosity class we hoped to minimise any non-LTE effects since one would expect stars with similar atmospheric properties to exhibit similar departures from equilibrium conditions. (The consequence of the differential technique is another reason why a more extensive grid of standard stars in the T_{eff}, log g *and* [M/H] domain should be compiled.)

Although the differential COG method is used to determine the abundances, its use requires the construction of stellar model atmospheres and implicit or explicit calculation of curves of growth. These can be undertaken using a number of analysis programmes (see Gustafsson et al 1975; WIDTH6 and ATLAS6 : see Kurucz 1970, 1979; LINES : Sneden 1973). Although many comparisons have been made between the derived abundances from various computer codes, I know of no complete comparison that has been made between the programmes themselves using a standard set of data and a standard grid of model atmospheres in the T_{eff}, log g and [M/H] domain. A similar comparison should also be made between the various spectrum synthesis codes (e.g. Bell 1970 and Sneden 1973), with the input dataset available in the literature. The use of the synthesis technique reflects the growing interest in not only the analysis of stars (see Gustafsson 1981) but of the total light output of galaxies (e.g. population synthesis calculations, see Pickles 1985).

3. RESULTS

The first set of results which I would like to discuss are the CNO abundances of CS, in which I will compare the old disk (OD) giants with previous analyses of principally young disk (YD) giants (Lambert and Ries

1981 [LR81]; Kjaergaard et al 1981 [KGWH]) and dwarfs (Clegg et al) 1981 [CLT].

When analysing the abundances of giant stars one is not necessarily obtaining information about the formation abundances for these objects but abundances which may have been altered by mixing of nuclear processed

TABLE II

Mean CNO abundances for F & G dwarf and G & K giant stars

Mean differential abundance[a]	Young disk (YD) dwarf CLT	Young disk (YD) giant LR81	Young disk (YD) giant CS	Old disk (OD) giant CS	(OD-YD) dwarf CLT	(OD-YD) giant CS
[C/X]	0.0	-0.24	-0.33	-0.08	0.0	+0.25
[N/X]	0.0	+0.37	+0.37	-0.05	0.0	-0.42
[O/X]	0.0	0.0	-0.09	+0.11	+0.2	+0.20
[X/H]	0.0	0.0	-0.13	-0.50	-0.4	-0.37

[a]X corresponds to either Fe (CLT and LR81) or Ti and Fe (CS)

material (see Lambert 1981). The F and G dwarf sample of CLT (and references therein) should indicate the formation CNO abundances for our comparison between old and disk giant stars. These are listed in Table II along with the relevant abundances from LR81 and CS. LR81 explained their results by the mixing of CN-processed material into the surface layers. Our old disk giant analysis provides quite a contrast. (See Figure 1)

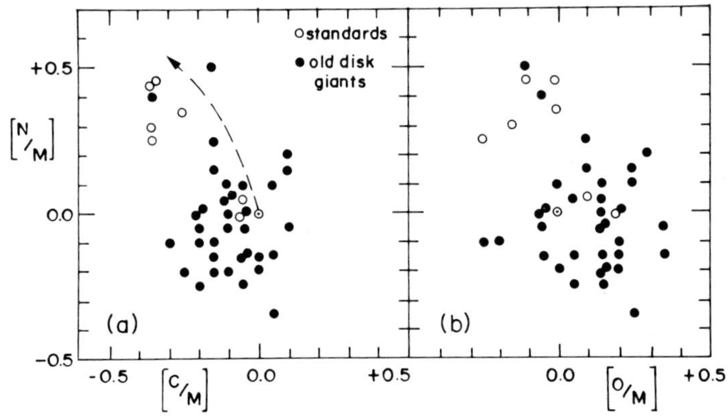

Figure 1. Carbon, nitrogen and oxygen abundances for old disk and young disk (standards) giant stars from CS. The solar abundances are shown by the usual solar symbol. The dashed line in (a) leading from the solar abundances to depleted carbon and enhanced nitrogen corresponds to the conversion of carbon to nitrogen by CN-processing with the conservation of the total amount of (C+N). The mean values of [C/M], [N/M] and [O/M] are given in Table II. M is the average metal abundance determined from Ti and Fe lines.

In particular our differential abundances (OD-YD) are shown in the final column of Table II, where they are compared with the abundances from CLT for a similar mean metallicity. One is able to see that the OD giants appear not to exhibit any of the effects of interior nuclear processing. [There was some indication of this effect in both LR81 and KGWH, in particular see Figures 13, 14 and 15 in the latter.] Some mixing of processed material may occur in the OD stars as some have low $^{12}C/^{13}C$ ratios (~10-20), which is well below the value (50-100) one would expect of the interstellar medium from which these objects formed.

One other domain in which we can illustrate our results is in terms of the kinematics of the stars, a point also emphasized by EGN. Figure 2 shows the difference between metal rich, [Fe/H] > -0.3 and metal poor [Fe/H] < -0.3 objects. A clear separation exists between the solar (C/N) ratio, high velocities, for the metal poor objects and low (typical of the YD stars) C/N ratio, low velocities, for the metal rich objects.

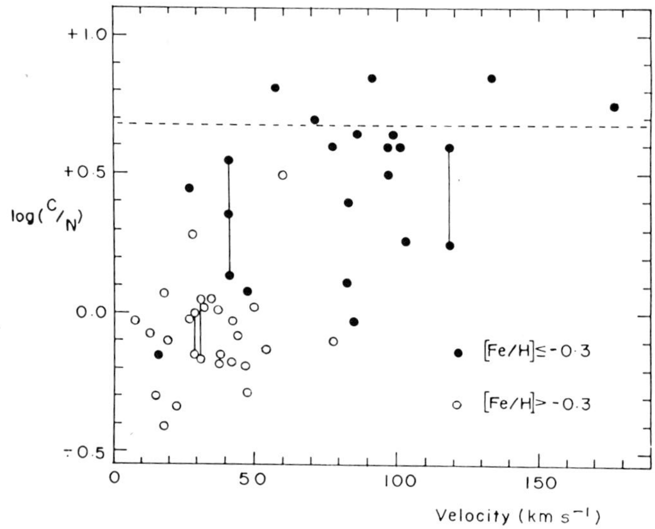

Figure 2. log (C/N) ratio versus total space velocity for a merged sample (from LR81, KGWH and CS) of giant stars. The dashed line corresponds to the solar value of C/N. The lines connecting some points correspond to observations of the same object by two or three of the groups mentioned above.

This quantitative abundance analysis adds another dimension to the photometric results of Hartwick and McClure (1980) where they compared the DDO CN index (41-42) for OD and YD objects, and illustrates the interplay between photometric and spectroscopic techniques.

So one can see that there is much to offer in this area (CNO abundance analyses) in terms of how stars in the different populations evolve.

I would now like to pass onto the next group of light elements (sodium to silicon). (Fluorine and neon have been bypassed since adequate stellar spectral information can not be obtained.) The dwarfs hold important keys to our understanding of galactic evolution, since for these stars their surface abundances are indicative of those they had at the time of their formation.

Begley (1985) has conducted a differential analysis between old and young disk giants. He found enhancement of both Na and Aℓ. (Open and filled circles respectively in Figure 3.)

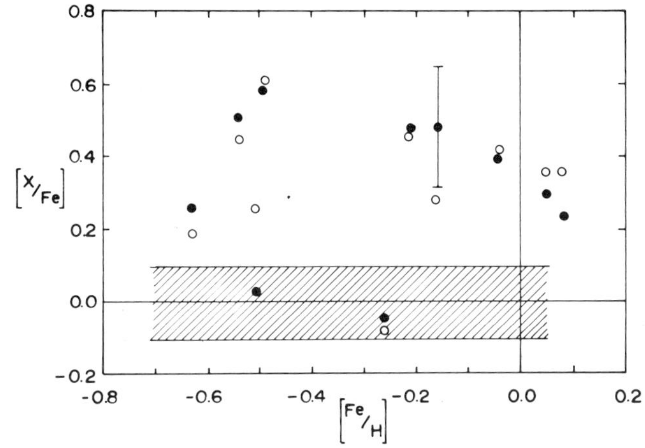

Figure 3. Sodium (open circles) and aluminium (filled circles) abundances, relative to iron for G and K *giant* stars belonging to the old and young disk populations (Begley 1985). The hatched region corresponds to the *spread* in sodium and aluminium abundances for F and G dwarf stars (EGN and TLB).

Begley discussed an interesting effect when one compared Na to Si abundances in G and K giants with their progenitors, the F and G dwarfs (EGN, TLB). Many analyses of late-type giants (see Begley 1985) also show enhancements of Na and Aℓ with respect to Fe whereas the dwarfs, from which they would be expected to have evolved, show no such enhancement (see Figure 3). (Similar enhancements of sodium in K giants were noted by Helfer and Wallerstein (1968).) Some departure from LTE in either the excitation or ionisation equilibria of Na and Aℓ is a possible explanation of this effect. Much more work in this area is required to follow up both the observational work (e.g. Ruland et al 1980 and Brown et al 1983) and theoretical non-LTE analyses (e.g. Kelch 1975 and Kelch and Milkey 1976). In particular these latter two papers predict that different energy levels (and hence different transitions) are affected in different ways by departures from LTE. The final solution for abundance analyses may require full non-LTE analyses for *all* species. Thus unless our understanding of stellar evolution is drastically incomplete, the enhancement of Na and Aℓ in the giants and supergiants relative to the dwarfs must be viewed with some caution and an atmospheric explanation (e.g. departures from LTE) may be more likely.

And so to the abundances of the α-rich elements Mg, Si and Ca (see EGN and TLB). I have combined their data (Figure 4) to give the

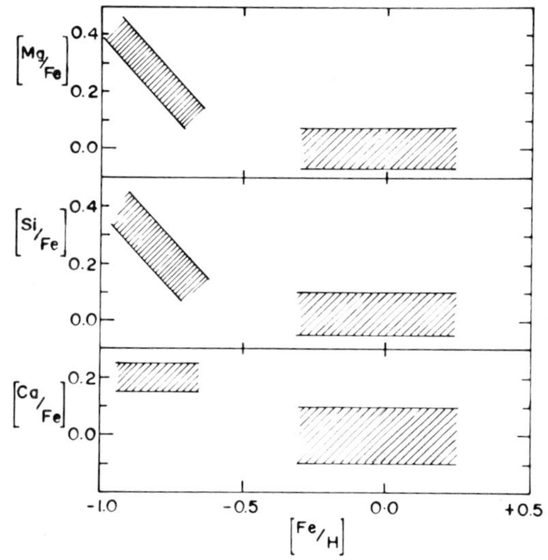

Figure 4. The spread of [Mg/Fe], [Si/Fe] and [Ca/Fe] abundances (from the data of EGN and TLB) as a function of overall metallicity.

schematic distribution of these abundances as a function of metallicity, [Fe/H]. This shows that for abundances $[Fe/H] < -0.5$, there is an enhancement, relative to iron, of these elements. This shows that there has been a change in the production of these α-rich elements relative to iron over the lifetime of the disk, based on the age-metallicity distribution of Twarog (1980b).

It would now seem appropriate to discuss some of the implications of these results.

4. IMPLICATIONS

As alluded to at the beginning of this talk the abundance analyses discussed here are referring to two different areas of evolution, namely stellar and galactic. I'll take each in turn.

The results for the OD giant stars (and for that matter those of the YD) indicate that there are different processes (or the same process to different extents) in action in the stars of different populations. Stars with different metallicity and yet at a similar position in the HR diagram should be expected to differ in their average mass and age. In such a way the OD stars are of lower average mass and have evolved much more slowly (than the YD stars) to now be at the same position in the HR diagram. In particular, it appears that the less massive, metal poor stars show little (or no) CNO-processed material in their outer layers. This raises a number of questions, the solutions to which I feel are beyond the scope of the current review and will require further theoretical modelling. For example, does the convective envelope in these less massive, metal poor stars not involve as much of the outer mass of the star (*cf* the more massive stars) so that the nuclear processed material is not reached by this envelope? Or, does shell burning occur closer to the centre of the star and hence the convective envelope doesn't reach the processed layers? (The $^{12}C/^{13}C$ results seem to indicate that there may be a little processed material mixed into the outer layers.) Anyway it appears that there is certainly some more work that can be done, both observationally and theoretically, on these types of objects. In addition, there may be some important consequences in terms of the origin of the CN anomalies (which are correlated with Na abundances) observed in the giant stars of many globular clusters (see Freeman and Norris 1981). The lack of any correlation between CN and Na (or Aℓ) in the disk (at least between the old and young disk stars) indicates that the environment in which the globular clusters evolved may have been dramatically different.

And so to the galactic evolution implications. EGN and TLB have both shown that the yields of the various α-rich elements can be explained by the predictions of explosive C nucleosynthesis (Arnett 1971) and massive star ($25M_\odot$) nucleosynthesis (Woosley and Weaver 1982). The Na and Aℓ results of EGN and TLB can also be explained by the work of Arnett (1971) and Pardo, Couch and Arnett (1974). The odd numbered nuclei species (i.e. Na and Aℓ) are produced less than the even numbered species (e.g. Mg) in the more metal poor stars. This is because the production of Na and Aℓ in explosive nucleosynthesis depends upon a source of neutrons, from ^{14}N, ^{18}O or ^{22}Ne, so that as more of these elements are produced (in the more metal rich objects) so more Na and Aℓ will be produced relative to adjacent species in the periodic table. This explains the flattening of the [Aℓ/Mg] versus [C+N+O/H] for [C+N+O/H] $\geqslant$ 0.5 (see Figure 4 of TLB).

Once again, continued observational work needs to be undertaken (as outlined by EGN) on large samples of F and G dwards. This should fill the [X/Fe] versus [Fe/H] diagrams between -1.0 dex and +0.3 dex in metallicity such that the extent of the enrichment of some species can be clearly seen and the transition from halo to disk abundances can be quantified. Kinematic information on *all* these objects is also essential.

All this data can then be fitted into a model for galactic evolution as has been shown by Twarog and Wheeler (1982) based on the data of Clegg, Lambert and Tomkin (1981). Twarog and Wheeler carefully considered which types of stars produced which elements (such that certain species are not under or over-produced), as well as the nature and type of the material falling into the disk, the star formation rate and the initial mass function at different epochs.

However, analyses should not be confined just to those elements considered here nor just to the "normal" stars, as some of the less abundant elements *and* more peculiar objects hold important keys linking various phases of stellar and galactic evolution. In particular, the use of small telescope time on these types of projects is an extremely useful task to increase the pool of abundance data.

Finally I would like to conclude my talk by once again drawing your attention to the work of Beatrice Tinsley (e.g. see Dearborn, Tinsley and Schramm 1978; Tinsley 1979), and in particular Figure 1 of Tinsley (1980) which illustrates the importance of fitting one's own place of research into the overall galactic picture. As Beatrice said, Tinsley (1980), one must treat 'different aspects of galactic evolution as pieces of a jigsaw puzzle that may someday be put together."

REFERENCES

Arnett, W.D. 1971, *Astrophys. J.*, 166, 153.

Begley, M.J. 1985, M.Sc. thesis, University of Canterbury, Christchurch, New Zealand.

Bell, R.A. 1970, *Monthly Notices Roy. Astron. Soc.*, 148, 25.

Brown, J.A., Tomkin, J., and Lambert, D.L. 1983, *Astrophys. J.* (Letters), 265, L93.

Cayrel de Strobel, G. 1976, in 'Abundance Effects in Classification' IAU Symposium No. 72, eds B. Hauck and P.C. Keenan (Dordrecht : Reidel), p29.

Cayrel de Strobel, G. 1983, *Pub. Astron. Soc. Pacific*, 95, 111.

Clegg, R.E.S., Lambert, D.L., and Tomkin, J. 1981, *Astrophys. J.*, 250, 262, [CLT].

Cottrell, P.L., and Sneden, C. 1986, *Astron. Astrophys.*, in press, [CS].

Dearborn, D., Tinsley, B.M., and Schramm, D.N. 1978, *Astrophys. J.*, 223, 557.

Edvardsson, B., Gustafsson, B., and Nissen, P.E. 1984, *The Messenger*, 38, 33, [EGN].

Freeman, K.C., and Norris, J. 1981, *Ann. Rev. Astron. Astrophys.*, 19, 319.

Greenstein, J.L. 1948, *Astrophys. J.*, 107, 151.

Gustafsson, B. 1981, *Astrophys. Space Science Library*, 88, 25.

Gustafsson, B. 1983, *Pub. Astron. Soc. Pacific*, 95, 101.

Gustafsson, B., Bell, R.A., Eriksson, K., and Nordlund, A. 1975, *Astron. Astrophys.*, 42, 407.

Hartwick, F.D.A., and McClure, R.D. 1980, *Astrophys. J.*, 235, 470.

Hearnshaw, J.B. 1972, *Mem. Roy. Astron. Soc.*, 77, 55.

Helfer, H.L., and Wallerstein, G. 1968, *Astrophys. J. Suppl.*, 16, 1.

Kelch, W.L. 1975, *Astrophys. J.*, 195, 679.

Kelch, W.L., and Milkey, R.W. 1976, *Astrophys. J.*, 208, 428.

Kjaergaard, P., Gustafsson, B., Walker, G.A.H., and Hultqvist, L. 1982, *Astron. Astrophys.*, 115, 145, [KGWH].

Kurucz, R.L. 1970, SAO Special Report No. 309.

Kurucz, R.L. 1979, *Astrophys. J. Suppl.*, 40, 1.

Laird, J.B. 1985a, *Astrophys. J.*, 289, 556.

Laird, J.B. 1985b, *Astrophys. J. Suppl.*, 57, 387.

Lambert, D.L. 1981, *Astrophys. Space Science Library*, 88, 115.

Lambert, D.L., and Ries, L.M. 1977, *Astrophys. J.*, 217, 508.

Lambert, D.L., and Ries, L.M. 1981, *Astrophys. J.*, 248, 228, [LR81].

Langer, G.E., Kraft, R.P., and Friel, E.D. 1985, *Pub. Astron. Soc. Pacific*, 97, 373.

Pardo, R.C., Couch, R.G., and Arnett, W.D. 1974, *Astrophys. J.*, 191, 711.
Peterson, R. 1981, *Astrophys. J.*, 244, 989.
Pickles, A.J. 1985, *Astrophys. J.*, 296, 340.
Ruland, F., Holweger, H., Griffin, R., Griffin, R., and Biehl, D. 1980, *Astron. Astrophys.* 92, 70.
Scalo, J.M. 1981, *Astrophys. Space Science Library*, 88, 77.
Sneden, C. 1973, Ph.D. thesis, The University of Texas at Austin.
Spite, M. 1965, *Ann. d'Astrophys.*, 31, 269.
Tinsley, B.M. 1979, *Astrophys. J.*, 229, 1046.
Tinsley, B.M. 1980, *Fundamentals Cosmic Physics*, 5, 287.
Tomkin, J., Lambert, D.L., and Balachandran, S. 1985, *Astrophys. J.*, 290, 289 [TLB].
Twarog, B.A. 1980a, *Astrophys. J. Suppl.*, 44, 1.
Twarog, B.A. 1980b, *Astrophys. J.*, 242, 242.
Twarog, B.A., and Wheeler, J.C. 1982, *Astrophys. J.*, 261, 636.
Wallerstein, G. 1962, *Astrophys. J. Suppl.*, 6, 407.
Woosley, S.E., and Weaver, T.A. 1982, in 'Essays in Nuclear Astrophys.', eds C.A. Barnes, D.D. Clayton, and D.N. Schramm (Cambridge : Cambridge University Press),p377.
Wright, K.O. 1947, *Pub. Dominion Astrophys. Obs.*, 8, 1.
Zinn, R. 1985, *Astrophys. J.*, 293, 424.

DISCUSSION

Garrison: I always worry a bit about comparisons of absolute observations derived using different techniques. How sensitive are the results for old and young disk stars to analysis methods? Are there comparisons for the same star between the Texan, Swedish and Australian groups?

Cottrell: Sneden's programme (LINES) has been used by many research workers (e.g. CS, TLB, SR81) so good absolute comparison should be possible. I don't know whether a direct comparison has been made between the Texan and Swedish programmes. However, I think the important thing one needs is a standard set of data (lines and equivalent widths) on some standard stars available to all users of analysis programmes.

Kumar: Have $^{12}C/^{13}C$ ratios been obtained with the échelle spectrograph and diode array detector which we saw at Mt John University Observatory?

Cottrell: No, but we expect to be able to obtain $^{12}C/^{13}C$ results similar to those which have been obtained at McDonald Observatory.

Hearnshaw: The main reason why different analyses of the same star give different results is that different observers have adopted different temperatures. The latter generally come from broad-band photometric observations, and so I would urge better calibration of the photometric data.

Mochnacki: There is a new temperature calibration in the uvby system by John Lester, Richard Gray and Bob Kurucz which will be out soon.

SPECTROSCOPIC OBSERVATIONS OF VISUAL DOUBLE STARS

A.H. Batten
Dominion Astrophysical Observatory
Herzberg Institute of Astrophysics
5071 W. Saanich Road
Victoria, B.C. V8X 4M6
Canada

ABSTRACT. Observations of visual binaries made with the 1.2-m telescope and coudé spectrograph at Victoria are described, with particular emphasis on their value as tests of evolutionary calculations.

1. INSTRUMENTS

Over a number of years, the writer and colleagues have used the 1.2-m telescope at Victoria and its coudé spectrograph, either photographically or in conjunction with a radial-velocity spectrometer (e.g. Griffin 1967) to observe visual binaries. With a telescope of a given aperture, a spectrometer reaches fainter (and more) stars than the photographic plate, at the cost of some of the information that a high-dispersion photographic record can give. Ideally, a spectrometer might be the principal instrument used in this work, but its results should be supplemented, whenever possible, by spectrograms or other observational records that permit the examination of individual line profiles.

2. STARS OBSERVED

Several different groups of visual binaries may usefully be observed spectroscopically. The most interesting are pairs of stars of nearly equal luminosity for which there is a well-determined visual orbit of such a shape and orientation that large differences in the radial velocities of the two components may be expected around nodal passage. Most pairs in this class will be so close that their images are indistinguishable from those of single stars at the spectrograph slit. A given binary, with a period of decades, is therefore likely to be of interest for about a couple of years, at most twice in an observer's active career. Nevertheless there are enough systems for a long-term programme that could be a useful part of a small observatory's work. The Catalogue by Dommanget and Nys (1982) is of considerable value in selecting systems for observation, although the necessary computations may easily be made directly from the orbital elements.

Although radial-velocity differences of the order of 100 km s^{-1}

J. B. Hearnshaw and P. L. Cottrell (eds.), Instrumentation and Research Programmes for Small Telescopes, 425–427.

occasionally occur, even between the components of quite long-period binaries, differences of the order of 20 km s^{-1} to 30 km s^{-1} are much more common. Most systems observed must have both components of late spectral type, if the spectra are to be resolved. Observational selection already favours this kind of system. We have found that velocity differences of about 20 km s^{-1} can be accurately measured with our spectrometer and pairs of about sixth magnitude or brighter can be observed with a photographic dispersion of 2.5 $Å mm^{-1}$ at even smaller velocity separations.

3. LIKELY RESULTS AND THEIR USEFULNESS

The principal results from these observations are the parallax and total mass of each system. Quite accurate parallaxes, beyond the range of trigonometrical measurement, can be obtained from only a few measurements (in principle, one) of radial-velocity difference. The total mass of a system can also be found accurately, although mass-ratios, and therefore individual masses, are usually harder to determine because of uncertainty in the systemic velocity. Popper (1980) lamented that "the very great amount of difficult observing [of visual binaries] for more than 150 years" has resulted in no more than a dozen or so masses being known even to within 20%. The kind of observation described here is one way of making use of that 150 years' work and increasing the number of systems for which both masses and luminosities are well known.

Of particular interest, however, is the surprising number of visual binaries that contain components not on the main sequence. Amongst nine visual binaries (including two interferometric ones) studied at Victoria, five or six contain at least one component above the main sequence. For these components, we know mass, luminosity and temperature with a fair accuracy (e.g. Batten et al. 1982). In time, it should be possible to amass enough data to provide even more useful checks on evolutionary models of stars than are obtained now from observations of star clusters.

Two or three groups are now making this kind of observation in the northern hemisphere. The rate of progress could be substantially increased if a suitable telescope in the southern hemisphere were used for similar work.

4. REFERENCES

Batten, A.H., Fletcher, J.M., Fisher, W.A., McClure, R.D., Morbey, C.L., Griffin, R.F. and Scarfe, C.D. 1982, Pub. Astron. Soc. Pacific **94**, 860.

Dommanget J. and Nys, O. Second Catalogue d'Ephémérides...1982 Comm. Obs. R. Belgique Série B, No. 124.

Griffin, R.F. 1967, Astrophs.J. **148**, 465.

Popper, D.M. 1980, Ann. Rev. Astron. Astrophys. **18**, 115.

DISCUSSION

Evans: Did you observe only those pairs with good visual orbits?

Batten: Yes, of course, I should have made that clear.

Kurtz: 1) Why can't you see the secondary in 70 Oph better since the masses of the two stars are similar? 2) What is the orbital eccentricity of OΣ 341?

Batten: 1) Immediately before and after nodal passage, the two components of 70 Oph cannot be resolved on the slithead at Victoria. Also the brighter component dominates the spectrum. 2) The orbital eccentricity of OΣ 341 is 0.96.

Garrison: There are other ways in which visual binaries are useful as tests of stellar evolution theory. Corbally has studied very close pairs (1"-5") with our 60cm in Chile using MK classification techniques. His results are published in an Ap. J. Suppl. and in the IAU Symposium Obs. Tests of Stellar Evolution Theory. He was about to put constraints on Demarque's latest models.

Batten: Chris Corbally's work is, of course, important, but orbital pairs do give you extra information: the masses of the components.

Evans: The Texas group has been working on triple systems from which you can extract much information. For example, one can get the equivalent of a visual orbit, using speckle or occultation measurements. We have done some of the binaries in the Hyades in this way.

Batten: Yes, other groups, including yours are doing similar work; but there is still no-one working in the southern hemisphere - where there are a lot of good stars to do.

PHOTOMETRIC AND SPECTROSCOPIC STUDY OF δ^2 LYRAE

Gustav A. Bakos
Department of Physics
University of Waterloo
Waterloo, Ont. N21 3G1
Canada

and

Jozef Tremko
Skalnate Pleso Observatory
Slovak Academy of Sciences
059 60 Tatranska Lomnica
Czechoslovakia

ABSTRACT. The bright star δ^2 Lyr has been observed photoelectrically in the V and B pass bands for three years. The star is a long period variable with an amplitude of 0.2 mag and a period of 83 days. Light changes are coupled with small range velocity changes. On some well-exposed spectrograms the emission features of the H and K lines of CaII appear.

1. INTRODUCTION

The fourth magnitude star δ^2 Lyr of spectral type M4 II has for long been suspected of variability. In the late fifties Biskupski (1960) observed the star visually, however he did not derive the type of variability nor the period. Eggen (1977) classified the star as a variable belonging to young disk population stars and derived its absolute bolometric magnitude as -5.75.

2. OBSERVATIONS

δ^2 Lyr has been observed in Waterloo since 1982 and more recently at the Skalnate Pleso Observatory in the V and B colour bands. The variable was compared with the star HD 176051 and checked against the star HD 175635. Both stars are of spectral type G. The period of the light variations has been derived from four epochs of minima and four epochs of maxima by the method of least squares. The average period has been adopted, P = 82.7736 days, as the period of light variations. The above period has been used to derive a mean light curve as a function of phase. The light curve exhibits three maxima, a high maximum at phase 0.15, a broad, lower maximum at phase 0.65 and third, low maximum at phase 0.92. The amplitude of light variations is 0.25 mag.

The colour index, Δ(B-V) changes by 0.1 mag between maxima and minima of the light curve in the sense that the colour index is bluest at the time of minimum brightness of the star. This is true for the minima associated with the first maximum.

J. B. Hearnshaw and P. L. Cottrell (eds.), Instrumentation and Research Programmes for Small Telescopes, 429–430.

Between 1969 and 1983 eleven spectrograms were obtained at David Dunlap Observatory having a dispersion of 12 A/mm. Although the range of radial velocities is only 5 kms^{-1} there is a correlation with the light curve in the sense that both curves reach a maximum at phase 0.15. The minimum is at phase 0.65.

On three well-exposed spectrograms the emission of the H and K lines appears. The radial velocities of the emission features K_2V and K_2R as well as of the absorption component K_3 have been measured. These velocities are respectively -43, -7 and +33 kms^{-1}.

3. DISCUSSION

The light curves so far obtained do not present a complete picture of light variations of this star. However, it can be said that the light curves do not repeat themselves rigorously and that the derived period represents an average period of light variations. The amplitude of the light curves is also variable.

The colour index, Δ(B-V), changes by about 0.1 mag between the maximum and minimum light of the star. Since the surface temperature of an M4 II star is about 3300°K the change of the colour index by the observed value would produce a change of about 150°K in the surface temperature of the star.

The radial velocity data indicate that the maximum velocity is reached at the time of the light maximum. If it is assumed that the velocity variations are produced by pulsations of the star then the most positive velocity (i.e. a contraction of the star) is associated with an increase of brightness, as indicated by the change in the colour index. The second and the third maximum of the light curve must have a different cause, for instance, shock waves as suggested by Wallerstein (1977).

Finally, a few comments about the emission features of the H and K lines. The emission features indicate an expanding chromosphere of the star while the deep absorption feature represents a circumstellar envelope (Deutsch, 1960). It is formed by matter ejected from the photosphere with a velocity close to the velocity of escape.

In summary, it has been shown that δ^2 Lyr is a small amplitude long period variable, easily accessible to small telescopes, in need of frequent observations extending over a time interval of years.

The first author would like to thank Dr. J. Donald Fernie, Director of David Dunlap Observatory for the use of the observing facilities. This research was supported by a grant from the Natural Sciences and Engineering Research Council of Canada.

REFERENCES

Biskupski, A. 1960, *Urania*, 31, 146.
Deutsch, A.J. 1960, *Stellar Atmospheres*, VI, 543. Univ. of Chicago Press.
Eggen, O.J. 1977, *Astrophys. J.* 213, 767.
Wallerstein, G. 1977, *J. Roy. Astron. Soc. Can.* 71, 298.

10 μm HETERODYNE SPECTROSCOPY AT TELESCOPES WITH 1 m APERTURE

H. Rothermel, U. Schrey, H.U. Käufl, S. Drapatz,
Max-Planck-Institut für Physik und Astrophysik,
Institut für Extraterrestrische Physik,
8046 Garching bei München, FRG.

ABSTRACT. Heterodyne spectroscopy combines high instrumental transmission with virtually unlimited spectral resolution. For 250 K bodies like Mars and Venus resolution is limited to 10^7 (= 30 m/s) because of photon statistics whereas for solar measurements the cm/s velocity range, desirable for study of solar oscillations, is feasible. Since a single heterodyne receiver is conserving its etendu, smaller collecting area results in larger field of view. The signal remains unchanged as long as the source is large enough to fill the beam. For Mars, Venus and the Sun heterodyne observation with 1 m aperture is competitive.

1. PRINCIPLE OF OPERATION

A heterodyne spectrometer converts by superposition with laser radiation an infrared spectrum down into the radio frequency (RF) range; where it is analysed with electrical rather than with optical filters. The RF heterodyne signal produced in the photomixing process is a function of both signal power P_S and laser power P_L:

$$\bar{i^2} = 2\eta\ (e/h\nu)^2\ P_S\ P_L$$

where η = quantum efficiency, $h\nu$ = photon energy. With sufficient laser power (~ 1 mW) the heterodyne signal can be increased to a level where other sources of noise such as preamplifier and RF electronics no longer are important. The heterodyne system works now in the quantum noise limited regime where it approaches the sensitivity of a hypothetical photon counter working at 10 μm wavelength with a quantum efficiency of 0.6. In addition a multiplex advantage is accomplished by use of a filter spectrometer analyzing the spectrum with 100 to 200 filters in parallel.

Heterodyne systems[1] are conserving the product of beam solid angle Ω and effective telescope area A_e:

$$A_e \cdot \Omega = \lambda^2$$

where λ = wavelength. This relation indicates that a heterodyne spectrometer exploits the telescope at the diffraction limit (about 2 arcsec. for λ = 10 μm and 1 m aperture).

J. B. Hearnshaw and P. L. Cottrell (eds.), Instrumentation and Research Programmes for Small Telescopes, 431–433.

2. INSTRUMENT

The spectrometer[2] working between 9.5 and 12.5 μm wavelength is based on a sealed-off CO_2 laser, a Hg Cd Te photovoltaic detector and a 200 channel filter spectrometer. In contrast to competitive instruments it is portable and operates at the Cassegrain focus. It was used first with a 1 m reflector on the Gornergrat (3200 m elevation in Switzerland). Five different isotopic combinations of CO_2 laser gas present about 250 discrete laser frequencies. This provides a sufficient number of coincidences with lines of other interesting molecules such as NH_3, SiH_4, C_2H_2, C_2H_4, OH and H_3O^+.

3. MEASUREMENTS

A heterodyne spectrum covers only one, in rare cases two molecular lines. These lines are perfectly resolved and have precise wavelength and flux calibration. Below is an example of a heterodyne spectrum from the Martian atmosphere[3]. The continuum is emitted by the Martian surface. The weak absorption on the left is $^{12}C^{16}O^{18}O$, the saturated absorption profile is normal CO_2. Adjusting synthetic line profiles results in temperature and pressure as a function of altitude. Furthermore isotopic ratios are derived with a precision competitive to in situ measurement by the Viking landers. The narrow emission line in the centre of the CO_2 absorption is a fluorescence produced by sunlight in the Martian atmosphere about 70km above the surface. From the line width or (better) comparing integrated fluxes in the emission for different rotational transitions, temperature mapping in the high atmosphere of Mars and Venus is possible. These investigations are beyond the capability of space probes. In future comets, outer planets and planetary nebulae will be investigated by heterodyne spectroscopy for thermal condition, isotopic abundances and expansion velocities.

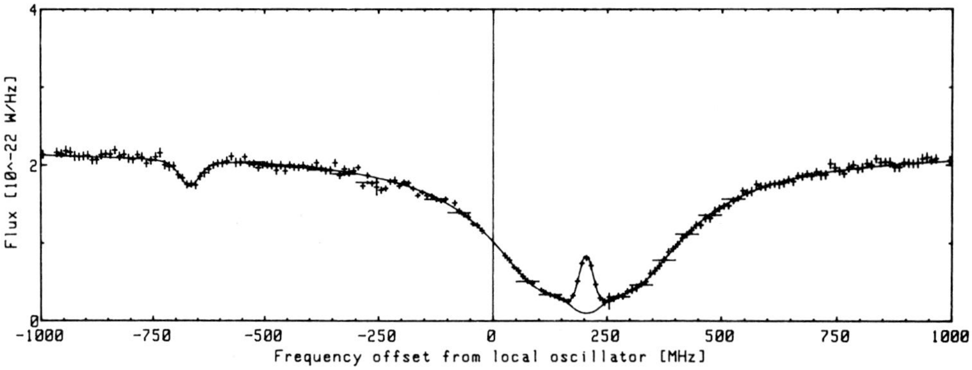

Line profiles of isotopic and normal CO_2 on Mars.

REFERENCES

1. A.E. Siegman, Proc. IEEE 54, 1350 (1966)
2. H. Rothermel, H.U. Käufl, Y. Yu, Astron. Astrophys. 126, 387 (1983)
3. U. Schrey, H. Rothermel, H.U. Käufl, S. Drapatz, Astron. Astrophys. in press (1985)

DISCUSSION

A'Hearn: How long an integration time is required to obtain a spectrum of Mars at one laser line?

Rothermel: About 10 minutes.

Genet: Is this telescope dedicated to this type of work?

Rothermel: No, it is not dedicated to IR work.

III. SPECTROSCOPIC RESEARCH PROGRAMMES

b) Poster papers

THE FIRST EIGHT YEARS OF RADIAL VELOCITY STUDIES AT FICK OBSERVATORY

W. I. Beavers and J. J. Eitter
E. W. Fick Observatory
Iowa State University
Ames, Iowa 50011
U.S.A.

ABSTRACT. The stellar radial velocity (RV) observations collected during the first eight years (1976-1984) of operation of the Fick Observatory photoelectric RV spectrometer have been tabulated for publication (Beavers and Eitter, 1986). Included are 16,000 observations of over 2,000 late-type stars. For these observations at the coudé focus of the 61 cm telescope the velocity zero-point is virtually the same as that for the early Lick radial velocity survey (Campbell and Moore, 1928), i.e. Lick-Fick = -0.08±0.08 km/sec. The zero-point comparison for DAO spectrometer measurements is DAO-Fick = -0.54±0.05 km/sec. During the last four years of this study the error of the "high quality dip" observations is ±0.8 km/sec. Thirty-three of these stars have been selected for use as future bright ($m < 7.0$) velocity standards at Fick Observatory.

1. FICK RV STUDIES

The stellar radial velocity observations made at I.S.U.'s Erwin E. Fick Observatory during the period 1976-1984 have recently been collected into a summary for publication (Beavers and Eitter, 1986). This work contains approximately 16,000 measurements made with the 61 cm coudé telescope and photoelectric spectrometer (Beavers and Eitter, 1977). The instrument employs a mask (solar spectrum) in the focal plane of a high dispersion (2.6 A/mm) spectrometer in a configuration inspired by that at Cambridge Observatory (Griffin, 1967).

Analysis of this large data set yields velocity zero-point comparisons with several other radial velocity instruments or collections of measurements. It also reveals a gradual reduction in the observational error for the Fick system to the present value ≈ ±0.8 km/sec for highest quality observations. From within this data sample a list of "constant velocity" stars has been selected to cover all right ascensions (see Table I). These stars are being used to determine the nightly zero-point drift of the Fick instrument. A few gaps are being filled by stars whose constancy, though not yet proven, will be monitored.

NSF support for the Fick Observatory stellar radial velocity

J. B. Hearnshaw and P. L. Cottrell (eds.), Instrumentation and Research Programmes for Small Telescopes, 437–438.

project through grants GP 6083, AST 76-10140, INT 82-05814, and AST 82-05956 is gratefully acknowledged.

TABLE I

FICK OBSERVATORY VELOCITY REFERENCE STARS

HR	RV(KM/SEC)	HR	RV(KM/SEC)
0022	-19.8	5854	2.7
0188	13.4	5947	-32.1
0253	-22.8	6056	-20.0
0402	16.8	6299	-56.2
0603	-10.9	6418	-25.3
0617	-14.5	6603	-12.2
0951	23.2	6973	36.0
1231	61.4[a]	7192	-17.7
1601	14.5	7602	-39.7
2016	45.2	7635	-34.0
2696	-27.7	7753	15.8
2990	3.6	8313	-22.6
3748	- 4.4	8414	6.8
3779	18.6	8498	- 9.0
4377	- 9.3	8795	- 5.2
4932	-14.1	8916	6.2
4954	-15.9		

[a]Velocity Constancy not yet proven

2. REFERENCES

Beavers, W. I., and Eitter, J. J. 1977, Pub. A.S.P. **89**, 733.
Beavers, W. I., and Eitter, J. J. 1986, (submitted to Ap. J. Suppl.).
Campbell, W. W., and Moore, J. H. 1928, Pub. Lick Obs. **16**.
Griffin, R. F. 1967, Ap. J. **148**, 465.

FAGS - A FAST ASTRONOMICAL GRATING SPECTROMETER

B. Denby, R. Dalglish, V. Meadows and K.N.R. Taylor
Astrophysics and Optics Department
University of New South Wales
Kensington, NSW, 2033, Australia.

ABSTRACT. A fast astronomical spectrometer is described using commercially available photographic lenses and a CCD detector. The size and weight of the instrument have been kept to a minimum to allow it to be used on telescopes as small as 0.6m.

1. INTRODUCTION

Despite the wide range of photometric facilities in the Australasian region for studying stellar objects of magnitudes less than about $V = 14$ using small ($\sim$1m) telescopes there are essentially no spectrometers available for carrying out low or intermediate spectral resolution studies of the same objects using the same telescopes. Only the AAT and the MSSSO 74" have versatile spectrometers which can provide resolutions appropriate for spectral classification. Apart from the obvious problems of oversubscription on these telescopes many of the objects in this brightness range would be embarrassingly bright for these spectrometers. At the other extreme, the MSSSO echelle system has a faint object limit of $V \cong 8$ on the 1m telescope at Siding Spring. In consequence objects in the range $8 < V < 14$ are inadequately provided for in spectral studies.

This is unfortunate, since in many fields observations of stars in this brightness range can provide valuable data for an understanding of the phenomena involved. From our own interests we can cite the need for spectral observations in studies of T- and R- association regions, but equally, the need exists in studies of variable stars, HII regions, emission line stars and so on.

2. DESIGN CONSIDERATIONS

To optimise its usefulness, the instrument was designed to be within the engineering and size limitations of a 0.6m telescope. This meant restricting both the mass and the maximum dimensions. Observationally we attempted to gain maximum spectral resolution while reaching limiting

J. B. Hearnshaw and P. L. Cottrell (eds.), Instrumentation and Research Programmes for Small Telescopes, 439–440.

magnitude stars in reasonable exposures of the order of 1000 secs.

The need for maximum possible speed dictated the use of a CCD array detector. The resulting optics, restricted by the field scales and focal ratios of 0.6 and 1.0m telescopes, and the need to match the stellar image and CCD pixel sizes were best satisfied by the use of high quality photographic lenses.

To obtain spectra adequate for classification purposes and at the same time have available sufficient resolution to observe the line profiles of objects such as Be stars, a dispersion of 4Å/pixel was set as a target. Using a Thomson -CSF 7802 linear array, with 1024 13 μm pixels implies a dispersion of approximately 300 Å/mm and an available spectral range of 4000Å. With a short wavelength CCD cut-off of slightly less than 4000Å two centre wavelengths can be used to cover the range to the long wavelength cut-off at 1.05 μm. A Jobyn Yvon 300g/mm grating provide this dispersion in conjunction with a Nikon graphic arts 485 mm F/9 collimator lens and a Canon 85 mm F/1.2 camera lens.

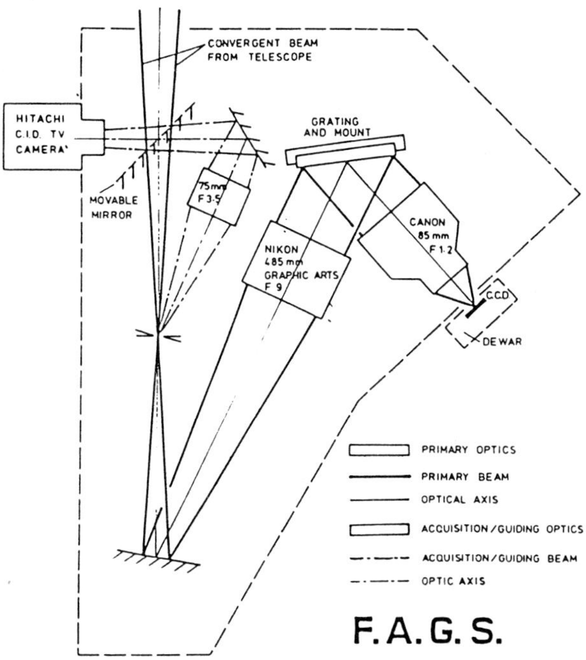

Figure 1. Ray Diagram of the Spectrometer

The final optical assembly is shown in Fig. 1. The maximum dimension (length) is only 68cm, the remaining dimensions lie within a circle of radius 30cm and hence within the lateral dimensions of a 0.6m telescope tube. Object acquisition is achieved using a solid state C.I.D. T.V. camera and diagonal mirror, while guiding is carried out by off-axis refocussing onto the stellar image on the front of the slits.

The data collection system uses dedicated peripheral hardware with its own memory and bus system. It is organised as a 24-bit, 1K (1024) memory block with its own parallel arithmetic processor. The memory block is arranged however, such that it can be accessed from the host microcomputer in single 8 bit words; this shared access to the memory is totally transparent for the computer and the data collection system. The system can be controlled with simple software, to add or subtract the data stream and to control the spectrometer choppers and shutters. All aspects of control and process monitoring are achieved via conventional I/O ports in the computer's address space.

The final cost, including the VVD and its signal processing hardware, but excluding the Apple computer and C.I.D. T.V. camera was approximately $3000.

MULTIPLE STELLAR SYSTEMS

A. Duquennoy and M. Mayor
Observatoire de Genève
CH-1290 Sauverny-Versoix
Switzerland

A spectroscopic survey of visual binaries with known orbital elements has been carried out with the radial velocity scanner CORAVEL at the Haute-Provence Observatory, since 1977, (Baranne, Mayor, Poncet, 1979). This survey of more than 100 visual systems, selected from Dommanget's catalogue (1967) (see also a new edition 1982) was first devoted to the determination of stellar masses. Several multiple systems were detected and have permitted also a study of the structure of triple systems. We have detected and measured in particular a class of triple systems with radial velocity variations of small amplitude. Taking advantage of the high resolution and high signal-to-noise ratio accessible with the cross-correlation technique, such small amplitude radial velocity curves are sometimes derived only through the change of width and shape of the cross-correlation function. Let us recall that the cc-function of a SB2 (or SB3) system is only the weighted sum of the individual cc-functions (Mayor, 1985). This property of the cross-correlation combined with the linearity of the detector allow a very simple analysis of blended dips. The full width at half depth of the cross-correlation dip is about FWHD = 16 km/s (in absence of noticeable rotation). Analysis of blended systems allows a good determination of the two individual velocities if the difference $|V_{r_1} - V_{r_2}|$ is equal or larger than about 0.15 *FWHD (about 2 km/s).

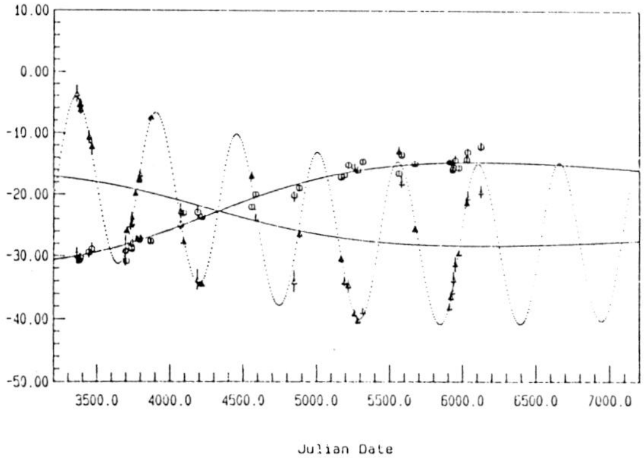

Fig.1 : HD 214608 a triple system with frequent small velocity differences between the two components.
A : G3V + K3V:
B : F8V
P_{Aa-Ab} = 1.5 year = P_{short}
P_{A-B} = 30 years = P_{long}

J. B. Hearnshaw and P. L. Cottrell (eds.), Instrumentation and Research Programmes for Small Telescopes, 441–442.

A 1-meter telescope, dedicated only to radial velocity measurements, gave us favorable conditions to discover and measure such systems with multiple periods ranging from days to years and sometimes very eccentric orbits. HD 214608 is a good example of a triple system with $|V_{r_1} - V_{r_2}|$ almost always less than FWHD (see fig.1). The long period (visual) is 30.0 years and the short period (spectroscopic) 1.5 years.

Considering the position of representative values for triple systems in a log P_{long}-log P_{short} plane figure 2, we are inclined to see a bimodal distribution. If most triple systems have a highly-hierarchical structure with spectroscopic period (P_{short}) much smaller than visual period (P_{long}), a non-negligible number of systems show only a factor of 10 to 100, between the two periods. Selecting a domain of P_{long} (for example $17y < P_{long} < 170y$) we can clearly see the bimodal distribution of P_{short}. (A bimodal distribution also exists for $P_{long} > 170$ years, the two periods in such systems only being determined by visual observations). The gap between the two modes (fig.3) of the distribution cannot be explained by observational bias. Such a bimodal distribution describing the geometry of triple systems can result from initial conditions or dynamical evolution.

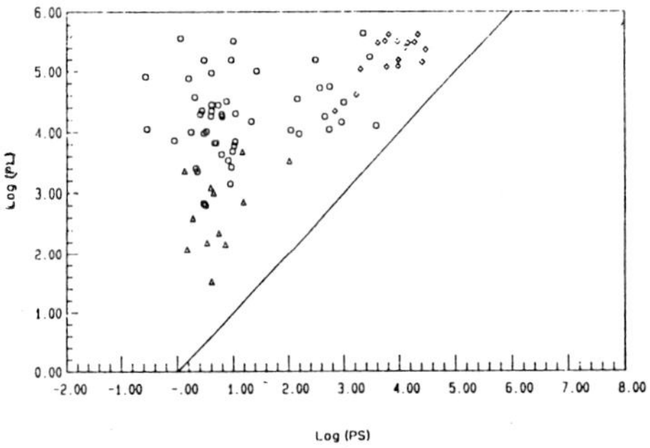

Fig.2 : Distribution of multiple stellar systems in a log P_{long} - log P_{short} plane. Δ=spectroscopic systems, O=spectroscopic-visual systems, ◇=visual systems.

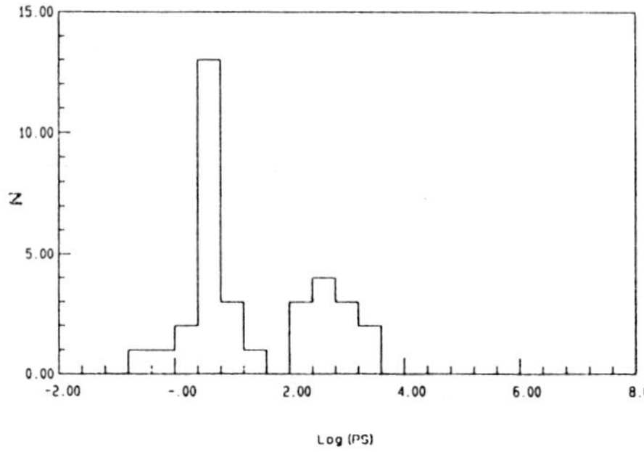

Fig.3 : In triple stellar systems having $17y<P_{long}<170y$, the distribution of short periods is bimodal.

Bibliography :

Baranne, A., Mayor, M., Poncet, J.L., 1979, *Vistas in Astronomy*, 23, 279

Dommanget, J., Nys, O., 1967, *Comm. Obs. Roy. Belgique*, Ser. B Nº 15

Dommanget, J., Nys, O., 1982, *Comm. Obs. Roy. Belgique*, Ser. B Nº 124

Mayor, M., 1985, "Cross-correlation spectroscopy" *in Stellar Radial Velocities, IAU coll. 88*, Ed. A.G.Davis Philip and D.W. Latham (Davis Press)

SPECTROSCOPIC SURVEILLANCE OF THE VARIABLE HERBIG Ae STAR AB AUR

U. Finkenzeller
Astronomy Department,
University of California,
Berkeley

and

J.G.V. Schiffer, H. Mandel
Landessternwarte,
Heidelberg-Königstuhl

AB Aur = HD 31293 = BD + $30^0 741$ = MWC 93 = HRC 78n is one of the brightest members of a class of pre-main sequence objects introduced by G. Herbig in 1960. The star is characterized by a spectral type of A0Ve, $L/L_\odot = 70$, $m_v = 7.2$ mag, $v \sin i = 140$ km/s, and d = 140 pc. Assuming a stellar radius of a main sequence dwarf $R/R_\odot = 3.0$, one can estimate the rotational period to be on the order of 28 hours if we see the system edge on. Previous work (Praderie et al. (1985) and Finkenzeller (1983)) has focused on high resolution spectroscopy and shown that AB Aur is subject to short term variations. Finkenzeller's observations have indicated that the emission features can disappear completely within a day. Praderie et al. found evidence of a rotational modulation (45 ± 6 hours) of the MgII λ 2795 and FeII UV lines which is interpreted as due to a non-axisymmetric wind interacting with the outer atmosphere.

Since AB Aur is a typical representative of the Herbig Ae and Be type stars, reasonably bright, and easily accessible we have initiated a spectroscopic surveillance program to monitor optical line variations. We were guided by the wish to achieve a temporally well resolved and contiguous data set in order to investigate short term variability ($t \leq 1h$) and to separate a superposed rotational modulation. It was hoped that the analysis of these data would yield an activity period which would be interesting to compare with the rotational period as deduced from photospheric lines.

All observations were carried out with the Boller and Chivens spectrograph at the 72 cm Waltz reflector of Landessternwarte Heidelberg-Königstuhl. The spectrograph can be used in conjunction with a two or three stage magnetically focused EMI image tube. (Work is under way to replace this detector by a linear diode array). The intensified spectrograph of the Landessternwarte is an ideal instrument to perform the surveillance project outlined above.

We confined our studies to H_β. Typical exposure times were on the order of 20 minutes with the two stage image tube. The spectra have a resolution of R ≈ 2000 at 22 Å/mm and were recorded on IIaO plates. After digitising the data with a Grant microdensitometer the reduction followed standard procedures. A subset of the final

J. B. Hearnshaw and P. L. Cottrell (eds.), Instrumentation and Research Programmes for Small Telescopes, 443–445.

spectrograms is displayed in fig. 1. In order to have a direct measure of the degree of activity, the residual emission strength has been determined by subtracting an undisturbed photospheric line profile, based on a Kurucz model. The variation of this value, corresponding to the spectrograms of fig. 1, is given in fig. 2.

Clearly, AB Aur did vary appreciably during the time covered by the program. A markedly on/off event as observed by Finkenzeller (1983) in February 1981 could not be seen. Changes can be verified down to scales of about an hour. It would be premature, however, to assign any period to the data values available at this point.

The presented observations should be considered as a pilot study to monitor fast rotators of intermediate mass among the pre-main sequence stars. These first results are encouraging and we intend to continue the program.

References

Finkenzeller, U. 1983, *Astron. Astrophys.* 124, 157
Herbig, G.H. 1960, *Astrophys. J. Suppl. Ser.* 4, 337
Praderie, F., Simon, T., Catala, C., Boesgard, A.M., 1985, *Astrophys. J.*, preprint

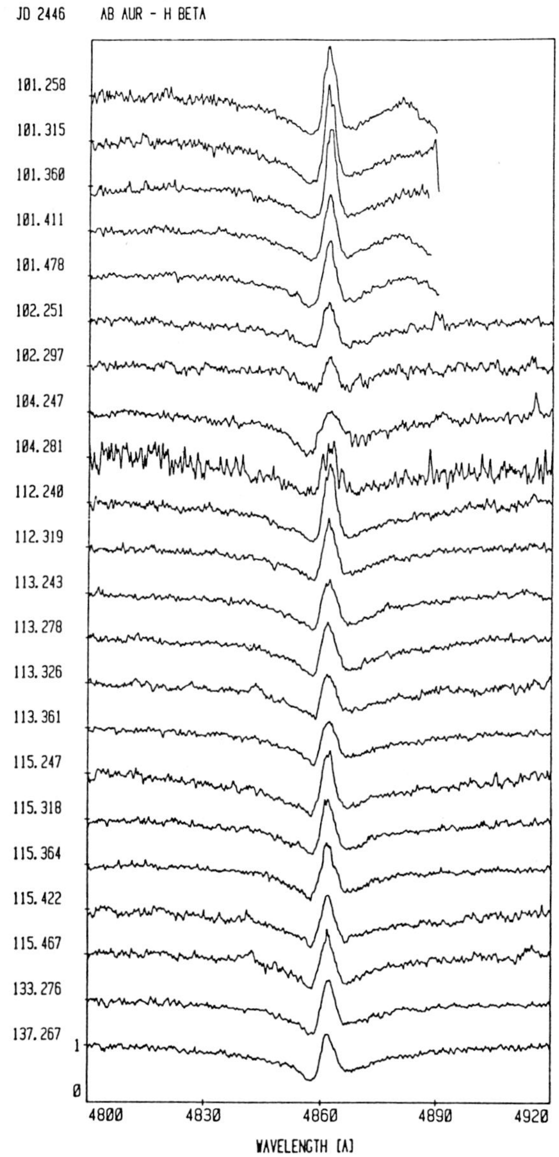

Fig.1 (right). Line profile variations of the Herbig Ae star AB Aur at $H\beta$. The numbers on the ordinate give J.D. 2446... Appreciable changes occurred within a fraction of a day.

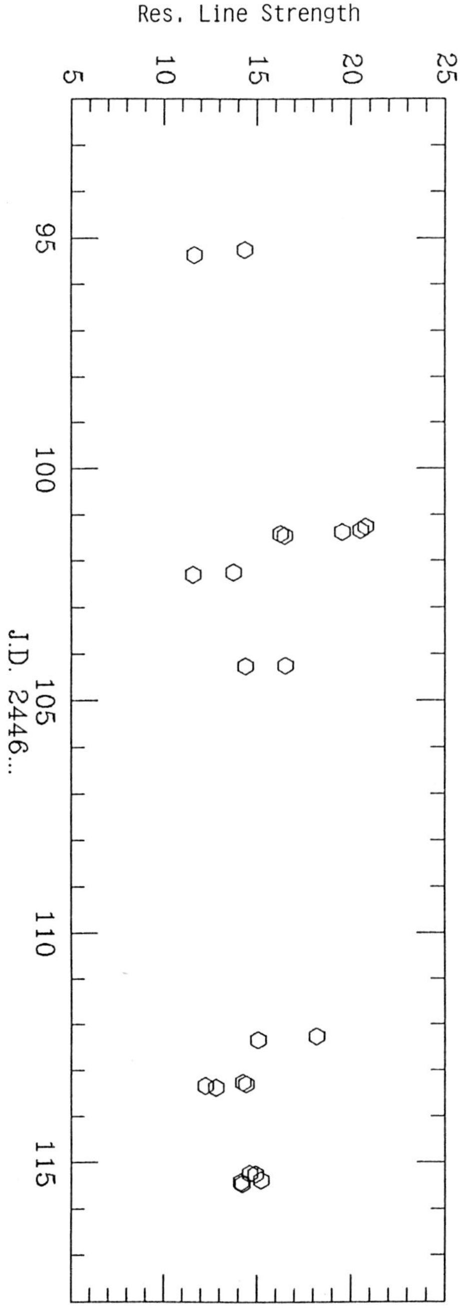

Fig.2 (left). Residual emission line strength after subtraction of an undisturbed photospheric line profile for the spectrograms shown in fig.1.

RADIAL VELOCITY SPECTROMETERS ON THE DOMINION ASTROPHYSICAL OBSERVATORY 1.2m AND 0.4m TELESCOPES

J.M. Fletcher and Robert D. McClure
Dominion Astrophysical Observatory
Victoria, B.C. Canada

1. 1.2m COUDE SPECTROMETER

A photoelectric radial velocity spectrometer has been in operation for several years at the coudé focus of the 1.2m telescope of the Dominion Astrophysical Observatory (DAO). The design of this instrument has been described by Fletcher *et al.* (1982) and by McClure *et al.* (1984). These papers discuss the similarities and differences between this spectrometer and others of its type such as that constructed by Griffin and Gunn (1974) at Palomar. The spectrum mask in the present instrument consists of more than 700 transparent slots on an opaque background, representing the spectrum of a K giant star, plus a few slots to coincide with the argon lines in a cadmium-argon discharge tube for use in obtaining comparison arc velocities. By progressively tilting the slots an appropriate amount along the mask and scanning by moving the mask relative to the stellar spectrum at 45° to the dispersion it is possible to retain a spectral match at non-zero velocities. The masks are produced using a standard measuring engine modified with stepping motors controlled by a computer to position and tilt the slots. Since guiding errors are our major source of error we hope to make a significant improvement in this area in the following manner. By inserting an image rotating prism behind the slit a 180° rotation of the slit will occur when the prism is rotated. Although this has not yet been tested, it is hoped that by making this rotation every few scans the effects of guiding errors due to uneven illumination of the slit and collimator will be vastly reduced.

One of the main uses of the 1.2m radial velocity spectrometer has been to accurately monitor the velocities of carbon and related stars such as Ba II and CH stars. The significance of this work has been discussed recently by McClure (1985); to summarize, the Ba II and CH stars appear to be almost all spectroscopic binaries, whereas the R type carbon stars appear to exhibit a normal frequency of multiplicity. Numerous other spectroscopic binary programs are also being pursued by various observers, the instrument being especially suited for such work where an approximate velocity can be predicted beforehand so that an inordinate range of velocity need not be scanned. Another research area for which the instrument is being used extensively is in cluster velocity dispersion and membership studies. Numerous velocities have

J. B. Hearnshaw and P. L. Cottrell (eds.), Instrumentation and Research Programmes for Small Telescopes, 447–448.

been obtained in the old open clusters NGC 188 and NGC 7789 and in the globular clusters M 71 and M 3 among others.

2. 0.4m SPECTROMETER

A second photoelectric radial velocity spectrometer has recently been constructed at the DAO for operation at the cassegrain focus of the 0.4m telescope. This telescope has a small off-axis spectrograph designed by Richardson and Brealey (1973), which has been modified to incorporate a Pechan roof prism which can be scanned behind the slit. This has the effect of scanning the spectrum back and forth at the focus of the spectrograph where a stationary spectrum mask is placed. The prism is scanned at 45° to the spectrum dispersion and the tilt of the slots in the mask again provides a match to the spectral dispersion as in the coudé instrument. The mask includes over 300 lines in the wavelength region 415-485 nm based on the spectrum of a K giant star, plus a few argon lines to match a cadmium-argon comparison spectrum. The light is collected at the photocathode of a photomultiplier tube, the grating being imaged by a simple fabry lens. The modifications to the spectrograph to make this velocity spectrometer were accomplished by the purchase of a few components such as a photomultiplier tube chamber, accurate bearings to hold the scanning screw, and the Pechan prism, and by a modest amount of machining in the observatory instrument shop. The control of the scanning and photon counting is done with an Apple II computer.

This velocity scanner has recently been tried on the 0.4m telescope with very promising results, but it is not yet in use. The accuracy appears to be about 2 km s^{-1} if care is taken to obtain comparison velocities before and after each stellar observation. The instrument is inconvenient to use in this form, however, because flexure in the spectrograph moves the zero point of the velocities significantly with position in the sky. As with photographic velocity measurements, therefore, an observation of the comparison spectrum must be made before and after the stellar observation. To eliminate many observations of the comparison and reduce guiding errors, we have been experimenting with feeding the spectrograph by an optical fiber so that the spectrograph can be placed in a stationary position nearby. We do not have extensive observations, but from preliminary measurements it appears that the end result will be an improvement in the accuracy of the velocities to about 1 km s^{-1}.

REFERENCES

Fletcher, J.M., Harris, H.C., McClure, R.D., and Scarfe, C.D., 1982, Publ. Astron. Soc. Pacific **94**, 1017.

Griffin, R.F., and Gunn, J.E., 1974, Astrophys. J. **191**, 545.

McClure, R.D., 1985, J. Roy. Astron. Soc. Canada, December issue.

McClure, R.D., Fletcher, J.M., Grundmann, W.A., and Richardson, E.H., 1985, in Stellar Radial Velocities, IAU Colloq. 88, eds. A.G.D. Philip and D.W. Latham (L. Davis Press, Schenectady), p 49.

Richardson, E.H., and Brealey, G.A., 1973, J. Roy. Astron. Soc. Canada **67**, 165.

TEMPERATURE AND BRIGHTNESS VARIATIONS ON BETELGEUSE

John E. Gaustad
Department of Astronomy
Swarthmore College
Swarthmore Pennsylvania 19081
U.S.A.

ABSTRACT. Changes in TiO band strengths correlate well with the brightness changes of α Orionis, thus supporting the hypothesis of Schwarzschild that the irregular luminosity variations of red giants are due to temperature changes in a few extremely large convective elements on their surface.

Schwarzschild (1975) suggested that because of the large depth of the outer convection zone in red giants, the dominant convective element diameter is comparable to the radius of the star. This was confirmed quantitatively by Antia, Chitre, and Narasimha (1984). Thus, the surface may contain only a few granules, and changes in the character of just one might produce changes in the integrated light of the star. It is well-known that the spectral type of α Ori is variable. To investigate whether this variation correlates with the brightness changes, I have used a 61-cm telescope and scanning spectrophotometer to obtain high-resolution (0.3Å) spectra of α Orionis at frequent intervals during the period March 1984 to April 1985. I determined the temperature from the strength of the R- and Q-branch bandheads of the β(0,0) system of TiO, measuring the equivalent width of the absorption beneath a pseudo-continuum drawn between the relative maxima at 5596Å and 5609Å.

For such a low temperature star, the V and B magnitude effective wavelengths lie on the short-wavelength side of the Planck function maximum, where the intensity varies rapidly with temperature. Therefore a fairly accurate calibration of the spectra is required. To make such a calibration, I obtained spectra at approximately the same resolution of several red giants, ranging in spectral type from M1.5 to M4.0, using the stellar spectrograph and Reticon detector of the McMath telescope at the National Solar Observatory on Kitt Peak. The spectral types (Wing 1978) were converted to effective temperature using the calibration given in Table 4 of Ridgway et al. (1980). This procedure is similar to that used by Ramsey (1981).

The effective temperature for α Orionis was determined for each of the 28 nights of observation. Means calculated for each successive 64-day

J. B. Hearnshaw and P. L. Cottrell (eds.), Instrumentation and Research Programmes for Small Telescopes, 449–450.

period are plotted in the Figure, along with B-magnitudes of α Orionis as observed by E. Guinan (1985) at Villanova University over the same period. The magnitude and temperature scales have been matched so that a black body of constant area undergoing a temperature change indicated by the right scale would show the change in intensity at 4500 Å indicated on the left scale. Thus the data can be directly compared in terms of the following question: can the brightness changes on α Orionis be explained solely by a change in temperature, with no change in surface area? The temperature is indeed lower when the luminosity is lower, and by about the right amount, giving some weak support to Schwarzschild's hypothesis that the irregular light variations of red giants are due to temperature changes of only a few supergranules on their surface.

REFERENCES

Antia, H.M., Chitre, S.M., and Narasimha, D. 1984, Ap.J. 282, 574.
Guinan, E.F. 1985, private communication.
Ramsey, L.W. 1981, Astron. J. 86, 557.
Ridgway, S.T., Joyce, R.R., White, N.M., and Wing R.F. 1980, Ap.J. 235, 127.
Schwarzschild, M. 1975, Ap.J. 195, 137.
Wing, R.F. 1978, "Spectral Classifications and Color Temperatures for 280 Bright Stars in the Range K4-M8" (unpublished report).

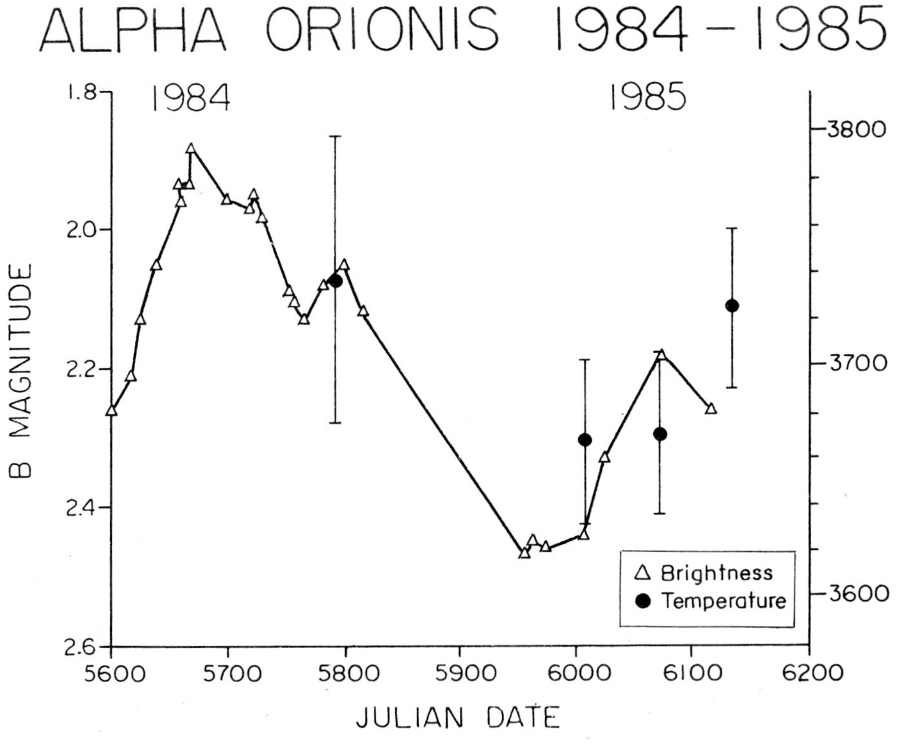

AN OFF-THE-SHELF RETICON SYSTEM

R.K. Honeycutt
Indiana University
Bloomington, Indiana 47401
U.S.A.

and

B.F. Peery
Howard University
Washington, DC 20059
U.S.A.

ABSTRACT. EG&G Princeton Applied Research produces a Reticon detector system intended for laboratory applications. However, current versions of the system are well-suited to astronomical observations at observatories with modest financial and technical resources.

The detector inputs are intensified by microchannel plates, fiber-optically coupled to a 1024-pixel Reticon. Controller and readout electronics, and software for data pre-processing and graphical display are also available off-the-shelf, enabling the observer to put the system to use without a major development effort.

The Optical Multichannel Analyser is a linear silicon-diode array detector system produced by Princeton Applied Research, Inc., of Princeton, New Jersey, U.S.A. The OMA has found application in laboratories, especially in spectroscopic studies of fast chemical reactions, but has been utilized by few astronomers for astronomical spectroscopy. The system is available as an off-the-shelf unit with all required electronics and software. It would seem to merit scrutiny by smaller observatories as a developed reticon system requiring relatively modest technical and financial resources to be put to use straightforwardly. We present here our preliminary evaluations of recent intensified versions of the PAR reticon as astronomical detectors.

The model 1420 detector is a 1024-pixel silicon diode array intensified by a proximity focussed microchannel plate and phosphor screen, fiber-optically coupled to the reticon. The size of the microchannel plate illuminates only either 512 or 700 pixels. A plane photocathode injects photoelectrons into the microchannel plate of model 1420. In the more recent model 1422 detector, photoelectrons are electrostatically focussed upon the microchannel plate. The microchannel plate and phosphor screen are large enough to illuminate all 1024 pixels of the reticon. Either detector can be powered, controlled and computer-interfaced by PAR model 1461 interface electronics package and model 1463 controller boards.

Table I presents a number of parameters of the PAR Reticon System.

The PAR Reticon Detectors incorporate a thermoelectric cooling

J. B. Hearnshaw and P. L. Cottrell (eds.), Instrumentation and Research Programmes for Small Telescopes, 451–452.

system that can achieve cooling to about -40 deg C. Even at that temperature, however, the dark current may still be too large for some applications. Table II presents extrapolated numbers of noise counts (less signal shot noise) at -40 deg C and at dry ice temperature (-70 deg C). We have subjected a sample assembly to temperatures near -150 deg C without impairment of performance. We conclude that dry ice cooling is unlikely to be hazardous to these detectors.

This investigation has been supported by NASA Grant NAGW-545.

Table I Parameters of PAR Models 1420 and 1422 Detectors

Geometric Distortion	2%
Number of Pixels	1024
Pixel Size	25 micrometers x 2.5 mm
Read out Noise	1 Electron/Pixel
Dark Current (at 5 deg C)	900 Counts/sec
Approximate Cost, (Electronics and Detector)	$US 31,000 (Model 1420) $US 25,000 (Model 1422)

Table II Predicted noise counts (less signal shot noise)

Integration time (sec):	-40 deg C	Counts -70 deg C	-100 deg C
1	3.2	1.2	1.0
10	10	2.5	1.1
100	32	7.4	1.9
1,000	102	23	5.3
10,000	322	73	17

STELLAR SPECTROPHOTOMETRY WITH A MICROCOMPUTER BASED, INTENSIFIED, SILICON VIDICON

S. Jeffers and T. Stiff,
Physics Dept. and Centre for Research in Experimental Space Science,
York University, 4700 Keele St.,
North York, Ontario, Canada, M3J 1P3.

1. INTRODUCTION

We have developed several detector systems for use in observational astronomy. These include a rapid scanning spectrometer (Jeffers and Weller, 1973), a rapid chopping spectrophotometer (Stiff and Jeffers, 1978) and an intensified, silicon vidicon detector for use either for spectrophotometry (Jeffers, Stiff and Weller, 1983) or direct imaging (Jeffers, 1982). These detector systems have been used for rapid variability studies of emission line strengths and profiles in Wolf-Rayet and Of stars (Weller and Jeffers, 1979; Jeffers, Stiff and Weller, 1985) and for low dispersion spectrophotometry of these stars (Jeffers and Weller, 1985). Here we report on the spectrophotometric performance of the intensified, silicon vidicon detector (RCA 4804H) when used with a low dispersion spectrograph.

2. SPECTROPHOTOMETRIC CALIBRATIONS

In order to recover the stellar, spectral energy distribution above the earth's atmosphere it is necessary to correct for the effects of (I) differential extinction in the Earth's atmosphere and (II) the wavelength dependence of the instrumental sensitivity.

2.1 Determination of wavelength dependence of extinction in the Earth's atmosphere

The extinction star used was HR7202, B5V, mv = 5.96. On July 27/28, 1985 this star was observed at eight different air masses through a range from 1.03 to 2.28. The integration time was typically 72 seconds giving a S/N ratio $\sim$ 150 at maximum. The spectrophotometric standard star HR6779 (B 9.5 V, mv = 3.83) was also observed (integration time 49.5 secs, S/N = 200 at maximum signal). The error in the measurement of individual extinction coefficients is $\sim \pm 5\%$. A plot of extinction coefficients vs $^1/\lambda^4$ was made and compared with that for pure Rayleigh scattering (R = .01085 $\pm$ .0021 mags/air mass/$^1/\lambda^4$ (van den Bergh and

J. B. Hearnshaw and P. L. Cottrell (eds.), Instrumentation and Research Programmes for Small Telescopes, 453–455.

Henry, 1962) quoted for the nearby David Dunlap Observatory.) The maximum discrepance between the two is ~ 35%. Clearly the assumption of Rayleigh scattering is a poor one at least for an urban atmosphere where there may be an appreciable contribution to extinction from aerosols.

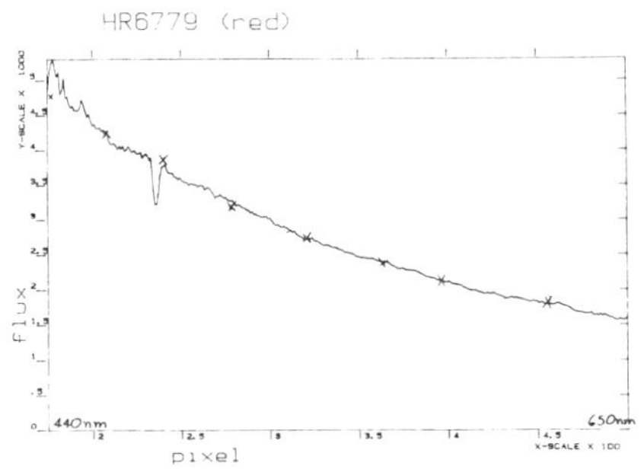

Figure 1. Reduced spectrum of HR6770. Crosses are from Breger's spectrophotometry

2.2 Instrumental Response Function

The instrumental response was determined from extinction corrected data for HR7202 which is a spectrophotometric standard star (Breger, 1976). Each extinction corrected data set was divided by a fit determined by Lagrangian interpolation to the published spectrophotometric data to yield the instrumental response. A mean instrumental response was computed and used in the reduction of the data for HR6779.

3. CONCLUSIONS

A low dispersion spectrophotometer incorporating an intensified silicon vidicon detector is being used for astronomical spectrophotometry. To recover accurate spectrophotometric data (± 1%) over the range of sensitivity of the detector, careful determinations of the wavelength dependence of atmospheric extinction must be made. Dependence on an assumed value of a Rayleigh coefficient leads to substantial errors.

REFERENCES

Breger, M., 1976, *Ap. J. Supp.* 32, 7-87.

Jeffers, S. and Weller, W.G., 1985, *Astronomy and Astrophysics Supplement Series*, 61, 173-183.

Jeffers, S., Stiff, T. and Weller, W.G., 1985, *Astron. J.*, 90, No. 9, 1852-1854.

Jeffers, S., 1982, *J. Roy. Astron. Can.*, 76, #1.
Jeffers, S., Stiff, T. and Weller, W.G., 1983, *Proceedings of Symposium on Instrumentation in Astronomy V*, 445, 199-203.
Jeffers, S. and Weller, W.G., 1973, *Proceedingof Symposium on Astronomical observations with television type sensors*, Eds. J.W. Glaspey and G.A.H. Walker.
Stiff, T. and Jeffers, S., 1978, *Applied Optics*, 17, 1811-1814.
van den Bergh, S. and Henry, R.C., 1962, Pub, DDO, 2, 281.
Weller, W.G. and Jeffers, S., 1979, *Proceeding of Symposium on Mass Loss and Evolution of O-type stars*, 27-30, Eds. P.S. Conti and C.W.H. de Loore.

A MICROCOMPUTER CONTROLLED CCD H-ALPHA SPECTROMETER

David A. McDavid
Division of Earth and Physical Sciences
University of Texas at San Antonio
San Antonio, Texas 78285

ABSTRACT. A miniaturized spectrometer designed for mounting at the Cassegrain focus of a 40 cm telescope has been built and laboratory tested with promising results. The optical system consists of a slit, spherical mirror, and reflection grating in a Littrow configuration. The detector is a 128 element linear CCD cooled by an evacuated microminiature nitrogen refrigerator and interfaced to a microcomputer using direct memory access. Laboratory spectra of a mercury vapor discharge tube have been obtained and demonstrate 1 Angstrom resolution with low noise. The spectral coverage of about 100 Angstroms is suitable for measurement of stellar H-alpha line profiles, and it is anticipated that the instrument can be used to monitor bright Be stars.

1. OPTICS

Since compactness and light weight are necessary for an instrument to be carried by a small telescope, a Littrow mount, in which a single mirror is used as both collimator and camera, was chosen. A spherical mirror of 60 mm diameter and 200 mm focal length accepts the full f/12 beam of the 40 cm Cassegrain telescope and brings the full diffracted beam to focus at the detector. A plane reflection grating 30 mm square with 600 lines/mm and 650 nm blaze wavelength is used at an angle of about 11 degrees to give a reciprocal linear dispersion of approximately 80 Angstroms/mm. The entrance slit is 13 microns wide on aluminum coated glass, tilted slightly so that a diagonal mirror can reflect an image by way of a relay lens to the guide eyepiece mounted at the side of the chassis. The slit width was chosen to cover one pixel width at the detector, resulting in 1 Angstrom resolution. It subtends about 0.5 arcsec on the sky. All of the optical elements are held in place by adjustable brackets machined from quarter inch aluminum plate. The spectrometer chassis is a box 12" long, 7" wide, and 5" deep with a 10" x 12" telescope mounting flange, also made from quarter inch aluminum plate.

J. B. Hearnshaw and P. L. Cottrell (eds.), Instrumentation and Research Programmes for Small Telescopes, 457–458.

2. ELECTRONICS

A Texas Instruments virtual phase 128 element CCD linear array is used as a detector. The pixels are 12.7 microns square, so a spectral interval of about 128 Angstroms is covered. Operation of the chip is supported by a factory supplied driver board mounted on one inside wall of the chassis. An onboard interface card designed around a National Semiconductor ADC 0820 8 bit half flash A/D converter digitizes and transfers the data to a Compaq microcomputer at 250 kHz through a parallel port using direct memory access. Memory refresh is suspended during the transfer so that no pixel values are lost. Exposure timing is done by a programmable interval timer chip connected to the clock output provided on the TI driver board. The TTL power supply is taken from the computer bus, and the driver board is powered by a 16 volt dual power supply mounted in the spectrometer chassis.

3. REFRIGERATION

To eliminate thermal noise, the CCD is cooled to -100 degrees Celsius by a microminiature refrigerator from MMR Technologies. This consists of a small vacuum chamber containing a glass strip on which the chip is mounted. Pressurized nitrogen gas expands through capillaries in the glass and liquefies beneath the chip, after which it recirculates to cool the incoming gas before venting at atmospheric pressure. The temperature can be controlled by a resistive heater embedded in the glass strip. The chamber is continuously evacuated while the instrument is on the telescope, requiring a high capacity vacuum pump because of the length of the connecting hose.

4. PRELIMINARY TEST RESULTS

In the laboratory, the spectrometer very cleanly resolved the 577-579 nm mercury doublet from a vapor discharge tube through a foil pinhole focused on the entrance slit, demonstrating the estimated resolution of 1 Angstrom per pixel. It was only recently adjusted for the H-alpha line and mounted on the telescope, and the results have been very discouraging. The electronics and refrigeration systems work perfectly well, but attempts to locate the sensitive point on the slit using Jupiter, the Moon, and even the bright star Sirius have failed completely. At the deadline time for this writing, the original slit has been replaced by a much wider one (150 microns, or about 6 arcsec) in a compromise of resolution for illumination. Regardless of slit width, photometric accuracy will be difficult because the linear detector array is so narrow. One solution to this problem might be running the star up and down along the length of the slit during exposure. Such a motion could be made quite regular since the telescope itself is computer controlled. Unfortunately, testing has been impossible because of a long period of cloudy weather.

A LONG-TERM RADIAL VELOCITY PROGRAM OF HIGH ACCURACY WITH A SMALL TELESCOPE

R. S. McMillan, P. H. Smith, J. E. Frecker, W. J. Merline, and M. L. Perry
Lunar & Planetary Laboratory, University of Arizona
Tucson, Arizona 85721 U.S.A.

We are making accurate observations of the change in Doppler shift of stellar absorption lines. The purpose is to detect the oscillatory reflex motion due to planets orbiting stars. The scrambling of incident light by an optical fiber and the stability of wavelength calibration by a Fabry-Perot etalon provide immunity to systematic errors. Selecting several echelle diffraction orders in the vicinity of 4250-4600 A, which are imaged on a CCD, about 350 points on the profile of the stellar spectrum are sampled by successive orders of interferometric transmission through the etalon. At 4300 A each interference order is 47 milliangstroms wide and the sample points are 0.64 A apart, causing distinct, widely-spaced monochromatic images of the entrance aperture to be formed in the focal plane of the camera. Changes in Doppler shift modify the relative intensities of these images, according to the slope of the spectral profile at each point sampled. To simplify operation and enhance sensitivity, the instrument is being operated as a null-measurement accelerometer, responding only to changes in radial velocity. With an argon emission line lamp the interferometer is calibrated to two parts in 100 million; this corresponds to ± 6 meters/sec in Doppler shift. These calibrations show instrumental variations of ± 27 meters/sec on a time scale of months; observations of stars are corrected for such changes. The internal repeatability of observations of the differential Doppler shift of Arcturus (K1 IIIb; B=1.19) is ± 6 meters/sec for each exposure of 600 square meter-seconds. These exposures are obtained in 15-20 minutes with a 0.9-meter telescope. The external repeatability (day-to-day differential accuracy) of nightly averages of stellar observations is ± 20 meters/second.

Frecker, J. E., Gehrels, T., McMillan, R. S., Merline, W. J., Perry, M. L., Scotti, J. V., and Smith, P. H. 1984. "A CCD system for direct and spectroscopic images", in Proceedings of the NASA/SDSU Workshop on Improvements in Photometry, San Diego, CA, 18-19 June; W. Borucki and A. Young, eds., NASA CP-2350, 137-151.

McMillan, R. S., Smith, P. H., Frecker, J. E., Merline, W. J., and Perry, M. L. 1985, "The LPL radial accelerometer", in Proceedings of

J. B. Hearnshaw and P. L. Cottrell (eds.), Instrumentation and Research Programmes for Small Telescopes, 459–460.

I.A.U. Colloquium No. 88, Stellar Radial Velocities, A. G. Davis Philip and D. W. Latham, eds. (New York: L. Davis Press), 63-86.

McMillan, R. S., Smith, P. H., Frecker, J. E., Perry, M. L., Castillo, N. D., and Merline, W. J. 1983. "A high accuracy radial velocity spectrometer", in Technical Digest of the Joint AAS/OSA Topical Meeting on Information Processing in Astronomy and Optics, June 23-24, St. Paul, MN (privately published by the Optical Society of America).

Merline, W. J., 1985. "Radial Velocity Information in Solar-Type Spectra", in Proceedings of I.A.U. Colloquium No. 88, op. cit., 87-98.

THE DDO PHOTON-COUNTING SPECTROMETER: A WELL-DRESSED SHECTOGRAPH

S. W. Mochnacki, S. Chew, W. Kunowski, F. Hawker, K. Kamper, D. Blyth, L. Zerafa and A. Platzer

David Dunlap Observatory, University of Toronto, Box 360, Richmond Hill, Ontario L4C 4Y6, Canada

ABSTRACT. We have built a Shectman-type photon-counting spectrometer based on the Latham-Geary "Z-Machine" design. The system is innovative in that the entire system rides on the telescope and communicates to the outside world via standard serial lines. Microprocessor architecture, memory controllers and new interconnect and packaging technologies have been employed. The system is portable and can be easily connected to almost any computer. It is suitable for calibrated spectrophotometry on a smaller telescope as well as for conventional spectroscopy.

1. INTRODUCTION

In the ten years since the prototype was developed (Shectman and Hiltner 1976), a number of "Shectographs" have been built and put to use on medium to large telescopes. The articles of Latham (1982), Allen et al. (1983) and Shectman (1983) describe the evolution of the design, at first using longer one-dimensional Reticon arrays and later using a two-dimensional CCD as the sensor element. At the DDO, we needed a system for one-dimensional Cassegrain spectroscopy on our 1.9 metre telescope, with sky subtraction. We chose tha CfA design (Latham 1982) for the degree of refinement of its electronics and because Drs. Latham and Geary of CfA were relatively nearby and very willing to help us.

2. SPECIFICATIONS

Our particular requirements were: (a) to detect spectra at a resolution of 25 microns or better, over a detector length of 25 mm or more, of stars as faint as 15th magnitude; (b) to measure radial velocities to a precision of ± 1 km s^{-1} at a reciprocal dispersion of 16 A mm^{-1}; and (c) to simultaneously detect the object and the sky through equivalent apertures or slits. These requirements could be met by a "Shectograph" system with the following specifications:

Image Tubes	2 x VARO 8605 40 mm "First Generation" diodes + 1 x VARO 3603 25 mm "Second Generation" MCP tube.
Quantum Efficiency	8% at 4000 A, 14% at 5000 A, 4.5% at 8000 A.

J. B. Hearnshaw and P. L. Cottrell (eds.), Instrumentation and Research Programmes for Small Telescopes, 461–462.

Usable Range	3800 – 8600 A (quantum efficiency > 1%).
Operating Temp.	-35°C, thermoelectric cooling.
Sensor Element	2 x Reticon RL1024-SF, each diode 25μmx2.5mm
Clock Frequency	0.5 MHz, Reticons read out in parallel.
Digitization	8 bit A/D, with frame subtraction.
Event Detection	1/4 diode event centring. Adjustable discrimination.
System Resolution	4 pixels (1 diode,25 microns), design goal 3 pixels.
Output Format	2 x 4096 pixels (A and B sides).
Coincidence Error	1% at 0.5 events sec^{-1} $pixel^{-1}$.
Max. Count Rates	2 events sec^{-1} per pixel, <u>or</u> 4000 events sec^{-1} per side.

3. DESIGN DETAILS

We have followed closely the well-proven design of Latham and Geary, with innovations in certain areas. We have packaged the entire photon counting electronics system into a single box which rides on the telescope. The on-board 6809-based microcomputer is commanded via a serial line using a standard terminal, and the data is sent to a host computer via another serial line. Analogue components are mounted on custom-made printed circuit boards, but the digital components are mounted and interconnected on BICC-Vero Eurocards using Speedwire technology. The cards are connected with DIN pin connectors and daisy-chained wire-wrap. These new interconnect schemes require care to achieve good contact and not break the wires. Nearly all cards are spared. We have aimed for modularity, compactness and reliability.

The intensifier head is potted in RTV and contained in a rather bulky housing thermoelectrically cooled to -35°C. The Reticons are also cooled, a necessary refinement given the large dark current of the S series chips. All optical couplings are custom fibre optic boules.

A digital encoding scheme for data transmission has been devised: the part of a number in the range 0-63 is represented using an ASCII code between 32 and 95; higher order segments (e.g. 64-2047, 2048-65536 etc.) use ASCII codes 96-127. An ASCII character in the range 32-95 therefore signals the least significant portion of a new number in the data stream. In practice, only about 70% as many bytes of data need be sent this way than if pure binary 16-bit data were sent.

This system can be used in the manner of the KPNO Intensified Reticon Scanner, which is used on a fairly large spectrograph with apertures on a 1-metre telescope and gives calibrated fluxes.

REFERENCES

Allen, R.G., Cromwell, R.H.,Liebert, J.W., Macklin, H. and Stockman, H.S. 1983, in <u>Instrumentation in Astronomy V</u>, ed. D. Crawford, Proc. SPIE <u>445</u>, 168.

Latham, D.W. 1982, in <u>Instrumentation for Large Telescopes</u>, IAU Coll. No. 67, ed. C. Humphries, p.259.

Shectman, S. 1983, in <u>Instrumentation in Astronomy V</u>, ed. D. Crawford, Proc. SPIE <u>445</u>, 128.

Shectman, S. and Hiltner, W.A. 1976, Pub. A. S. P., <u>88</u>, 960.

SPECTRAL CLASSIFICATION OF ECLIPSING BINARIES

George W. Wolf and Janet T. Kern

Department of Physics and Astronomy
Southwest Missouri State University
Springfield, Missouri 65804 U.S.A.

ABSTRACT. Approximately 375 classification spectra of 130 Southern Hemisphere eclipsing binary stars were obtained between 1978 and 1982 at Mt. John University Observatory, New Zealand using the 0.6 meter reflector, and at Cerro Tololo Inter-American Observatory, Chile using the 0.4, 0.6, 0.9, and 1.0 meter telescopes. The spectra have been classified by one of us (GWW) using a grid of standards obtained on the various spectrographs at each of the observatories. Since many of the spectra were taken during primary and secondary minima, it has been possible in many cases to classify separately each component in the binaries.

1. INTRODUCTION

New and improved computer modeling techniques used in the study of the nature and evolution of eclipsing binary stars have emphasized the need for accurate observational data on these systems. Among the most important information necessary for the detailed analyses are the temperature and luminosity of one, or preferably both, of the companions in a system under study. To ascertain these properties, the techniques of spectroscopy and standardized photometry are required.

To obtain a broad base of such information for researchers in this area, Hilditch and Hill (1975) have published a uvby photometric survey, and Hill et al. (1975) an MK spectrographic survey of eclipsing binaries observable from the Northern Hemisphere. To fill the need for information on Southern Hemisphere systems, the present authors have undertaken similar, but modified, uvbyβ photometric and spectrographic surveys. The results of the former survey have already been published by Wolf and Kern (1983). We report here on the completion of the first phase of the MK spectrographic survey.

2. THE OBSERVATIONS

Approximately 375 classification spectra of 130 Southern Hemisphere

J. B. Hearnshaw and P. L. Cottrell (eds.), Instrumentation and Research Programmes for Small Telescopes, 463–464.

eclipsing binary stars were obtained between 1978 and 1982 at Mt. John Observatory in New Zealand using the 0.6 meter reflector, and at Cerro Tololo Inter-American Observatory in Chile using the 0.4,0.6,0.9, and 1.0 meter telescopes. The reciprocal dispersions of the spectra were 62 A/mm for all the Mt. John spectra, and 91,91,62, and 43 A/mm, respectively for the Cerro Tololo spectra. Although only small telescopes were used for this survey, it was possible to obtain spectra of stars down to 10th magnitude in reasonable amounts of time (to avoid phase spread) on hydrogen hyper-sensitized IIIa-J or IIa-O photographic plates.

As in the photometric survey, the present study included specific phase coverage, usually at the minima and quadratures, for all binaries. The intention with this was to attempt to separate the spectral types of the binary members. Prior to each observing run, ephemerides were calculated for photometric phases 0.00, 0.25, 0.50, and 0.75 for all stars on the program using the best available times of minimum and periods from the Eclipsing Binary Card Catalog of the University of Florida, or from recent IBVS circulars. Events were combined, sorted by date and time, and arranged into nightly observing lists.

The spectra have been classified by one of us (GWW) using a fine grid of MK standards obtained on the various spectrographs at each of the observatories. Since many of the spectra were taken during primary and secondary minima, it has been possible in many cases to classify separately each component in the binaries. A comparison of this study with the results of the uvbyβ photometry on the same stars is presently underway. Preliminary results indicate a close correlation in the temperatures determined from each technique. The actual results of this analysis and the MK spectral classifications will be published elsewhere.

ACKNOWLEDGEMENTS

The authors would like to thank Dr. John Hearnshaw, Prof. A.G. McLellan and staff for the generous support and allocation of observing time at Mt. John University Observatory during the period of this observing project. We also wish to thank the University of Florida for use of their spectrograph at Mt. John and for use of the Eclipsing Binary Card Catalog. In addition, thanks are also given to the staff of CTIO, particularly Drs. John Graham and Victor M. Blanco, for their help and allocation of observing time.

REFERENCES

Hilditch, R.W. and Hill, G. 1975, Mem. R. Astr. Soc. 79, 101

Hill, G., Hilditch, R.W., Younger, F., and Fisher, W.A. 1975, Mem. R. Astr. Soc. 79, 131

Wolf, G.W. and Kern, J.T. 1983, Astroph. J. Suppl. 52, 429

DISCUSSION OF POSTER PAPERS : SECTION III

Beavers and Eitter - The first eight years of radial velocity studies...

Burki: Could you make a comment on the zero-point comparisons you made between your radial velocities and those observed by other groups? What about the IAU standards?

Beavers: Yes, we compared our zero-point with that of other observatories. We looked at stars in common, some of which were IAU standards. The number of stars in common varied from institution to institution. Our zero-point is similar to that of the Lick Observatory which is close to some of the IAU standards.

Latham: At the IAU General Assembly in Delhi a working group was formed to work on setting up a new set of standards of stellar radial velocities. We plan to observe about 15 stars extensively at three or four observatories, with an absolute zero-point calibration against minor planets. We think it may be possible to achieve a precision of 0.1 km s^{-1} in the mean velocities of the new standards and in the absolute zero-point.

Gray: I believe there will be difficulties in attaining a grid of radial velocities accurate to 100 ms^{-1}, because at this level the asymmetries, caused by stellar granulation, come in. As a result, the measured radial velocity will depend on the technique used to measure it. Internal-consistency errors and radial-velocity variations, of course, can be pushed to this level and beyond. The problem will not be cured by modelling the solar case in isolation because stellar granulation varies with spectral type and luminosity class, and it may also depend on age. The granulation asymmetries depend on rotation velocity, and there are smaller dependences with excitation potential of the line and line strength. Stellar granulation may be a real limiting factor for radial velocity measurement, but then, stellar granulation is a fascinating area of astrophysical study for some of us.

Latham: I agree that the accuracy of the zero-point determination may be limited by astrophysical effects. By using minor planets the zero-point will be referenced to the sun, so perhaps these effects can be modelled theoretically.

Denby et al - FAGS : a fast Astronomical Grating Spectrograph

Graham: On what telescopes do you plan to use this instrument?

J. B. Hearnshaw and P. L. Cottrell (eds.), Instrumentation and Research Programmes for Small Telescopes, 465–472.

Denby: On the 24" and 40" at Siding Spring Observatory. At the moment we have it on the 24" at Mount John.

Latham: Can you describe the wavelength sensitivity and readout noice of your CCD?

Denby: The readout noise hasn't been determined yet. We are only in the very early stages of commissioning the instrument and it has not been cooled yet. Spectral sensitivity drops off rapidly below λ360nm. In the red it extends to ~1μm.

Hiltner: I want to emphasise that it is important to eliminate all unnecessary reflections in the design of optical instruments. Each mirror "costs" approximately 2 hours of observing of each 10 hour night.

Duquennoy and Mayor - Multiple stellar systems

Scarfe: In the cases where the spectra are never fully resolved, how do you determine the parameters of the fitting functions used to resolve the dips?

Burki: We cannot obtain all the parameters of the dips, but we can obtain, in some cases, the separation using a mathematical treatment developed by Duquennoy.

Latham: The survey has tended to pick out companions with fairly short periods and low velocity amplitudes. This is a good recipe for a low mass companion. What is the lowest mass found this way?

Burki: I don't know.

Garrison: I want to raise two questions (complaints) about this type of spectrograph: 1) early-type stars are not being done and 2) there are no archives, so one can't go back and look at variations of spectra over time. We have been getting lots of good temporal information of shell stars from old DDO, DAO, Mount Wilson and Michigan plates.

Latham: There are techniques which are now being developed (e.g. Abt at Kitt Peak), using solid state detectors, to obtain radial velocities for early-type stars. It's crucial to do good flat-fielding so the continuum is straightened out across broad, strong lines, e.g. Hβ. Comparing the wings gives radial velocities to ~3 km s^{-1}.

Scarfe: I have used a mask (at DAO) based on Procyon to measure A stars with low rotation velocities.

Latham: With the CfA stellar velocity systems we do make archival records of the spectra. However, these are not the best spectra for other applications. First, the signal-to-noise ratio tends to be in the range 3 to 5, and it is hard to measure individual lines. Second, the observations are taken for radial velocities, often with lots of moon light or twilight contamination. Thus the spectra are poorly suited for spectrophotometry. I agree that it is helpful to have the spectra, but I worry that they are not suitable for other applications.

Djorgovski: The trouble with these types of spectrographs is that one wastes a lot of light with the mask. One should take real spectra and use Fourier techniques on it to extract velocities or whatever you want.

Fletcher: You don't lose a lot of light with the mask, maybe 30%. Also the cost of this type of instrument is much less than a Shectman-type system.

Hearnshaw: There is no radial velocity information in the continuum, so why not throw it away?

Finkenzeller et al - Spectroscopic surveillance of the variable... .

Garrison: Are you intending to follow up this work?

Finkenzeller: Yes, using CCD spectrographs at Lick and at Berkeley. We will do a short-term monitoring program on T Tauri stars, which have periods of 4-8 days for which one can obtain better rotation information than for the higher mass, faster rotators.

Rucinski: Fred Vrba, Steve Strom and other people have started photometry of a number of T Tauri stars using a CCD photometer at Naval Observatory. This should provide photometric rotation periods for T Tauri stars in NGC 2264 or, possibly, in the Taurus Cloud.

Garrison: Herbst has bought 5 nights/month (for a 2 year period - six months still to go) at the University of Toronto's field station in Chile to do Hα photometry of T Tauri stars.

Mochnacki: Is there any evidence of binaries in these stars?

Finkenzeller: No, only a few appear to be double.

Latham: More will be uncovered because these objects are very difficult to study spectroscopically. This could hide a lot of the binaries.

Finkenzeller: Walker (at Santa Cruz) has set up a long term program to monitor radial velocity changes in T Tauri stars.

Fletcher and McClure - Radial velocity spectrometers on the DAO... .

Mochnacki: Is the spectrometer on the 0.4m telescope a coudé or cassegrain?

Fletcher: A bit of both! Its optically fiber fed from the cassegrain focus. We want to reduce the amount of time doing comparison spectra (before and after each stellar observation) because of flexure in the cassegrain spectrograph.

Latham: What do you think the limiting magnitude will be on the 0.4m telescope?

Fletcher: From some of the tests we have done we can probably get to about 14 mag, but maybe 12 mag will be more realistic under normal observing conditions.

Garrison: Don't you have trouble with degradation of the focal ratio when using the optical fiber?

Fletcher: We have taken care of this with "extra optical elements". We don't lose too much overall because we don't have to observe as many standards.

Latham: When you said you were losing a fair amount of light with the fiber you actually mean a lot!

Fletcher: We are now getting about 1/3 of the light through the fiber but we expect to improve that to about 1/2 with better alignment of the optical fiber. We may also be able to get around other problems, e.g., guiding errors, by having the spectrograph fixed, and thus improve the efficiency of our observing technique.

Hearnshaw: What is the diameter of your fiber?

Fletcher: About 50μm.

Gaustad - Temperature and brightness variations on Betelgeuse

Gaustad: Indeed, this is a very inefficient system at present. This was a pilot project to see if I could do anything useful on stars with an instrument designed for solar spectroscopy. The detector system will soon be replaced with a Reticon, and then I should be able to reach 6th magnitude stars at 0.1Å resolution with a S/N = 100 in 1 hour.

Rowe: Has anyone made a series of speckle observations of Betelgeuse which might show rotation?

Gaustad: Not to my knowledge, I have seen nothing published since the original speckle observations which are now some 8 years old. I am not sure why this technique has not been used more.

Honeycutt and Peery - An off-the-shelf Reticon system

Gaustad: What is the effective quantum efficiency of your system?

Peery: We haven't measured it yet. However, I would like to add that one of the reasons we bought the detector was that is has been used extensively, and successfully, by chemists in the laboratory.

Levato: We have used an OMA II system with a small spectrograph manufactured by Ridell-Spotz Scientific Instruments. The system works quite well. We used it for monitoring Be spectra but we did not lower the temperature below -40°C. The efficiency in the red is ~60%.

Jugaku: With your system what kind of research programs do you have in mind?

Peery: My research intersts have been and will continue to be in high dispersion spectroscopy (abundance work and isotope ratios) but we expect to undertake routine monitoring projects as well.

Jeffers and Stiff - Stellar spectrophotometry with a micro-computer... .

Graham: Is the fact that you are situated within the city limits of Toronto the reason why you take such care with your extinction measurements?

Jeffers: We were interested to find that even in an urban site pure Rayleigh scattering was the best fit to the data.

Weller: I have measured the extinction as a function of wavelength with the SIT vidicon. Over most of the wavelengths covered Rayleigh scattering was a good approximation but breaks in the curve of k(λ) vs λ are seen, caused for example by the onset of the ozone continuum absorption.

McDavid - A microcomputer controlled CCD Hα spectrometer

Garrison: Where is your observatory?

McDavid: About 15 miles north west of San Antonio in the Texas hill country.

McMillan et al - A long-term radial velocity program of high accuracy...

Garrison: What is the limiting magnitude of the system? What is the cost?

Wisniewski: About 4-5 mag, and they expect to go fainter. The cost of the instrument is high, but it is very elegant.

Barwig: 1) To what extent is the accuracy of your instrument influenced by guiding errors? 2) What's the core diameter of the fibre used?

Wisniewski: 1) No influence. 2) 125 μm.

Kurtz: Is the instrument portable? That is, can you put it on an aircraft and take it to another observatory?

Wisniewski: Yes.

Mochnacki et al - The DDO photon counting spectrometer : a well... .

Scarfe: Could you explain what you mean by "well-dressed"?

Mochnacki: It refers to the way we built it. Although we followed the Latham-Geary design we made a number of innovations. We want to mount the whole instrument on the telescope and talk to the outside world via RS232 serial lines. We also have an on-board microprocessor. We also use speed-wire technology.

Levato: Can you comment about the cost of the system?

Mochnacki: We had a grant for ~$CAN90,000 which included labour costs.

<u>Wolf and Kern</u> - Spectral classification of eclipsing binaries

Batten: A homogeneous list of spectral types, such as is given in this paper, is very valuable and will be of much use to future compilers of catalogues of spectroscopic binary orbits.

Graham: Is the difference you give in spectral type for a given system significant?

Wolf: Most of the variations, especially those from one minimum to another are believed to be real. Those variations in spectral type of spectra at maximum may be due to effects of the combined spectra. One subclass variation may be the scatter in the classification process; 2 or more subclasses variation is probably real.

Garrison: Corbally is publishing an atlas of visual double stars, showing individual spectra and combined spectra, so that one can see the effects of contamination.

Mochnacki: Have you or anyone else classified all the binary stars in Popper's lists that are used in the mass-radius-luminosity relation?

Wolf: I did not specifically classify the stars in Popper's or other surveys, although many of those stars may have been on the program anyway.

Garrison: There is another list, compiled by Hiltner, Schild and myself, of 63 OB eclipsing binary stars for southern hemisphere observers which could form a starting point for a project with small telescopes. These stars come from composite MK dispersion spectra which we noticed when looking at 1100 OB stars in order to understand galactic structure. Follow-up photometry also confirmed their binary nature. Masses for these stars would be extremely useful, since there are very few masses for these types of stars.

Scarfe: Could you please remind me what your limiting magnitude is, and in which band?

Wolf: The magnitude limit for the survey was 10th magnitude in V. For the later spectral types it was probably closer to 9th magnitude in V.

SUMMARY

SUMMARY: OPPORTUNITIES FOR RESEARCH WITH SMALL TELESCOPES

J. A. Graham
Carnegie Institution of Washington
Department of Terrestrial Magnetism
Washington, D. C.
U. S. A.

Our meeting has been unusual in that we have covered so many fields of present-day astronomy and astrophysics. The common link has been the potential of small telescopes for making substantial advances in our knowledge of the observable universe. Attention has been focussed not just on those programmes which can be done best with small telescopes but we have also highlighted many which can only be done with small telescopes.

At this IAU symposium, we followed the by now almost established format of having a number of invited review papers with the bulk of the contributed papers in poster format. The invited papers served to bring us up-to-date on what had been done in the various research areas. However, most of the new results appeared at the poster sessions. Our three poster discussion periods worked especially well through the efforts of the excellent chairmen, Dave Latham, Brian Warner and Bob Garrison. There were several lively discussions with question and answer exchanges which compelled many of us to go back for a second look at many of the posters. I particularly liked the practice followed by many of photographing posters so that the new ideas displayed there could be carried back to the people at home who were not able to be with us in New Zealand.

One's main job in attempting a summary such as this is to try and sense the main themes running through the meeting. Perhaps a major one was that astronomy is a science which is in no way problem limited at present. Rather the limit seems to be determined by the number of people who are able to undertake research in the different fields. In his splendid overview which started off proceedings, Brian Warner reminded us that, with small telescopes, discoveries can and have been made which opened up whole new research areas. He gave several examples and many of us felt we could have added to his list.

There is an abundance of new instrumentation available. New detector technology was reviewed by Dave Latham. Some of it is expensive both in terms of money and time and should not be undertaken without careful

J. B. Hearnshaw and P. L. Cottrell (eds.), Instrumentation and Research Programmes for Small Telescopes, 475–478.

planning. However, the benefits are enormous. With CCD detectors, for example, photometry and spectroscopy of faint objects are no longer just in the realm of large telescopes at a few institutions. However, their use does require a major effort. Why do we do it? It is very much the old spirit of exploration as we try to push back the boundaries of knowledge a little bit further. Our standards are high! As Latham reminded us, "If you can satisfy an astronomer you can satisfy anyone."

With modest, conventional instrumentation, progress can still be made. However, the path is not so direct. Here, the effort sometimes has to be put into programmes which cover a long period of careful observation often extending over many years. The results can be of fundamental value and can equally well determine directions of future research. Hugo Levato summed it up nicely when he said, "A modest budget does not have to mean a modest institution." It is the scientific creativity which counts.

In the instrumentation line, many new telescopes of the 0.5-1.5 m class are coming into use. Several were described in the posters. We all welcomed the University of Canterbury 1 m telescope and look forward to its inauguration during 1986. Automatic telescopes have come among us, this time to stay. Russ Genet and others showed what fine work the amateurs have been doing here. Paul Payne described the automatic patrol camera being developed at the University of New South Wales.

A number of new instruments were described which took advantage of new technology. I was especially impressed by the photometers and spectrographs which incorporate the relatively cheap but very efficient fibre-optic feeds which are now available. A. K. Saxena described a new polarimeter with a CCD detector which he is building at the Indian Institute of Astrophysics. Costs and effort in getting a CCD system into operation on a small telescope were described by Bryan Laubscher of the University of New Mexico while Ben Peery told us about the off-the-shelf Reticon system he had obtained for Howard University. With the accurate linear detectors now available, the distinction between spectroscopy and photometry is indeed becoming diffuse. Several radial velocity spectrophotometers, for example, were described at the meeting which are producing velocities to an accuracy better than 1 km s^{-1}.

Many new observing opportunities were described. A prominent theme, reiterated again and again by participants, was the tremendous amount of work still needing to be done on the apparently brightest stars which are difficult or impossible to observe with large telescopes. Information about diameters, multiplicity and even the shapes of stars comes from the study of occultations by the moon. After David Evans's review, Wayne Osborn pointed out to me that this is a particularly good field in which to get involved if one is attached to a physics department. Not only is the relatively high-tech equipment usually available but also there are students freshly acquainted with the concepts of signal and noise; important ones in this work.

With telescopes which can be set rapidly and accurately, large survey programs can be carried out to give good statistical samples of many thousands of stars. B. Twarog of the University of Kansas showed us how photometry of such samples can give important information on the history of galactic star formation. W. Gliese of Heidelberg reminded us several times about how incomplete our knowledge is of the 1000-2000 nearest stars.

Alistair Walker from Cape Town talked about the methods of CCD photometry. With this technique, one can produce better photometry for stars of 12th mag and fainter than by any other means. Photometry done with small telescopes complements and essentially ties down photometric work carried out with large telescopes.

All through Tuesday, Thursday and Friday we were reminded of the importance, particularly in the southern hemisphere, of observation of binary stars. Visual binaries, spectroscopic binaries, eclipsing binaries and contact binaries all competed in their own way for our attention in the talks by Batten, Plavec and Rucinski. Much of this can be done with small telescopes. Photometry, spectroscopy and radial velocities are all needed. Small, dedicated telescopes can undertake fundamental programs. Brad Wood, speaking about eclipsing binaries, reminded us, "If you want to achieve immortality, observe times of minimum and publish them." His statement can be readily adapted for application to other fields although a certain amount of the appropriate insight and intuition is necessary.

In this symposium, we have talked mostly about stars but valuable work can be done with small telescopes on solar system objects (A'Hearn, Schober, Millis) and on external galaxies (Djogovski, Phillips). Telescopes with short focal lengths must be used, for example, to get representative spectra of extended objects like comets and nearby galaxies.

I am deeply conscious of not being able to cover all the highlights of this meeting which has embraced so much. For instance, I have not reviewed the many opportunities for MK spectral classification and the development of the MK process as related by Hugo Levato and Bob Garrison. However in the available time, I would like to reiterate two subjects brought up during the meeting by W. Wisniewski and M. Zeilik. First, it is very clear that, with the expansion of space astronomy, ground-based support and follow-up is going to be needed urgently. We should be opening up more small telescopes, certainly not closing them. Arthur Page reminded us that here on cloud-bound Earth, any one small telescope may be in a unique position to obtain further coverage of a special event observed from space. Then, as Mike Zeilik told us, we have to worry about funding. Government money will remain tight unless we can get out and demonstrate the importance of astronomy to the people who control its distribution. Especially in the United States, astronomers have been successful in the past in getting help from industrial groups and in turn helping them to push their techniques to the limit. In the long run, they, as well as we, have benefited. As I see it, more countries could follow this track. It is this characteristic of forever pushing towards greater limits which has made astronomy such an exciting field in which to work for so many of us.

INDEX OF SUBJECTS